de Gruyter Series in Nonlinear Analysis and Applications 7

Mikhail Kamenskii
Valeri Obukhovskii
Pietro Zecca

Condensing Multivalued Maps and Semilinear Differential Inclusions in Banach Spaces

Walter de Gruyter · Berlin · New York 2001

Authors

Mikhail Kamenskii
Faculty of Mathematics
University of Voronezh
Universitetskaya pl. 1
394693 Voronezh
Russia

Valeri Obukhovskii
Faculty of Mathematics
University of Voronezh
Universitetskaya pl. 1
394693 Voronezh
Russia

Pietro Zecca
Dipartimento di Energetica
Università degli studi di Firenze
Via di S. Marta, 3
50139 Firenze
Italy

Mathematics Subject Classification 2000: 47-02, 34-02; 47H04, 47H09, 47H10, 47H11, 34A60, 34C25, 34C29, 34G25, 49J24, 49J27, 49J53, 54C60, 54C65, 54H25, 55M20, 55M25

Keywords: Multivalued map, Condensing map, Fixed point, Differential inclusion, Measure of noncompactness, Cauchy problem, Periodic solution

♾ Printed on acid-free paper which falls within the guidelines of the ANSI to ensure permanence and durability.

Library of Congress Cataloging-in-Publication Data

Kamenskii, Mikhail, 1950–
Condensing multivalued maps and semilinear differential inclusions in Banach spaces / by Mikhail Kamenskii, Valeri Obukhovskii, Pietro Zecca.
p. cm. – (De Gruyter series in nonlinear analysis and applications, ISSN 0941-813X ; 7)
Includes bibliographical references.
ISBN 3-11-016989-4
1. Set-valued maps. 2. Differential inclusions. 3. Banach spaces. I. Obukhovskii, Valeri, 1947– . II. Zecca, P. (Pietro). III. Title. IV. Series.
QA611.3 .K36 2001
515.2–dc21 2001017247

Die Deutsche Bibliothek – Cataloging-in-Publication Data

Kamenskii, Mikhail:
Condensing multivalued maps and semilinear differential inclusions in Banach spaces / Mikhail Kamenskii ; Valeri Obukhovskii ; Pietro Zecca. – Berlin ; New York : de Gruyter, 2001
(De Gruyter series in nonlinear analysis and applications ; 7)
ISBN 3-11-016989-4

ISSN 0941-813 X

Printed in Germany.
Typesetting using the authors' TEX files: I. Zimmermann, Freiburg.
Printing and binding: WB-Druck GmbH & Co., Rieden/Allgäu.
Cover design: Thomas Bonnie, Hamburg.

Introduction

Two key phrases, "condensing multivalued maps" and "semilinear differential inclusions", completely characterize this book.

In fact, our aim is twofold. First, we give a self-contained presentation of the theory of condensing multivalued maps; secondly we apply this theory to investigate semilinear differential inclusions in Banach spaces.

To fulfill the first objective, after a presentation of the basic properties of multivalued maps, we focus our attention on the notion of measure of noncompactness and on the development of a degree theory for condensing multimaps. For the second objective we restrict ourselves to the case when the linear part of the inclusion generates a (not necessarily compact) strongly continuous semigroup, and we study both the Cauchy problem and the periodic case. For the periodic case the principal properties of the integral and translation multioperators are investigated.

The plan of the book is as follows.

In order to make the book self-contained, we devote the first chapter to a detailed description of the fundamental, general properties of multimaps. In particular, we discuss different types of continuity for multimaps and various operations on multimaps (including those which have no "single-valued" analogs). We devote particular attention to the problem of the existence of single-valued selections and approximation for multimaps. The last part of the first chapter contains a detailed investigation of measurable multimaps including a justification of the notion of superposition multioperator which is used constantly when dealing with differential inclusions.

In the second chapter, after describing the concept of measure of noncompactness, we present a number of examples of different measures of noncompactness in spaces of continuous functions. Furthermore, we consider special classes of multimaps which are condensing with respect to various measures of noncompactness in specific infinite-dimensional spaces. In subsequent chapters we show how these particular multimaps provide a very convenient machinery for treating several different problems involving differential inclusions and control systems in infinite-dimensional spaces.

The third chapter is devoted to the construction of a topological degree theory for both convex-valued and nonconvex-valued condensing multimaps. To make this theory universal and applicable to different classes of condensing multimaps we implement it for a general class of fundamentally restrictible multimaps. We also describe some properties of the fixed point sets of condensing multimaps and we apply the degree theory to the problem of solvability of systems of inclusions with condensing multivalued operators. We apply our results to some optimal control problems.

In the fourth chapter we first review some properties of linear operators in Banach spaces generating strongly continuous semigroups and describe some properties of measures of noncompactness in function spaces paying special attention on the conditions under which it is possible to estimate measure of noncompactness of multivalued integrals. We conclude the chapter with the study of a special class of condensing semigroups and giving an example of a condensing semigroup appearing in a nonlinear transmission line problem. We want to point out that, in spite of the independent character of Chapters 1–4, we develop the theory having in mind applications to the study of semilinear differential inclusions in infinite-dimensional Banach spaces. In this setting, in fact, we have to deal with noncompact multioperators for which the classical topological degree theory is not suitable.

The fifth chapter begins with a detailed description of the properties of the integral multioperator, which plays a key role in the study of semilinear differential inclusions. The most important of these properties is the fact that this multioperator is condensing with respect to a special measure of noncompactness. On this basis we are able to give local and global existence theorems for a semilinear differential inclusion in a Banach space. We give two examples of semilinear differential inclusions satisfying the assumptions of these theorems: the first example arises in a mathematical model of a transmission line with nonlinear parameters, the second deals with a hybrid system with dry friction in an anisotropic environment. The compactness of the solutions set, being an additional product of our approach, makes it possible to give an application to the optimization problem for a general feedback control system. In this chapter we also investigate the topological index of the solutions set, study the continuous dependence of the solutions set on parameters and initial data, and describe also the topological structure of this set. All these results play a key role in the construction of the translation multioperator. We describe also the averaging principle for the Cauchy problem for a semilinear differential inclusion with a small positive parameter. In the last section of this chapter, using the continuous selection theorem for a multimap with decomposable values, we prove an existence theorem for a semilinear differential inclusion whose linear part generates a noncompact semigroup and whose multivalued nonlinearity is lower semicontinuous and nonconvex-valued. We obtain also an analog of the classical Kneser property, proving that the integral funnel of such an inclusion is a connected set.

The sixth and last chapter is devoted mainly to the investigation of periodic problems for systems governed by semilinear differential inclusions. Here two different operator methods are used: the method of the integral multioperator and the method of the translation multioperator along the solutions of the inclusion. In both cases it is necessary to deal with the difficult problem that the multioperator is condensing with respect to an appropriate measure of noncompactness. The application of both methods allows us to present new existence results for periodic problems for differential inclusions and control systems governed by them.

Having explained the title and the plan of the book, we would like to stress again that our main goal is, from one side, to give a self-contained introduction to the contemporary theory of noncompact multivalued maps, and, from the other, to demonstrate in the second half of the monograph some applications of our approach to various problems involving differential inclusions and control systems. Here we are able to present some new results and this represents the main difference between our book and other recent monographs completely or partially devoted to the theory of differential inclusions (e.g. [16], [18], [19], [20], [56], [73], [86], [103], [104], [119], [141], [152], [197], [210]).

Finally, the first two authors would like to thank the Russian Foundation for Basic Research for constant backing as well as the Italian CNR (Consiglio Nazionale delle Ricerche) and the University of Florence for supporting their visits to Italy. V. Obukhovskii is much obliged to the Italian Ministry of Foreign Affairs and to the Landau Network-Centro Volta for a fellowship which permitted his work on the book. M. Kamenskii appreciates the contribution of the University of Rouen, France, where he wrote some parts of the book during his stay as Visiting Professor.

We all are grateful to our wives for their understanding and support.

Particular thanks go to Signora Mirta Stampella, who deciphered our handwriting and so carefully typed the manuscript, and to Professor Jack W. Macki for improving English language. At last we have the pleasure to express our thanks to the editors of the Walter de Gruyter Publishing House for constructive cooperation.

Firenze and Voronezh, February 2001

Mikhail Kamenskii
Valeri Obukhovskii
Pietro Zecca

Contents

Chapter 1

Multivalued maps: general properties

A multivalued map (multimap) F of a set X into a set Y is a correspondence which associates to every $x \in X$ a non-empty subset $F(x) \subseteq Y$, called the *value of* x. Denoting by $P(Y)$ the collection of all non-empty subsets of Y we write this correspondence as

$$F : X \to P(Y)$$

Sometimes we use the symbols $x \multimap F(x)$ or $F : X \multimap Y$. Throughout the book we denote multimaps by capital letters.

If $A \subseteq X$, then the set $F(A) = \bigcup_{x \in A} F(x)$ is called *the image of* A under F.

The set $\Gamma_F \subseteq X \times Y$, defined by $\Gamma_F = \{(x, y) : x \in X, y \in F(x)\}$ is *the graph* of F.

The notion of multimap arises naturally in various branches of modern mathematics, such as mathematical economics, theory of games, convex analysis, etc.

Multimaps play a significant role in the description of processes in control theory since the presence of controls provides an intrinsic multivalence in the evolution of the system. As an illustration, consider the following simple examples.

Let $\mathbb{R}^n$ be the state space of a control system and $\mathbb{R}^m$ the space in which controls take their values. The dynamics of the system is determined by a map $f : \mathbb{R} \times \mathbb{R}^n \times \mathbb{R}^m \to \mathbb{R}^n$; the feedback condition indicating the restriction on the choice of control, at a given moment, under a given state of the system, can be expressed by the relation $u(t) \in U(t, x(t))$, where $U : \mathbb{R} \times \mathbb{R}^n \to P(\mathbb{R}^m)$ is the feedback multimap. Then the description of the control system can be given as:

$$\begin{cases} x'(t) = f(t, x(t), u(t)) \\ u(t) \in U(t, x(t)). \end{cases}$$

Under general conditions specified in the sequel, the description of the system can be presented in the form of the differential inclusion

$$x' \in F(t, x(t))$$

where $F : \mathbb{R} \times \mathbb{R}^n \to P(\mathbb{R}^m)$ is defined as $F(t, x) = f(t, x, U(t, x))$.

In the investigation of a control system an important part is played by the multioperator of translation along trajectories, i.e., the multimap $Q : [0, \infty) \times X \to P(X)$, assigning to every initial state x from the phase space X the set $Q(t, x)$ of all possible reachable states of the system at the time $t \geq 0$. The translation multioperator plays a key role in problems of periodicity and controllability, and will be studied in Chapter 6.

1.1 Continuity of multimaps

The classical characterizations of continuity of a single-valued map split into different concepts when generalized to multifunctions. We introduce, first, the following sets.

Let X and Y be topological spaces, $F : X \to P(Y)$ be a multimap and $D \subseteq Y$ be a set. The *small preimage* $F_+^{-1}(D)$ and the *complete preimage* $F_-^{-1}(D)$ of a set D are, respectively, the sets

$$F_+^{-1}(D) = \{x \in X : F(x) \subset D\} \quad \text{and} \quad F_-^{-1}(D) = \{x \in X : F(x) \cap D \neq \emptyset\}.$$

Definition 1.1.1. A multimap $F : X \to P(Y)$ is *upper semicontinuous at the point* $x \in X$ if, for every open set $W \subseteq Y$ such that $F(x) \subset W$, there exists a neighborhood $V(x)$ of x with the property that $F(V(x)) \subset W$.

A multimap is *upper semicontinuous* (u.s.c.) if it is upper semicontinuous at every point $x \in X$.

Theorem 1.1.1. *The following conditions are equivalent:*

(a) *the multimap F is u.s.c.;*

(b) *the set $F_+^{-1}(W)$ is open for every open set $W \subset Y$;*

(c) *the set $F_-^{-1}(Q)$ is closed for every closed set $Q \subset Y$;*

(d) $F_-^{-1}(\overline{D}) \supseteq \overline{F_-^{-1}(D)}$ *for every set $D \subset Y$.*

As usual $\overline{A}$ denotes the closure of the set A.

Example 1.1.1. The multimap $F : [0, 1] \to P([0, 1])$, defined as

$$F(x) = \begin{cases} [0, 1/2], & \text{if } x \neq 1/2, \\ [0, 1], & \text{if } x = 1/2, \end{cases}$$

is u.s.c.

Definition 1.1.2. A multimap $F : X \to P(Y)$ is *lower semicontinuous at the point* $x \in X$ if for every open set $W \subseteq Y$ such that $F(x) \cap W \neq \emptyset$ there exists a neighborhood $V(x)$ of x with the property that $F(x') \cap W \neq \emptyset$ for all $x' \in V(x)$.

A multimap is *lower semicontinuous* (l.s.c.) if it is lower semicontinuous at every point $x \in X$.

The following dual of Theorem 1.1.1 gives some criteria for lower semicontinuity.

Theorem 1.1.2. *The following conditions are equivalent:*

(a) *the multimap F is l.s.c.;*

(b) *the set $F_-^{-1}(W)$ is open for every open set $W \subset Y$;*

(c) *the set* $F_+^{-1}(Q)$ *is closed for every closed set* $Q \subset Y$;

(d) *if the system of open sets* $\{W_j\}_{j\in J}$ *forms a base for the topology of* Y, *then every set* $F_-^{-1}(W_j)$ *is open;*

(e) $F_+^{-1}(\overline{D}) \supseteq \overline{F_+^{-1}(D)}$ *for every set* $D \subset Y$;

(f) $F(\overline{A}) \subseteq \overline{F(A)}$ *for every set* $A \subset X$;

(g) *for any* $x \in X$, *if* $\{x_\alpha\} \subset X$ *is a generalized sequence,* $x_\alpha \to x$, *then for every* $y \in F(x)$ *there exists a generalized sequence* $\{y_\alpha\} \subset Y$, $y_\alpha \in F(x_\alpha)$, $y_\alpha \to y$.

Note that when X and Y are metric spaces, in condition g) we can take sequences instead of generalized sequences.

Example 1.1.2. a) The multimap $F : [0, 1] \to P([0, 1])$, defined as

$$F(x) = \begin{cases} [0, 1], & \text{if } x \neq 1/2, \\ [0, 1/2], & \text{if } x = 1/2, \end{cases}$$

is l.s.c.

b) Let E be a Banach space, $C([a, b]; E)$ denotes the space of continuous functions $x : [a, b] \to E$, equipped with the usual norm $\| x \| = \sup_{t\in[a,b]} \| x(t) \|$. For an arbitrary nonempty subset $\Omega \subset C([a, b]; E)$ the multimap $Q : [a, b] \to P(E)$, defined as

$$t \multimap Q(t) = \Omega(t) := \{y(t) : y \in \Omega\}$$

is l.s.c.. This is easily verified using Theorem 1.1.2 (g).

Definition 1.1.3. A multimap F which is both upper and lower semicontinuous is said to be *continuous.*

Example 1.1.3. Let $f_1, f_2 : [0, 1] \to \mathbb{R}$ be continuous functions such that $f_1(x) \leq f_2(x)$ for all $x \in [0, 1]$. Then the multimap $F : [0, 1] \to P(\mathbb{R})$, $F(x) = [f_1(x), f_2(x)]$ is continuous.

We consider one more important class of multimaps.

Definition 1.1.4. A multimap F is said to be *closed* if its graph Γ_F is a closed subset of the space $X \times Y$.

Theorem 1.1.3. *The following properties are equivalent:*

(a) *the multimap* F *is closed;*

(b) *for any pair* $x \in X$, $y \in Y$ *such that* $y \notin F(x)$ *there exist neighborhoods* $V(x)$ *of* x *and* $W(y)$ *of* y *such that* $F(V(x)) \cap W(y) = \emptyset$;

(c) *for any generalized sequences* $\{x_\alpha\} \subset X$, $\{y_\alpha\} \subset Y$, *if* $x_\alpha \to x$, *and* $y_\alpha \in F(x_\alpha)$, $y_\alpha \to y$ *then* $y \in F(x)$.

In the last condition sequences can be used when X and Y are metric spaces.

Example 1.1.4. Let $f : Y \to X$ be a continuous surjective map between topological spaces and let X be Hausdorff. Then the inverse multimap $F : X \to P(Y)$, $F(x) = f^{-1}(x)$ is closed.

We introduce some notation. Let Y be a topological space. (Recall that $\emptyset \notin P(Y)$.)

$$C(Y) = \{D \in P(Y) : D \text{ is closed}\};$$

$$K(Y) = \{D \in P(Y) : D \text{ is compact}\};$$

If Y is a topological vector space we introduce:

$$Pv(Y) = \{D \in P(Y) : D \text{ is convex}\};$$

$$Cv(Y) = Pv(Y) \cap C(Y) = \{D \in P(Y) : D \text{ is closed and convex}\};$$

$$Kv(Y) = Pv(Y) \cap K(Y) = \{D \in P(Y) : D \text{ is compact and convex}\}.$$

When a multimap F maps into the collections $C(Y)$, $K(Y)$, or $Pv(Y)$, we will say that F has closed, compact or convex values respectively.

From the definition it is clear that a closed multimap has closed values.

Closed and upper semicontinuous multimaps are a short distance apart. The relation between them is clarified by the following statements.

Theorem 1.1.4. *Let X be a topological space, Y a regular topological space and $F : X \to C(Y)$ an u.s.c. multimap. Then F is closed.*

Proof. Let $y \in Y$, $y \notin F(x)$. Since Y is regular there exist an open neighborhood $W(y)$ of the point y and an open neighborhood W_1 of the set $F(x)$ such that $W(y) \cap W_1 = \emptyset$. Let $V(x)$ be a neighborhood of x such that $F(V(x)) \subset W_1$. Then $F(V(x)) \cap W(y) = \emptyset$ and the statement follows from Theorem 1.1.3 part (b). □

Remark 1.1.1. It is clear from the proof that when F has compact values the condition of regularity on Y can be replaced by the weaker condition that Y is a Hausdorff space.

To formulate a sufficient condition for a closed multimap to be u.s.c. we need the following definitions.

Definition 1.1.5. A multimap $F : X \to P(Y)$ is

(a) *compact* if its range $F(X)$ is relatively compact in Y, i.e., $\overline{F(X)}$ is compact in Y;

(b) *locally compact* if every point $x \in X$ has a neighborhood $V(x)$ such that the restriction of F to $V(x)$ is compact;

(c) *quasicompact* if its restriction to any compact subset $A \subset X$ is compact.

It is clear that a) $\Longrightarrow$ b) $\Longrightarrow$ c).

Theorem 1.1.5. *Let $F : X \to K(Y)$ be a closed locally compact multimap. Then F is u.s.c.*

Proof. Let $x \in X$, W an open neighborhood of the set $F(x)$ and $V(x)$ an open neighborhood of x such that the restriction of F to $V(x)$ is compact. Suppose that the set $Q = \overline{F(V(x))} \setminus W$ is nonempty. Since F is closed, for any $y \in Q$, there exist neighborhoods $\widetilde{W}(y)$ of y and $V_y(x)$ of x such that $F(V_y(x)) \cap \widetilde{W}(y) = \emptyset$. By the compactness of Q we can extract a finite subcover $\widetilde{W}(y_1), \widetilde{W}(y_2), \ldots, \widetilde{W}(y_n)$. Then if we consider the open neighborhood of x defined by $\widetilde{V}(x) = V(x) \cap (\bigcap_{i=1}^{n} V_{y_i}(x))$ we have $F(\widetilde{V}(x)) \subset W$. □

Example 1.1.5. The condition of local compactness is essential. The multimap $F : [-1, 1] \to K(\mathbb{R})$,

$$F(x) = \begin{cases} \{1/x\}, & \text{if } x \neq 0, \\ \{0\}, & \text{if } x = 0, \end{cases}$$

is closed but is not u.s.c. at $x = 0$.

Let us consider some properties of closed and u.s.c. multimaps.

Theorem 1.1.6. *Let $F : X \to C(Y)$ be a closed multimap. If $A \subset X$ is a compact set then its image $F(A)$ is a closed subset of Y.*

Proof. The case $F(A) = Y$ is trivial. Let $y \in Y \setminus F(A)$. For any $x \in A$ let $V(x)$ and $W_x(y)$ be neighborhoods of x and y such that $F(V(x)) \cap W_x(y) = \emptyset$. If $V(x_1), V(x_2), \ldots, V(x_n)$ are neighborhoods forming a finite cover of A then $W(y) = \bigcap_{i=1}^{n} W_{x_i}(y)$ is a neighborhood of y such that $W(y) \cap F(A) = \emptyset$. □

Theorem 1.1.7. *Let $F : X \to K(Y)$ be an u.s.c. multimap. If $A \subset X$ is a compact set then its image $F(A)$ is a compact subset of Y.*

Proof. Let $\{W_j\}_{j \in J}$ be an open cover of the set $F(A)$. At each point $x \in A$ the value $F(x)$ can be covered by a finite collection of sets $W_{j_1}, W_{j_2}, \ldots, W_{j_{n(x)}}$. We denote $W_x = \bigcup_{i=1}^{n(x)} W_{j_i}$. The sets $F_+^{-1}(W_x)$, $x \in A$, form an open cover of A. If we select a finite subcover $F_+^{-1}(W_{x_1}), F_+^{-1}(W_{x_2}), \ldots, F_+^{-1}(W_{x_m})$, then the sets $W_{x_1}, W_{x_2}, \ldots, W_{x_m}$ form an open cover of the set $F(A)$. □

Remark 1.1.2. Upper semicontinuity is essential in this theorem. In fact, for the closed multimap F in Example 1.1.5 we have $F([-1, 1]) = \mathbb{R}$.

We mention also the following property of u.s.c. multimaps which can be easily verified.

Proposition 1.1.1. *Let $F : X \to P(Y)$ be an u.s.c. multimap. If $A \subset X$ is a connected set and $F(x)$ is connected for every $x \in A$ then the image $F(A)$ is a connected subset of Y.*

When a multimap acts into a metric space (Y, d) one can obtain convenient characterizations for the types of continuity described above. For $A \subset Y$, denote by $W_\varepsilon(A)$ the *ε-neighborhood* of the set A, $W_\varepsilon(A) = \{y \in Y : d(y, A) < \varepsilon\}$, where $d(y, A) \underset{\text{def}}{=} \inf_{x \in A} d(x, y)$ is the distance of y from A.

Theorem 1.1.8. *Let X be a topological space. (Y, d) a metric space.*

(i) *A multimap $F : X \to K(Y)$ is u.s.c. at a point $x \in X$ if and only if for every $\varepsilon > 0$ there exists a neighborhood $V(x)$ such that $F(x') \subset W_\varepsilon(F(x))$ for every $x' \in V(x)$;*

(ii) *A multimap $F : X \to K(Y)$ is l.s.c. at a point $x \in X$ if and only if for every $\varepsilon > 0$ there exists a neighborhood $V(x)$ such that $F(x) \subset W_\varepsilon(F(x'))$ for every $x' \in V(x)$.*

Proof. (i) (a) Note that $W_\varepsilon(F(x))$ is an open neighborhood of the set $F(x)$ and apply Definition 1.1.1.

(i) (b) Let W be an open neighborhood of the set $F(x)$. Since $F(x)$ is a compact set there exists $\varepsilon > 0$ such that $W_\varepsilon(F(x)) \subset W$, hence there exists a neighborhood $V(x)$ such that $F(V(x)) \subset W_\varepsilon(F(x)) \subset W$.

(ii) (a) Take $\varepsilon > 0$ and let $y_1, y_2, \dots, y_n$ be points of the set $F(x)$ such that the collection of balls $\{B_{\varepsilon/2}(y_i)\}$, $i = 1, \dots, n$ forms an open cover of $F(x)$. Since F is l.s.c., for every i there exists an open neighborhood $V_i(x)$ of a point x such that from $x' \in V_i(x)$ it follows that $F(x') \cap B_{\varepsilon/2}(y_i) \neq \emptyset$. Then $x' \in V(x) = \bigcap_{i=1}^n V_i(x)$ implies $W_\varepsilon(F(x')) \supset B_{\varepsilon/2}(y_i)$ for all $i = 1, \dots, n$, and hence the neighborhood $V(x)$ is the desired one.

(ii) (b) Let W be an open set such that $F(x) \cap W \neq \emptyset$. Let $y \in F(x) \cap W$ an arbitrary point and $\varepsilon > 0$ be such that $B_\varepsilon(y) \subset W$. If $V(x)$ is an open neighborhood of x such that $x' \in V(x)$ implies $F(x) \subset W_\varepsilon(F(x'))$ then $F(x') \cap B_\varepsilon(y) \neq \emptyset$ for $x' \in V(x)$ proving that F is l.s.c. at x. □

Let $\mathrm{Cb}(Y)$ denote the collection of all nonempty closed bounded subsets of Y. The function $h : \mathrm{Cb}(Y) \times \mathrm{Cb}(Y) \to \mathbb{R}_+$ defined by

$$h(A, B) = \inf\{\varepsilon : A \subset W_\varepsilon(B),\ B \subset W_\varepsilon(A)\}$$

is a metric on the set $\mathrm{Cb}(Y)$. It is called the *Hausdorff metric*.

Theorem 1.1.9. *A multimap $F : X \to K(Y)$ is continuous if and only if it is continuous as a single-valued map from X to* $(\mathrm{Cb}(Y), h)$.

The proof follows immediately from Theorem 1.1.8.

Example 1.1.6. If E is a Banach space and $\Omega \subset C([a, b]; E)$ is a nonempty compact subset then the multimap $Q : [a, b] \to K(E)$, $Q(t) = \Omega(t) = \{y(t) : y \in \Omega\}$ is continuous. In fact, using the equicontinuity of the set Ω we can apply Theorem 1.1.8 (i) to verify that Q is u.s.c. From Example 1.1.2 (b) we know that Q is l.s.c.

If (Y, d) is a separable metric space we can obtain a useful characterization of lower and upper semicontinuity of multimaps.

Let $\{r_n\}_{n=1}^{\infty}$ be a countable dense subset of Y. Given a multimap $F : X \to P(Y)$, define the functions $\{\varphi_n\}_{n=1}^{\infty}$, $\varphi_n : X \to \mathbb{R}_+$ as $\varphi_n(x) = d(r_n, F(x))$.

Theorem 1.1.10. *A multimap $F : X \to P(Y)$ is l.s.c. if and only if all functions φ_n are upper semicontinuous (in the single-valued sense).*

Proof. For any $a > 0$ and $n \in \mathbb{N}$, the sets $\{x \in X : \varphi_n(x) < a\}$ coincide with the sets $F_-^{-1}(B_a(r_n))$. It remains to note that the balls with centers at the points r_n form a base for the topology of the space Y and to apply Theorem 1.1.2 (d). □

Before we can prove an analogous result for the u.s.c. case we need the following.

Lemma 1.1.1. *Let $F : X \to P(Y)$ be a multimap, $Q \subset Y$ a closed set. Let a sequence of sets $\{Q_m\}_{m=1}^{\infty}$ satisfy the relations $Q \subset Q_m \subset V_{\varepsilon_m}(Q)$ for any sequence $\{\varepsilon_m\}_{m=1}^{\infty}$, $\varepsilon_m > 0$, $\varepsilon_m \to 0$. Then*

(a) $F_+^{-1}(Q) = \bigcap_{m=1}^{\infty} F_+^{-1}(Q_m)$;

(b) $F_-^{-1}(Q) = \bigcap_{m=1}^{\infty} F_m^{-1}(Q_m)$ *if F has compact values.*

Proof. The inclusions $F_+^{-1}(Q) \subset \bigcap_{m=1}^{\infty} F_+^{-1}(Q_m)$, $F_-^{-1}(Q) \subset \bigcap_{m=1}^{\infty} F_-^{-1}(Q_m)$ are obvious.

(a) if $x \notin F_+^{-1}(Q)$ then there exists $y \in F(x)$ such that $y \notin Q$. But then $\varepsilon_m < d(y, Q)$ implies $x \notin F_+^{-1}(Q_m)$.

(b) if $x \notin F_-^{-1}(Q)$ then $F(x) \cap Q = \emptyset$. Since the set $F(x)$ is compact, there exists $\varepsilon > 0$ such that $W_\varepsilon(F(x)) \cap Q = \emptyset$. Then if $\varepsilon_m < \varepsilon$ we have $x \notin F_-^{-1}(Q_m)$. □

Theorem 1.1.11. (i) *If a multimap $F : X \to K(Y)$ is u.s.c. then the functions φ_n, defined above, are lower semicontinuous (in the single valued sense).*

(ii) *If F is compact then the lower semicontinuity of all the functions φ_n implies the upper semicontinuity of F.*

Proof. (i) If a multimap F is u.s.c. then for any $a > 0$ and $n \in \mathbb{N}$ the sets $\{x \in X : \varphi_n(x) > a\} = F_+^{-1}(X \backslash \overline{B}_a(r_n))$ are open.

(ii) Since F is compact it is sufficient to show that the set $F_-^{-1}(K)$ is closed for every compact set $K \subset Y$. Given a sequence $\{\varepsilon_m\}_{m=1}^{\infty}$, $\varepsilon_m > 0$, $\varepsilon_m \to 0$, consider finite covers of K by closed balls of radius ε_m centered in points taken from the set $\{r_n\}_{n=1}^{\infty}$:

$$K \subset K_m \underset{\text{def}}{=} \bigcup_{i=1}^{q(m)} \overline{B}_{\varepsilon_m}(r_{n_{m,i}})$$

For every m the set

$$F_-^{-1}(K_m) = \bigcup_{i=1}^{q(m)} F_-^{-1}(\overline{B}_{\varepsilon_m}(r_{n_{m,i}})) = \bigcup_{i=1}^{q(m)} \{x \in X : \varphi_{n_{m,i}}(x) \le \varepsilon_m\}$$

is closed. Applying Lemma 1.1.1 (b) we obtain

$$F_-^{-1}(K) = \bigcap_{m=1}^{\infty} F_-^{-1}(K_m)$$

proving that the set $F_-^{-1}(K)$ is closed. □

When X is a metric space we obtain the following refinement of the upper semicontinuity condition given in Theorem 1.1.5.

Theorem 1.1.12. *Let X and Y be metric spaces and $F : X \to K(Y)$ a closed quasicompact multimap. Then F is u.s.c.*

Proof. Let $x \in X$ be a point, W be an open neighborhood of the set $F(x)$. If F is not u.s.c. at x then there exists a sequence $\{x_n\} \subset X$, $x_n \to x$ such that we can take $y_n \in F(x_n) \backslash W$ for each $n \in \mathbb{N}$. The quasicompactness condition implies that we can assume, without loss of generality that $y_n \to y \notin W$, contradicting $y \in F(x)$. □

1.2 Operations on multimaps. Selections and approximations

1.2.1 Union and intersection of multimaps

The variety of operations that can be defined on multimaps is intrinsically richer than for single-valued maps. Operations as union, intersection and others have no "single-valued" analogs. In this section we investigate continuity properties of operations on multimaps.

Let X and Y be topological spaces, $\{F_j\}_{j \in J}$, $F_j : X \to P(Y)$ a family of multimaps.

Definition 1.2.1. The multimap $\bigcup_{j\in J} F_j : X \to P(Y)$,

$$\Big(\bigcup_{j\in J} F_j\Big)(x) = \bigcup_{j\in J} F_j(x)$$

is called the *union of the multimaps* F_j.

Definition 1.2.2. Let $\bigcap_{j\in J} F_j(x) \neq \emptyset$ for every $x \in X$. The multimap $\bigcap_{j\in J} F_j : X \to P(Y)$,

$$\Big(\bigcap_{j\in J} F_j\Big)(x) = \bigcap_{j\in J} F_j(x)$$

is called the *intersection of multimaps* F_j.

Theorem 1.2.1.

(i) *Let F_j be u.s.c. multimaps. If the set of indices J is finite then the union $\bigcup_{j\in J} F_j : X \to P(Y)$ is u.s.c.;*

(ii) *Let F_j be l.s.c. multimaps. Then the union $\bigcup_{j\in J} F_j : X \to P(Y)$ is l.s.c.;*

(iii) *Let F_j be closed multimaps. If the set of indices J is finite, the union $\bigcup_{j\in J} F_j : X \to P(Y)$ is closed.*

Proof. Let $W \subset Y$ be an open set. Statements (i) and (ii) follow from the easily verified relations

$$\Big(\bigcup_{j\in J} F_j\Big)_+^{-1}(W) = \bigcap_{j\in J} (F_j)_+^{-1}(W)$$
$$\Big(\bigcup_{j\in J} F_j\Big)_-^{-1}(W) = \bigcup_{j\in J} (F_j)_-^{-1}(W)$$

and from Theorems 1.1.1 part (b) and 1.1.2 part (b).

To prove (iii) it is sufficient to note that the following relation holds for the graphs of multimaps,

$$\Gamma_{\bigcup_{j\in J} F_j} = \bigcup_{j\in J} \Gamma_{F_j} . \qquad \square$$

Let $\{F_j\}_{j\in J}$, $F_j : X \to C(Y)$ be a collection of multimaps such that $\bigcap_{j\in J} F_j(x) \neq \emptyset$ for every $x \in X$. We have the following result.

Theorem 1.2.2.

(i) *Let F_j be u.s.c. multimaps. If the set of indices J is finite and the space Y is normal then the intersection $\bigcap_{j\in J} F_j : X \to C(Y)$ is u.s.c.;*

(ii) *Let F_j be closed multimaps. Then the intersection $\bigcap_{j\in J} F_j : X \to C(Y)$ is closed.*

Proof. (i) For $x \in X$, let W be an open neighborhood of the set $F(x) = \bigcap_{j\in J} F_j(x)$. The closed subsets $F_j(x)\backslash W,\ j \in J$ do not intersect, then the normality of Y implies that there exist disjoint open sets $U_j,\ j \in J$ such that $F_j(x)\backslash W \subset U_j$. Then for every $j \in J$ we have $F_j(x) \subset W \cup (F_j(x)\backslash W) \subset W \cup U_j$. From the upper semicontinuity of multimaps F_j it follows that for every $j \in J$ there exists a neighborhood $V_j(x)$ of x such that $F_j(V_j(x)) \subset W \cup U_j$. For the neighborhood $V(x) = \bigcap_{j\in J} V_j(x)$ we have

$$F(V(x)) = \bigcap_{j\in J} F_j(V(x)) \subset \bigcap_{j\in J} F_j(V_j(x)) \subset \bigcap_{j\in J}(W \cup U_j) = W$$

proving the upper semicontinuity of F at x.

(ii) The validity of the statement follows directly from the fact that the graph of the multimap $\bigcap_{j\in J} F_j$ is the intersection of the graphs of the multimaps F_j. □

The following statement also holds.

Theorem 1.2.3. *Let $F_0 : X \to C(Y)$ be a closed multimap and $F_1 : X \to K(Y)$ an u.s.c. multimap. Suppose that $F_0(x) \cap F_1(x) \neq \emptyset$ for every $x \in X$. Then the intersection $F_0 \cap F_1 : X \to K(Y)$ is u.s.c.*

Proof. Given $x \in X$ let $W \subset Y$ be any open neighborhood of the set $(F_0 \cap F_1)(x)$. We will show that there exists an open neighborhood $V(x)$ of x such that $(F_0 \cap F_1)(V(x)) \subset W$.

When $F_1(x) \subset W$ the existence of such a neighborhood follows from the upper semicontinuity of F_1. If $K = F_1(x)\backslash W \neq \emptyset$, then the set K is compact and $K \cap F_0(x) = \emptyset$. As F_0 is a closed multimap, for any point $y \in K$ there exist neighborhoods $W(y) \subset Y$ of y and $V_y(x) \subset X$ of x such that $F_0(V_y(x)) \cap W(y) = \emptyset$ (see Theorem 1.1.3 (b)).

Let $\{W(y_i)\}_{i=1}^n$ be a finite cover of K taken from the neighborhoods $W(y)$, and $W(K) = \bigcup_{i=1}^n W(y_i)$. The open set $W \cup W(K)$ contains $F_1(x)$, hence there exists a neighborhood $V_1(x)$ of x such that $F_1(V_1(x)) \subset W \cup W(K)$. The neighborhood $V(x) = V_1(x) \cap V_{y_1}(x) \cap \ldots V_{y_n}(x)$ is the required one. In fact $F_0(V(x)) \cap W(K) = \emptyset$ and $F_1(V(x)) \subset W \cup W(K)$, therefore $(F_0 \cap F_1)(V(x)) \subset W$. □

Corollary 1.2.1. *Let $F : X \to K(Y)$ be an u.s.c. multimap, $C \subset Y$ a closed set and $F(x) \cap C \neq \emptyset$ for all $x \in X$. Then the multimap $\widetilde{F} : X \to K(Y)$, $\widetilde{F}(x) = F(x) \cap C$ is u.s.c.*

Proof. It is clear that the multimap $F_0 : X \to C(Y)$, $F_0(x) \equiv C$ is closed. Take $F_1 = F$ and apply the previous theorem. □

Corollary 1.2.2. *Let Y be a Hausdorff topological space, $\{F_j\}_{j\in J}$, $F_j : X \to K(Y)$ a collection of u.s.c. multimaps such that $\bigcap_{j\in J} F_j(x) \neq \emptyset$ for every $x \in X$. Then the intersection $\bigcap_{j\in J} F_j : X \to K(Y)$ is u.s.c..*

Proof. Let F_{j_0} be one of the multimaps in the family. Since all the multimaps F_j are closed (see Remark 1.1.1) the multimap $F_1 = \bigcap_{j\neq j_0, j\in J} F_j$ is also closed (see Theorem 1.2.2 (ii)). Take $F_0 = F_{j_0}$ and apply the theorem. □

The continuity properties of the intersection of l.s.c. multimaps are more complicated. The following example proves that in general such an intersection is not l.s.c.

Example 1.2.1. Consider the multimaps F_0, $F_1 : [0,\pi] \to Kv(\mathbb{R}^2)$ defined in the following way. The multimap F_0 is constant:

$$F_0(\varphi) = \{(y_0, y_1) : y_0^2 + y_1^2 \leq 1,\ y_1 \geq 0\},$$

the multimap F_1 is defined as

$$F_1(\varphi) = \{(y_0, y_1) : y_0 = \lambda \cos\varphi,\ y_1 = \lambda \sin\varphi,\ \lambda \in [-1, 1]\}.$$

The multimaps F_0, F_1 are l.s.c. (they are even continuous) but their intersection loses the lower semicontinuity property at the points 0 and π.

To clarify the conditions under which we can guarantee the lower semicontinuity of the intersection of multimaps the following notion is useful.

Definition 1.2.3. A multimap $F : X \to P(Y)$ is said to be *quasi-open* at a point $x \in X$ if $\operatorname{int} F(x) \neq \emptyset$ and for every $y \in \operatorname{int} F(x)$ there exist neighborhoods $W(y) \subset Y$ and $V(x) \subset X$ such that $W(y) \subset F(x')$ for all $x' \in V(x)$. A multimap F is said to be quasi-open provided it is quasi-open at every point $x \in X$.

It is easy to see that a multimap $F : X \to P(Y)$ such that $\operatorname{int} F(x) \neq \emptyset$ for every $x \in X$ is quasi-open if and only if the multimap $\operatorname{int} F : X \to P(Y)$ defined as $(\operatorname{int} F)(x) = \operatorname{int} F(x)$ has open graph $\Gamma_{\operatorname{int} F} \subset X \times Y$.

We have the following important characterization of a quasi-open multimap.

Theorem 1.2.4. *Let X be a topological space and Y a normed space. A multimap $F : X \to Cv(Y)$ is quasi-open at a point $x \in X$ if and only if $\operatorname{int} F(x) \neq \emptyset$ and F is l.s.c. at x.*

Proof. Let F be quasi-open at x, then $\operatorname{int} F(x) \neq \emptyset$. Let $W \subset Y$ be an open set such that $W \cap F(x) \neq \emptyset$. It easy to see that $W \cap \operatorname{int} F(x) \neq \emptyset$.

For an arbitrary $y \in W \cap \operatorname{int} F(x)$ let $W(y) \subset Y$ and $V(x) \subset X$ be neighborhoods such that $W(y) \subset F(x')$ for all $x' \in V(x)$. But $W \cap W(y) \neq \emptyset$ implies that $W \cap F(x') \neq \emptyset$ for all $x' \in V(x)$, giving the lower semicontinuity of F at x.

Conversely, let $\operatorname{int} F(x) \neq \emptyset$ and F be l.s.c. at $x \in X$. Let $y \in \operatorname{int} F(x)$ and $B_\delta(y) \subset F(x)$ for some $\delta > 0$. Take $0 < \delta_1 < \delta$. From Theorem 1.1.8 (ii) it follows that there exists a neighborhood $V(x)$ of x such that $F(x) \subset W_\eta(F(x'))$ for all $x' \in V(x)$ where $\eta = \delta - \delta_1$.

Let now $y' \in B_{\delta_1}(y)$ but $y' \notin F(x')$ for some $x' \in V(x)$. Then it is easy to see that the ball $B_\eta(y')$ will contain points whose distance from $F(x')$ is greater than η, contradicting the fact that $B_\eta(y') \subset B_\delta(y) \subset F(x) \subset W_\eta(F(x'))$. Therefore, $B_{\delta_1}(y) \subset F(x')$ for all $x' \in V(x)$. □

We now formulate a condition that guarantees the lower semicontinuity of the intersections of multimaps.

Theorem 1.2.5. *Let X and Y be topological spaces. Let $F_0 : X \to P(Y)$ be a multimap l.s.c. at $x_0 \in X$ and let $F_1 : X \to P(Y)$ be a multimap quasi-open at x_0, $F_0(x) \cap F_1(x) \neq \emptyset$ for every $x \in X$ and $F_0(x_0) \cap F_1(x_0) \subset \overline{F_0(x_0) \cap \operatorname{int} F_1(x_0)}$. Then the intersection $F_0 \cap F_1$ is l.s.c. at x_0.*

Proof. Let $W \subset Y$ be an open set such that $W \cap (F_0 \cap F_1)(x_0) \neq \emptyset$. From the assumptions it follows that there exists a point $y \in (F_0 \cap F_1)(x_0)$ which is an interior point of the set $F_1(x_0)$. Let $W(y)$ be a neighborhood of y such that $W(y) \subset W \cap F_1(x_0)$. From the quasi-openness of the multimap F_1 it follows that there exists a neighborhood $V_1(x_0)$ of the point x_0 such that $W(y) \subset F_1(x')$ for all $x' \in V_1(x_0)$.

Since $y \in F_0(x_0)$ and the multimap F_0 is l.s.c. there exists a neighborhood $V_0(x_0)$ of a point x_0 such that $F_0(x'') \cap W(y) \neq \emptyset$ for all $x'' \in V_0(x_0)$. Then for every $\widetilde{x} \in V(x_0) = V_0(x_0) \cap V_1(x_0)$ we have $(F_0 \cap F_1)(\widetilde{x}) \cap W(y) \neq \emptyset$, that is the lower semicontinuity of the multimap $F_0 \cap F_1$ at x_0. □

As a corollary, we obtain a sufficient condition for the lower semicontinuity of the intersection of l.s.c. multimaps.

Corollary 1.2.3. *Let X be a topological space, Y a normed space and $F_0, F_1 : X \to Cv(Y)$ l.s.c. multimaps. Assume that $F_0(x) \cap F_1(x) \neq \emptyset$ for all $x \in X$ and $F_0(x_0) \cap \operatorname{int} F_1(x_0) \neq \emptyset$ for some $x_0 \in X$. Then the intersection $F_0 \cap F_1 : X \to Cv(Y)$ is l.s.c. at x_0.*

Proof. From Theorem 1.2.4 it follows that the multimap F_1 is quasi-open at x_0. Let $y \in (F_0 \cap F_1)(x_0)$ be an arbitrary point and $\widetilde{y} \in F_0(x_0) \cap \operatorname{int} F_1(x_0)$.

It is clear that $[y, \widetilde{y}] \subset (F_0 \cap F_1)(x_0)$ and $(y, \widetilde{y}] \subset \operatorname{int} F_1(x_0)$. Hence the multimaps F_0, F_1 satisfy the assumptions of Theorem 1.2.5. □

We use the above results to give sufficient conditions for lower semicontinuity of some special types of multimaps appearing in control theory.

Theorem 1.2.6. *Let X be a topological space, Y and Z normed spaces. Assume that:*

(i) *$f : X \times Z \to Y$ is a map such that $f(x, \cdot) : Z \to Y$ is continuous and affine for every $x \in X$;*

(ii) *$T : X \to Cv(Y)$ is an arbitrary multimap and $U : X \to Cv(Z)$ is a l.s.c. multimap;*

(iii) *for every $x \in X$ there exist $u \in U(x)$ and neighborhoods $V(x) \subset X$ and $N(u) \subset Z$ such that $f(x', N(u)) \subset T(x')$ for all $x' \in V(x)$.*

Then the multimap $R : X \to Cv(Z)$, $R(x) = \{u \in U(x) : f(x, u) \in T(x)\}$ is l.s.c.

Proof. Consider the multimap $Q : X \to P(Z)$,

$$Q(x) = \{u \in Z : \exists \text{ neighbourhoods } V(x) \subset X,\ N(u) \subset Z \\ \text{such that } f(x', N(u)) \subset T(x')\ \forall x' \in V(x)\}.$$

From conditions (i) and (ii) it follows that the multimap Q has nonempty open convex values. It is clear also that Q is quasi-open.

If we consider the multimap $S : X \to Cv(Z)$, $S(x) = \{u \in Z : f(x, u) \in T(x)\}$ we have $S(x) = \overline{Q(x)}$ for all $x \in X$. In fact it is clear that $\overline{Q(x)} \subset S(x)$. If $u \in S(x)$ then for arbitrary $u' \in Q(x)$, $[u', u) \subset Q(x)$. Indeed, if $N(u')$ is a neighborhood of u' such that $f(x', N(u')) \subset T(x')$ for all $x' \in V(x)$, where $V(x)$ is a neighborhood of x and $\widetilde{u} = \lambda u' + (1 - \lambda)u$, $0 < \lambda \leq 1$ then $f(x', \widetilde{N}(\widetilde{u})) \subset T(x')$ where $\widetilde{N}(\widetilde{u}) = \lambda N(u') + (1 - \lambda)u$ for all $x' \in V(x)$. Hence $S(x) \subset \overline{Q(x)}$ and we obtain the desired equality. Therefore the multimap S is quasi-open and from condition (iii) it follows that $U(x) \cap \operatorname{int} S(x) \neq \emptyset$ for all $x \in X$. It remains to note that $R = U \cap S$ and apply Corollary 1.2.3. □

Corollary 1.2.4 (Cf. [16], Proposition 6.3.2). *Let X be a metric space, Y and Z normed spaces. Assume that:*

(i) *$f : X \times Z \to Y$ is a continuous map such that $f(x, \cdot) : Z \to Y$ is affine for every $x \in X$;*

(ii) *$T : X \to Cv(Y)$ and $U : X \to Cv(Z)$ are l.s.c. multimaps;*

(iii) *for every $x \in X$ there exists $u \in U(x)$ such that $f(x, u) \in \operatorname{int} T(x)$.*

Then the multimap $R : X \to Cv(Z)$ defined by $R(x) = \{u \in U(x) : f(x, u) \in T(x)\}$ is l.s.c.

Proof. It is sufficient to show that condition (iii) of Theorem 1.2.6 holds. Supposing the contrary, we can find points $x \in X$ and $u \in U(x)$ such that $f(x, u) \in \operatorname{int} T(x)$, and sequences $x_n \to x$, $u_n \to u$ with $f(x_n, u_n) \notin T(x_n)$.

From the continuity of f it follows that $(x_n, f(x_n, u_n)) \to (x, f(x, u)) \in \Gamma_{\operatorname{int} T}$. By Theorem 1.2.4, the multimap T is quasi-open and hence the graph $\Gamma_{\operatorname{int} T}$ is an open set. Therefore $f(x_n, u_n) \in \operatorname{int} T(x_n)$ for n large enough, a contradiction. □

For the sake of completeness we conclude with the following statement ([16], Proposition 6.3.1).

Theorem 1.2.7. *Let X be a metric space, Y and Z normed spaces. Assume that:*

(i) *$f : X \times Z \to Y$ is a continuous map such that $f(x, \cdot) : Z \to Y$ is affine for every $x \in X$;*

(ii) *$T : X \to Cv(Y)$ and $U : X \to Cv(Z)$ are l.s.c. multimaps;*

(iii) *for every $x \in X$ there exist $\gamma > 0, \delta > 0$ and $r > 0$ such that $\gamma B_Y \subset f(x', U(x') \cap rB_Z) - T(x')$ for all $x' \in B_\delta(x)$ where B_Y, B_Z denote the unit balls in the spaces Y and Z respectively.*

Then the multimap $R : X \to Cv(Z)$, $R(x) = \{u \in U(x) : f(x, u) \in T(x)\}$ is l.s.c.

Proof. Consider a point $u \in R(x)$ and a sequence $x_n \to x$. From condition (ii) and the lower semicontinuity criterion from Theorem 1.1.2 (g), it follows that there exist sequences $u_n \in U(x_n)$, $y_n \in T(x_n)$ such that $u_n \to u$ and $y_n \to f(x, u)$.

From the continuity of the map f it follows that $\varepsilon_n \underset{\text{def}}{=} \| y_n - f(x_n, u_n) \| \to 0$. If we set $\theta_n = \gamma/(\gamma + \varepsilon_n)$ then

$$\begin{aligned} \theta_n(y_n - f(x_n, u_n)) &\in \theta_n \varepsilon_n B_Y = (1 - \theta_n)\gamma B_Y \\ &\subset (1 - \theta_n)(f(x_n, U(x_n) \cap rB_Z) - T(x_n)) \end{aligned}$$

for all n large enough. Therefore there exist $\widetilde{u}_n \in U(x_n) \cap rB_z$ and $\widetilde{y}_n \in T(x_n)$ such that $\theta_n(y_n - f(x_n, u_n)) = (1 - \theta_n)(f(x_n, \widetilde{u}_n) - \widetilde{y}_n)$ and hence from (i) it follows that

$$f(x_n, \theta_n u_n + (1 - \theta_n)\widetilde{u}_n) = \theta_n y_n + (1 - \theta_n)\widetilde{y}_n \in T(x_n)$$

and therefore $v_n = \theta_n u_n + (1 - \theta_n)\widetilde{u}_n$ belongs to $R(x_n)$.

It remains to note that $\| u_n - v_n \| = (1 - \theta_n)\| u_n - \widetilde{u}_n \| \to 0$ since $\theta_n \to 1$, $\| \widetilde{u}_n \| \leq r$ and the sequence $\{u_n\}$ is bounded. Hence $v_n \to u$ and we can again apply Theorem 1.1.2(g) concluding the proof. □

1.2.2 Composition and Cartesian product of multimaps

We consider now some continuity properties for other set-theoretic operations over multimaps.

Definition 1.2.4. Let X, Y and Z be topological spaces , and $F_0 : X \to P(Y)$, $F_1 : Y \to P(Z)$ multimaps.

The multimap $F_1 \circ F_0 : X \to P(Z)$, defined as $(F_1 \circ F_0)(x) = F_1(F_0(x))$ is said to be the *composition* of the multimaps F_0 and F_1.

Theorem 1.2.8. *If the multimaps $F_0 : X \to P(Y)$, $F_1 : Y \to P(Z)$ are u.s.c. (l.s.c.) then their composition $F_1 \circ F_0 : X \to P(Z)$ is u.s.c. (respectively l.s.c.).*

Proof. Let $W \subset Z$ be an open set. It is easy to verify that

$$(F_1 \circ F_0)_+^{-1}(W) = (F_0)_+^{-1}((F_1)_+^{-1}(W))$$
$$(F_1 \circ F_0)_-^{-1}(W) = (F_0)_-^{-1}((F_1)_-^{-1}(W))$$

and the statement follows from Theorems 1.1.1 (b) and 1.1.2 (b). □

Theorem 1.2.9. *Let $F_0 : X \to K(Y)$ be an u.s.c. multimap and $F_1 : Y \to C(Z)$ a closed multimap. Then the composition $F_1 \circ F_0 : X \to C(Z)$ is a closed multimap.*

Proof. Let $z \in Z$ be such that $z \notin F_1 \circ F_0(x)$, $x \in X$. From Theorem 1.1.3 (b) it follows that for any point $y \in F_0(x)$ there exist neighborhoods $W_y(z)$ of z and $V(y)$ of y such that $F_1(V(y)) \cap W_y(z) = \emptyset$. Let $\{V(y_i)\}_{i=1}^n$ be a finite cover of the set $F_0(x)$. If now $U(x)$ is a neighborhood of a point x such that $F_0(U(x)) \subset \bigcup_{i=1}^n V(y_i)$ then $F_1 \circ F_0(U(x)) \cap \left(\bigcap_{i=1}^n W_{y_i}(z)\right) = \emptyset$ and the application of Theorem 1.1.3 (b) concludes the proof. □

Remark 1.2.1. The condition of upper semicontinuity for the map F_0 is essential. The following example shows that the composition of closed multimaps is not necessarily closed.

Example 1.2.2. The multimaps $F_0 : \mathbb{R} \to K(\mathbb{R})$ and $F_1 : \mathbb{R} \to K(\mathbb{R})$ defined by

$$F_0(x) = \begin{cases} \{\frac{1}{x}\}, & x \neq 0, \\ \{0\}, & x = 0, \end{cases}$$

$$F_1(x) = \begin{cases} \{\frac{1}{x}\}, & x \neq 0, \\ \{1\}, & x = 0, \end{cases}$$

are closed but not u.s.c. Their composition $F_1 \circ F_0 : \mathbb{R} \to K(\mathbb{R})$ given by

$$F_1 \circ F_0(x) = \begin{cases} \{x\}, & x \neq 0, \\ \{1\}, & x = 0, \end{cases}$$

is not closed.

We consider now the Cartesian product of multimaps.

Definition 1.2.5. Let X, Y and Z be topological spaces, $F_0 : X \to P(Y)$, $F_1 : X \to P(Z)$ multimaps.

The multimap $F_0 \times F_1 : X \to P(Y \times Z)$, defined as $(F_0 \times F_1)(x) = F_0(x) \times F_1(x)$ is said to be the *Cartesian product* of the multimaps F_0 and F_1.

Theorem 1.2.10. *If the multimaps $F_0 : X \to P(Y)$, $F_1 : X \to P(Z)$ are l.s.c., then their Cartesian product $F_0 \times F_1 : X \to P(Y \times Z)$ is l.s.c.*

Proof. Note that the sets $W_0 \times W_1$, where $W_0 \subset Y$ and $W_1 \subset Z$ are open sets, form a base for the topology of the space $X \times Z$. The statement follows from the relation

$$(F_0 \times F_1)_-^{-1}(W_0 \times W_1) = (F_0)_-^{-1}(W_0) \cap (F_1)_-^{-1}(W_1)$$

and Theorem 1.1.2 (d). □

Theorem 1.2.11. *If the multimaps $F_0 : X \to C(Y)$, $F_1 : X \to C(Z)$ are closed then their Cartesian product $F_0 \times F_1 : X \to C(Y \times Z)$ is closed.*

Proof. Let us consider generalized sequences $\{x_\alpha\} \subset X$, $\{v_\alpha\} \subset Y \times Z$ with $v_\alpha \in (F_0 \times F_1)(x_\alpha)$ such that $x_\alpha \to x$, $v_\alpha \to v$. Then $v_\alpha = y_\alpha \times z_\alpha$, where $y_\alpha \in F_0(x_\alpha)$, $z_\alpha \in F_1(x_\alpha)$. From the definition of the product topology, the convergence of $v_\alpha \to v = (y, z)$ implies the convergence of $y_\alpha \to y$ and $z_\alpha \to z$. From Theorem 1.1.3 (c) it follows that $y \in F_0(x)$, $z \in F_1(x)$, hence $v \in (F_0 \times F_1)(x)$ concluding the proof.□

To consider the upper semicontinuity of the Cartesian product of multimaps we need the compactness of their values.

Theorem 1.2.12. *If the multimaps $F_0 : X \to K(Y)$, $F_1 : X \to K(Z)$ are u.s.c. then their Cartesian product $F_0 \times F_1 : X \to K(Y \times Z)$ is u.s.c.*

Proof. The Tychonoff theorem implies that the multimap $F_0 \times F_1$ has compact values. For an arbitrary point $x \in X$, let $G \supset (F_0 \times F_1)(x)$ be an open subset of $X \times Y$. From the definition of product topology in $X \times Y$ it follows that for every $(y, z) \in (F_0 \times F_1)(x)$ there exist open sets $G_0(y, z) \subset Y$, $G_1(y, z) \subset Z$ such that $(y, z) \in G_0(y, z) \times G_1(y, z) \subset G$. For every $y \in F_0(x)$ consider the cover $\sum_y = \{G_1(y, \tilde{z}) : \tilde{z} \in F_1(x)\}$ of the set $F_1(x)$. Since the set $F_1(x)$ is compact we can extract a finite subcover $\sum_y' = \{G_1(y, \tilde{z}_1), G_1(y, \tilde{z}_2), \ldots, G_1(y, \tilde{z}_n)\}$ from $\sum_y$. The set $V(y) = \bigcap_{j=1}^n G_0(y, \tilde{z}_j)$ is a neighborhood of y in Y and the set $W_y = \bigcup_{j=i}^n G_1(y, \tilde{z}_j)$ is a neighborhood of $F_1(x)$ in Z and moreover $V(y) \times W_y \subset G$. The sets $V(y)$, $y \in F_0(x)$ form an open cover of the compact set $F_0(x)$. Choose a finite subcover $V(y_1), \ldots, V(y_m)$ and set $V = \bigcup_{i=1}^m V(y_i)$, $W = \bigcap_{i=1}^m W_{y_i}$. Then the set V is a neighborhood of $F_0(x)$ in Y and W is a neighborhood of $F_1(x)$ in Z and $V \times W \subset G$.

From the relation

$$(F_0 \times F_1)_+^{-1}(V \times W) = (F_0)_+^{-1}(V) \cap (F_1)_+^{-1}(W)$$

and the upper semicontinuity of the multimaps F_0 and F_1 it follows that there exists a neighborhood $U(x)$ of the point x such that $(F_0 \times F_1)(U(x)) \subset V \times W \subset G$ proving the upper semicontinuity of $F_0 \times F_1$. □

1.2.3 Algebraic operations on multimaps

We consider continuity properties of algebraic operations on multimaps.

Let X be a topological space, Y a topological vector space.

Definition 1.2.6. Let $F_0, F_1 : X \to P(Y)$ be multimaps. The multimap $F_0 + F_1 : X \to P(Y)$ defined by $(F_0 + F_1)(x) = F_0(x) + F_1(x)$ is said to be the *sum* of the multimaps F_0 and F_1.

Theorem 1.2.13. *If the multimaps $F_0, F_1 : X \to P(Y)$ are l.s.c. then the multimap $F_0 + F_1 : X \to P(Y)$ is l.s.c.*

Proof. The multimap $F_0 \times F_1 : X \to P(Y \times Y)$ is l.s.c. by Theorem 1.2.10. The single valued map $f : Y \times Y \to Y$, $f(u, v) = u + v$ is continuous. We have

$$F_0 + F_1 = f \circ (F_0 \times F_1)$$

and the conclusion follows from Theorem 1.2.8. □

Similar application of Theorems 1.2.12 and 1.2.8 gives the following result.

Theorem 1.2.14. *If the multimaps $F_0, F_1 : X \to K(Y)$ are u.s.c., then their sum $F_0 + F_1 : X \to K(Y)$ is u.s.c.*

Remark 1.2.2. Note that the assumption of compactness of the values of F_0 and F_1 is essential. In fact, consider the multimaps $F_0, F_1 : \mathbb{R}^2 \to P(\mathbb{R}^2)$ defined in the following way:

$$\begin{aligned} F_0((x_1, x_2)) &= \{(x_1, x_2)\} \\ F_1((x_1, x_2)) &= \{(y_1, y_2) : y_1 \cdot y_2 = 1,\ y_1 > 0\}. \end{aligned}$$

They are continuous, but their sum

$$(F_0 + F_1)((x_1, x_2)) = \{(z_1, z_2) : z_1 = x_1 + y_1,\ z_2 = x_2 + y_2,\ y_1 \cdot y_2 = 1,\ y_1 > 0\}$$

is not u.s.c. since $(F_0 + F_1)_+^{-1}(\mathbb{R}_+^2) = \overline{\mathbb{R}_+^2}$, where $\mathbb{R}_+^2 = \{(x_1, x_2) : x_1 > 0,\ x_2 > 0\}$ and $\overline{\mathbb{R}_+^2} = \{(x_1, x_2) : x_1 \geq 0,\ x_2 \geq 0\}$.

Definition 1.2.7. Let $F : X \to P(Y)$ be a multimap, $f : X \to \mathbb{R}$ a function. The multimap $f \cdot F : X \to P(Y)$, defined as $(f \cdot F)(x) = f(x) \cdot F(x)$ is said to be the *product* of f and F.

Theorem 1.2.15. *Let $F : X \to P(Y)$ be a l.s.c. multimap, $f : X \to \mathbb{R}$ a continuous function. Then the product multimap $f \cdot F : X \to P(Y)$ is l.s.c.*

Proof. The multimap $f \times F : X \to P(\mathbb{R} \times Y)$ is l.s.c. by Theorem 1.2.10. The map $\varphi : \mathbb{R} \times Y \to Y$ defined as $\varphi(r, y) = r \cdot y$ is continuous. Then the multimap $f \cdot F = \varphi \circ (f \times F)$ is l.s.c. by Theorem 1.2.8. □

The following statement can be proved with a similar application of Theorems 1.2.12 and 1.2.8.

Theorem 1.2.16. *If the multimap $F : X \to K(Y)$ is u.s.c. and the function $f : X \to \mathbb{R}$ is continuous, then their product $f \cdot F : X \to K(Y)$ is u.s.c.*

Definition 1.2.8. Let $F : X \to P(Y)$ be a multimap. The multimap $\overline{\text{co}}\, F : X \to Cv(Y)$, defined by $(\overline{\text{co}}\, F)(x) = \overline{\text{co}}(F(x))$, the closed convex hull of $F(x)$, is said to be the *convex closure* of the multimap F.

Theorem 1.2.17. *Let Y be a Banach space. If a multimap $F : X \to K(Y)$ is u.s.c. (l.s.c.) then its convex closure $\overline{\text{co}}\, F : X \to Kv(Y)$ is u.s.c. (l.s.c.).*

Proof. First, we note that the multimap $\overline{\text{co}}\, F$ has compact values by Mazur's Theorem. Let F be u.s.c.; consider a point $x \in X$ and let $\varepsilon > 0$. Then for any ε_1, $0 < \varepsilon_1 < \varepsilon$ there exists a neighborhood $V(x)$ of the point x such that $F(V(x)) \subset W_{\varepsilon_1}(F(x))$ (see Theorem 1.1.8 (i)). As $W_{\varepsilon_1}(F(x)) \subset W_{\varepsilon_1}(\overline{\text{co}}\, F(x))$ we have $\text{co}\, F(x') \subset W_{\varepsilon_1}(\overline{\text{co}}\, F(x))$ for every $x' \in V(x)$. Then $\overline{\text{co}}\, F(x') \subset W_{\varepsilon}(\overline{\text{co}}\, F(x))$ for every $x' \in V(x)$ proving the upper semicontinuity of the multimap $\overline{\text{co}}\, F$ by Theorem 1.1.8(i).

The lower semicontinuity of the multimap $\overline{\text{co}}\, F$ can be proved in a similar way applying Theorem 1.1.8(ii). □

The property of closedness of a multimap can be lost under the operation of convex closure, as the following example shows.

Example 1.2.3. The multimap $F : \mathbb{R} \to C(\mathbb{R})$, defined as

$$F(x) = \begin{cases} \{0\}, & x \le 0, \\ \{\pm 1/x\}, & x > 0, \end{cases}$$

is closed, but it is clear that its convex closure $\overline{\text{co}}\, F : \mathbb{R} \to Cv(\mathbb{R})$,

$$F(x) = \begin{cases} \{0\}, & x \le 0, \\ [-1/x, +1/x], & x > 0, \end{cases}$$

is not closed.

1.2.4 Selections and approximations

In this section we describe briefly some techniques for studying multimaps by means of single-valued selections and approximations.

The existence of selections and approximations is an enormous topic in the theory of multimaps.

We will restrict ourselves only to some basic facts.

Definition 1.2.9. Let X and Y be sets. A single valued map $f : X \to Y$ is said to be a *selection* of a multimap $F : X \to P(Y)$ if

$$f(x) \in F(x)$$

for all $x \in X$.

It is clear that only the problem of existence of selections with some regularity properties is of mathematical interest. The next result follows from the classical Michael continuous selection theorem (see [168]).

Theorem 1.2.18. *Let X be a metric space, Y a Banach space. Then every l.s.c. multimap $F : X \to Cv(Y)$ admits a continuous selection.*

The class of multimaps which admit a continuous selection is fairly narrow: as can be easily seen, it does not contain u.s.c. and closed multimaps. A simple example is the following "step" multimap,

$$F(x) = \begin{cases} \{0\}, & x < 0, \\ [0,1], & x = 0, \\ \{1\}, & x > 0. \end{cases}$$

In the upper semicontinuous case, however, it is possible to introduce the notion of single-valued approximation of a multimap.

Definition 1.2.10. Let (X, d) be a metric space, Y a normed space and $F : X \to P(Y)$ a multimap. A continuous map $f_\varepsilon : X \to Y$, $\varepsilon > 0$, is called an *ε-approximation* of F if:

(i) for every $x \in X$ there exists $x' \in X$ such that $d(x, x') < \varepsilon$ and

$$f_\varepsilon(x) \cup F(x) \subset W_\varepsilon(F(x'));$$

(ii) $f_\varepsilon(X) \subset \mathrm{co}\, F(X)$.

Note that condition (i) means that the graph Γ_{f_ε} is contained in the ε-neighborhood of the graph Γ_F.

Theorem 1.2.19. *Let (X, d) be a metric space, Y a normed space. Every u.s.c. multimap $F : X \to Cv(Y)$ has an ε-approximation $f_\varepsilon : X \to Y$, for any $\varepsilon > 0$.*

Proof. Fix $\varepsilon > 0$ and for each $x \in X$ let $\delta = \delta(x), 0 < \delta < \varepsilon$, be such that $F(B_{\delta(x)}(x)) \subset W_\varepsilon(F(x))$. For $\eta(x) = (1/4)\delta(x)$, consider the cover $\{B_{\eta(x)}(x)\}_{x \in X}$ of the space X and let $\{V_j\}_{j \in J}$ be a subordinated locally finite cover with $\{p_j\}_{j \in J}$ the corresponding partition of unity.

Now, choosing for each index $j \in J$, an arbitrary point $y_j \in F(V_j)$ define the map $f_\varepsilon : X \to Y$ by

$$f_\varepsilon(x) = \sum_{j \in J} p_j(x) y_j .$$

The map f_ε is the desired one. In fact, suppose $x \in X$ belongs to all members of the family $\{V_j\}_{j=1}^n$ from the cover $\{V_j\}_{j \in J}$. Every V_j, $j = 1, \dots, n$ is contained in some ball $B_{\eta(x_j)}(x_j)$, therefore $x \in \bigcap_{j=1}^n B_{\eta(x_j)}(x_j)$. Let k, $1 \le k \le n$ be such that $\eta_k = \max_{1 \le j \le n} \eta(x_j)$. Take $x' = x_k$, then we have $x_j \in B_{\eta_k}(x)$ and hence $x_j \in B_{2\eta_k}(x')$ for all $j = 1, \dots, n$. Therefore $B_{\eta(x_j)}(x_j) \subset B_{4\eta_k}(x')$, $j = 1, \dots, n$.

Then we conclude that

$$y_j \in F(V_j) \subset F(B_{\eta(x_j)}(x_j)) \subset F(B_{4\eta_k}(x')) \subset W_\varepsilon(F(x'))$$

for all $j = 1, \dots, n$ and the convexity of the set $W_\varepsilon(F(x'))$ implies that $f_\varepsilon(x) \in W_\varepsilon(F(x'))$. Since $x \in V_j$, $j = 1, \dots, n$, we also have that $F(x) \subset W_\varepsilon(F(x'))$. □

1.3 Measurable multimaps and the superposition multioperator

1.3.1 Measurable multimaps and multivalued integral

The notion of measurable multimap is a natural extension of the classical one and turns out to be very useful with numerous applications in different branches of analysis. At the present time a detailed description of properties of measurable multimaps can be found in a number of works (see e.g. [14], [18], [20], [41], [42], [56], [73], [118], [142], [199] and others). This means that we need not strive for the maximal generality in our presentation, giving our definitions in a form convenient for subsequent applications. In general, we will restrict ourselves to the case of multimaps defined on an interval of the real line equipped with Lebesgue measure, mapping into compact subsets of a Banach space.

So let $I \subset \mathbb{R}$ be a compact interval, μ a Lebesgue measure on I and E a Banach space.

Definition 1.3.1. A multifunction $F : I \to K(E)$ is said to be *measurable* if for every open subset $W \subset E$ the set $F_+^{-1}(W)$ is measurable.

It is clear that an equivalent definition is the measurability of the complete preimage $F_-^{-1}(Q)$ of every closed subset $Q \subset E$. The following statement gives two more equivalent definitions of the measurability of a multifunction.

Lemma 1.3.1. *A multifunction $F : I \to K(E)$ is measurable if and only if:* (i) *for every closed set $Q \subset E$ the small preimage $F_+^{-1}(Q)$ is measurable;* (ii) *for every open set $W \subset E$ the complete preimage $F_-^{-1}(W)$ is measurable.*

Proof. Follows from Lemma 1.1.1. □

Note that from the above definition and Lemma 1.3.1 it follows that every u.s.c. or l.s.c. multifunction is measurable.

To describe additional properties of measurable multifunctions we will need the following definitions.

Definition 1.3.2. A function $f : I \to E$ is said to be a *measurable selection* of a multifunction $F : I \to K(E)$ if f is measurable and

$$f(t) \in F(t) \quad \text{for } \mu\text{-a.e. } t \in I.$$

The set of all measurable selections of F will be denoted as S_F.

Definition 1.3.3. A countable family $\{f_n\}_{n=1}^{\infty} \subset S_F$ is said to be a *Castaing representation of F* if

$$\overline{\bigcup_{n=1}^{\infty} f(t)} = F(t)$$

for μ-a.e. $t \in I$.

We say that the multifunction $\widetilde{F} : I \to K(E)$ is a step multifunction if there exists a partition of I into a finite family of disjoint measurable subsets $\{I_j\}$, $\bigcup_j I_j = I$ such that $\widetilde{F}$ is constant on each I_j.

Definition 1.3.4. A multifunction $F : I \to K(E)$ is said to be *strongly measurable* if there exists a sequence $\{F_n\}_{n=1}^{\infty}$ of step multifunctions such that

$$h(F_n(t), F(t)) \to 0$$

as $n \to \infty$ for μ-a.e. $t \in I$ where h is the Hausdorff metric on $K(E)$ (see Section 1.1).

It is known that in the same way we can define the concept of strongly measurable function and, hence, of strongly measurable selection. Note that, in general, a measurable multifunction is not a strongly measurable one (see, e.g. [73]). But for compact-valued multifunctions acting into a separable Banach space these notions coincide. This becomes clear from the following proposition describing the main properties of measurable multifunctions.

Theorem 1.3.1. *Let E be a separable Banach space. For a multifunction $F : I \to K(E)$ the following conditions are equivalent:*

(a) *F is measurable;*

(b) *for every countable dense subset $\{x_n\}_{n=1}^{\infty}$ of E the functions $\{\varphi_n\}_{n=1}^{\infty}$,*

$$\varphi_n : I \to \mathbb{R}, \quad \varphi_n(t) = d(x_n, F(t))$$

are measurable;

(c) *F has a Castaing representation;*

(d) *F is strongly measurable;*

(e) *F is measurable as a single-valued map from I into a metric space $(K(E), h)$;*

(f) *F has the Lusin property: for every $\delta > 0$ there exists a closed subset $I_\delta \subset I$ such that $\mu(I \setminus I_\delta) \leq \delta$ and the restriction of F on I_δ is continuous.*

Proof. (a) $\Leftrightarrow$ (b). Note that the collection of balls $\{B_r(x_n)\}$ of rational radius r forms a countable base of the topology of the space E. Then, it follows that (a) is equivalent to the measurability of the preimage $F_{-}^{-1}(B_r(x_n))$ of every such ball. But this preimage is equal to the Lebesgue set

$$\Delta_{x_n}(r) = \{t \in I : \varphi_n(t) < r\}.$$

(b) $\Rightarrow$ (c). Let $N = \{x_n\}_{n=1}^{\infty}$ be a countable dense subset of E. Define the sequence of functions $\{\psi_k\}_{k=1}^{\infty}$, $\psi_k : I \to E$ by the following inductive process. Set

$$\psi_1(t) = x_i$$

if for a given $t \in I$, i is the minimum index such that

$$\varphi_i(t) = d(x_i, F(t)) \leq \frac{1}{2}.$$

If ψ_k is constructed then ψ_{k+1} is defined as

$$\psi_{k+1}(t) = x_i$$

if for a given $t \in I$, i is the least number such that

$$\varphi_i(t) \leq \frac{1}{2^{k+1}},$$

and

$$d(\psi_k(t), x_i) \leq \frac{1}{2^{k-1}}.$$

Note that the functions ψ_k are measurable. In fact, each of these functions have a countable number of values and the Lebesgue sets of the functions φ_i

$$\Delta_i(\alpha) = \{t \in I : \varphi_i(t) \leq \alpha\}$$

are measurable. The function ψ_1 is then measurable, since

$$\{t \in I : \psi_1(t) = x_i\} = \Delta_i\left(\frac{1}{2}\right) \setminus \bigcup_{p<i} \Delta_p\left(\frac{1}{2}\right)$$

and if we suppose that the function $\psi_k,\ k \geq 1$ is measurable, then

$$\begin{aligned}\{t \in I : \psi_{k+1}(t) = x_i\} = &\left\{\Delta_i\left(\frac{1}{2^{k+1}}\right) \cap \left[t \in I : d(\psi_k(t), x_i) \leq \frac{i}{2^{k-1}}\right]\right\} \\ &\setminus \left\{\bigcup_{p<i}\left[\Delta_p\left(\frac{1}{2^{k+1}}\right) \cap \left[t \in I : d(\psi_k(t), x_p) \leq \frac{1}{2^{k-1}}\right]\right]\right\}\end{aligned}$$

giving the measurability of ψ_{k+1}.

For every $t \in I$ we have

$$d(\psi_k(t), \psi_{k+1}(t)) \leq \frac{1}{2^{k-1}},$$

so the sequence $\{\psi_k\}_{k=1}^{\infty}$ converges uniformly to a measurable function $\psi_N : I \to E$.

From the relation

$$d(\psi_k(t), F(t)) \leq \frac{1}{2^{k+1}}$$

it follows that ψ_N is a measurable selection of F.

Therefore we can assign a measurable selection ψ_N to every countable dense subset $N \subset E$. Note that if for any $t \in I$ and $x \in F(t)$ we have

$$d(x, x_1) \leq \frac{1}{2^k}$$

then $\psi_k(t) = x_1$ and

$$d(x, \psi_N(t)) \leq \frac{1}{2^{k-2}} + \frac{1}{2^k}.$$

Now let $\{N_m\}_{m=1}^{\infty}$ be a sequence of countable dense subsets of E constructed in the following way. If $N_0 = \{x_1, \dots, x_i, \dots\}$ then $N_m = \{x_{m_1}, \dots, x_{m_r}, \dots\}$ where $x_{m_r} = x_{m+r}$. The sequence of functions $\{f_m\}_{m=1}^{\infty},\ f_m = f_{N_m}$ forms a Castaing representation of F.

In fact, if $t \in I,\ x \in F(t)$ and integer $k > 0$ are given then we can find a number m such that

$$d(x_{m_1}, x) \leq \frac{1}{2^k}$$

so,

$$d(x, \psi_{N_m}(t)) \leq \frac{1}{2^{k-2}} + \frac{1}{2^k}.$$

(c) $\Rightarrow$ (b). Let $x \in E$ be any point and $\{f_m\}_{m=1}^{\infty}$ a Castaing representation of F. Then

$$\varphi_x(t) = d(x, F(t)) = \inf_m d(x, f_m(t)),$$

so the function φ_x is measurable.

(d) $\Leftrightarrow$ (e) $\Leftrightarrow$ (f). It is known that the metric space $(K(E), h)$ is separable (see, e.g. [85]), then the equivalence is the direct consequence of the analogous properties of single-valued measurable functions (see, e.g. [204], Theorem 23_2 and Corollary of Theorem 33 of Chapter IV).

(f) $\Rightarrow$ (a). Let $Q \subset E$ be a closed set. For arbitrary $\delta > 0$ take a closed subset $I_\delta \subset I$ such that $\mu(I \setminus I_\delta) \leq \delta$ and the restriction of F to I_δ is continuous. The inverse image $F_-^{-1}(Q)$ consists of a closed set $F_-^{-1}(Q) \cap I_\delta$ and a set $F_-^{-1}(Q) \cap (I \setminus I_\delta)$ whose outer measure is less or equal to δ, therefore $F_-^{-1}(Q)$ is measurable.

(c) $\Rightarrow$ (e). Let $M \in K(E)$ and $\{f_n\}_{n=1}^{\infty}$ be a Castaing representation of F. Then

$$h(M, F(t)) = \max\left\{ \sup_n d(f_n(t), M),\ \sup_{x \in M} \inf_n \| x - f_n(t) \| \right\}$$

so the function $t \to h(M, F(t)),\ t \in I$ is measurable. □

Remark 1.3.1. From the proof it is clear that for a closed-valued multifunction $F : I \to C(E)$ conditions (a), (b) and (c) are equivalent too (see also [56]).

Corollary 1.3.1.

(a) *Let $\{F_j\}_{j \in J}$ be an at most countable family of measurable multifunctions $F_j : I \to K(E)$ such that $\bigcap_{j \in J} F_j \neq \emptyset$ for all $t \in I$. Then the intersection of these multifunctions $\bigcap_{j \in J} F_j : I \to K(E)$ is measurable;*

(b) *If $F_1, F_2 : I \to K(E)$ are measurable multifunctions then their Cartesian product $F_1 \times F_2 : I \to K(E \times E)$ and sum $F_1 + F_2 : I \to K(E)$ are measurable multifunctions;*

(c) *If $F : I \to K(E)$ is a measurable multifunction and $f : I \to \mathbb{R}$ is a measurable function then the product $f \cdot F : I \to K(E)$ is a measurable multifunction;*

(d) *If $F : I \to K(E)$ is a measurable multifunction then its convex closure $\overline{\text{co}}\, F : I \to Kv(E)$ is a measurable multifunction.*

Proof. All these statements can be easily deduced from the Lusin property (Theorem 1.3.1 (f)) and from the propositions on the continuity of the results of corresponding operations over multimaps (see Section 1.2). □

Corollary 1.3.2. *If $F_1, F_2 : I \to K(E)$ are measurable multifunctions then the deviation function $d_* : I \to \mathbb{R}_+$,*

$$d_*(t) = d_*(F_1(t), F_2(t)) = \sup_{y \in F_1(t)} d(y, F_2(t))$$

and the Hausdorff function $h : I \to \mathbb{R}_+$,

$$h(t) = h(F_1(t), F_2(t))$$

are measurable.

Proof. The measurability of the function d_* follows from the equality

$$d_*(t) = \sup_n [d(x_n, F_2(t)) - d(x_n, F_1(t))]$$

where $\{x_n\}_{n=1}^{\infty}$ is a countable dense subset of E, and from Theorem 1.3.1 (b). The relation

$$h(F_1(t), F_2(t)) = \max\{d_*(F_1(t), F_2(t)), d_*(F_2(t), F_1(t))\}$$

implies the measurability of the function $h(t)$. □

Lemma 1.3.2. *Let $\{F_j\}_{j=1}^{\infty}$, $F_j : I \to K(E)$ be a family of measurable multifunctions and $\Phi : I \to K(E)$ a multifunction such that*

$$\bigcup_{j=1}^{\infty} F(t) \subset \Phi(t)$$

for all $t \in I$. Then the multifunction $\overline{\bigcup_{j=1}^{\infty} F_j} : I \to K(E)$,

$$\left(\overline{\bigcup_{j=1}^{\infty} F_j}\right)(t) = \overline{\bigcup_{j=1}^{\infty} F_j(t)}$$

is measurable.

Proof. Let $W \subset E$ be an open set, then

$$\left(\overline{\bigcup_{j=1}^{\infty} F_j}\right)_-^{-1}(W) = \left(\bigcup_{j=1}^{\infty} F_j\right)_-^{-1}(W) = \bigcup_{j=1}^{\infty} [(F_j)_-^{-1}(W)]$$

is a measurable set and we can use Lemma 1.3.1 (ii). □

Let E be an arbitrary (not necessarily separable) Banach space. As a consequence of Theorem 1.3.1 we obtain

Lemma 1.3.3. *Let $F : I \to K(E)$ be a strongly measurable multifunction. Then F is measurable and has a Castaing representation.*

Proof. Let $\{F_n\}_{n=1}^{\infty}$ be a sequence of step multifunctions approximating the multifunction F. Then for μ-a.e. $t \in I$ the sets $F(t)$ are contained in the separable Banach space

$$E' = \overline{\text{sp}} \bigcup_{n=1}^{\infty} F_n(I)$$

and we can apply Theorem 1.3.1. □

We consider now the notion of multivalued integral. Let E be a Banach space and $F : I \to P(E)$ be a multifunction.

By the symbol of S_F^1 we will denote the set of all Bochner integrable selectors of the multifunction F, i.e.,

$$S_F^1 = \{f \in L^1(I; E) : f(t) \in F(t) \text{ for } \mu\text{-a.e. } t \in I\}.$$

If $S_F^1 \neq \emptyset$ then the multifunction F is called *integrable* and

$$\int_{\mathcal{I}} F(s)\, ds \underset{\text{def}}{=} \left\{ \int_{\mathcal{I}} f(s)\, ds : \; f \in S_F^1 \right\}$$

for any measurable set $\mathcal{I} \subseteq I$. Clearly if $F : I \to K(E)$ is strongly measurable and *integrably bounded*, i.e., there exists $\alpha \in L_+^1(I)$ such that

$$\| F(t) \| \underset{\text{def}}{=} \max\{\| y \| : y \in F(t)\} \leq \alpha(t)$$

for μ-a.e. $t \in I$, then F is integrable.

1.3.2 The Carathéodory conditions and the Filippov Implicit Function Lemma

Let E, E_0 be separable Banach spaces; I as in the previous section.

Definition 1.3.5. (a) A multimap $F : I \times E_0 \to K(E)$ verifies *the upper Carathéodory conditions* if

(i) for every $x \in E_0$ the multifunction $F(\cdot, x) : I \to K(E)$ is measurable;

(ii) for almost every $t \in I$ the multimap $F(t, \cdot) : E_0 \to K(E)$ is u.s.c.

(b) A multimap $F : I \times E_0 \to K(E)$ verifies *the Carathéodory conditions* if condition (i) holds and condition

(ii′) for almost every $t \in I$ the multimap $F(t, \cdot) : E_0 \to K(E)$ is continuous

is satisfied.

We consider some properties of multimaps satisfying these conditions.

First of all note that from Theorem 1.3.1 (e) and Theorem 1.1.9 it follows that a multimap $F : I \times E_0 \to K(E)$ satisfying the Carathéodory conditions can be considered as a single valued map into a metric space $(K(E), h)$ satisfying "single-valued" Carathéodory conditions. Therefore we can conclude (see, e.g. [118]) that the following Scorza–Dragoni property is satisfied.

Theorem 1.3.2. *Let the multimap $F : I \times E_0 \to K(E)$ satisfy the Carathéodory conditions. Then for every $\delta > 0$ there exists a closed subset $I_\delta \subset I$ such that $\mu(I \setminus I_\delta) \leq \delta$ and the restriction of F to $I_\delta \times E_0$ is continuous.*

We note that the analogous property for multimaps satisfying the upper Carathéodory conditions does not hold (see Example 1.3.1 below).

The following proposition is a generalization of a result proved in its initial form by A. F. Filippov [87]. This statement, known as the "Filippov Implicit Function Lemma" plays an important role in control theory.

Theorem 1.3.3. *Suppose that the multimap $F : I \times E_0 \to K(E)$ satisfies the Carathéodory conditions and that $U : I \to K(E_0)$ is a measurable multifunction. Let $g : I \to E$ be a measurable function such that*

$$g(t) \in F(t, U(t)) \quad \textit{for } \mu\text{-a.e. } t \in I.$$

Then there exists a function $u \in S_U$ such that

$$g(t) \in F(t, u(t)) \quad \textit{for } \mu\text{-a.e. } t \in I.$$

Before proving this theorem we state the following proposition which can be easily verified.

Lemma 1.3.4. *Let $F : I \times E_0 \to C(E)$ be a closed multimap, $g : I \to E$ a continuous map. Suppose that for every $t \in I$ there exists $x \in E_0$ such that*

$$g(t) \in F(t, x).$$

Then the multifunction $\Gamma : I \to C(E_0)$

$$\Gamma(t) = \{x \in E_0 : g(t) \in F(t, x)\}$$

is closed.

Proof of Theorem 1.3.3. Without loss of generality we can assume that the multimap F is continuous in the second argument and $g(t) \in F(t, U(t))$ for every $t \in I$. Consider the multifunction $\Gamma : I \to C(E_0)$,

$$\Gamma(t) = \{x \in E_0 : g(t) \in F(t, x)\}.$$

It is clear that the multifunction $\Phi = \Gamma \cap U : I \to K(E_0)$ is well defined and its measurable selection is the desired function $u(\cdot)$. Hence it is sufficient to demonstrate the measurability of the multifunction Φ.

To this aim, fix $\delta > 0$. By Theorem 1.3.2 there exists a closed subset $I_1 \subset I$ such that $\mu(I \setminus I_1) \leq \delta/2$ and the restriction of F to $I_1 \times E_0$ is continuous. The application of the Lusin property to the function $g(\cdot)$ and the multifunction $U(\cdot)$ gives the existence of a closed subset $I_2 \subset I$ such that $\mu(I \setminus I_2) \leq \delta/2$ and the restrictions of g and U to I_2 are continuous.

Let $I_\delta = I_1 \cap I_2$, then $\mu(I \setminus I_\delta) \leq \delta$ and from Theorem 1.1.4 it follows that the restriction of F on $I_\delta \times E_0$ is closed. Then, by Lemma 1.3.4, the multifunction Γ is closed on I_δ. Theorem 1.2.3 implies that the multifunction Φ is u.s.c. on I_δ and application of the Theorem 1.3.1 (f) completes the proof. □

1.3.3 The superposition multioperator

Every multimap $F : I \times E_0 \to P(E)$ induces a correspondence assigning to every multifunction $Q : I \to P(E_0)$ the multifunction $\Phi : I \to P(E)$ defined by the formula

$$\Phi(t) = F(t, Q(t)).$$

Our aim now is to describe some properties of this correspondence.

Theorem 1.3.4. *If a multimap $F : I \times E_0 \to K(E)$ satisfies the Carathéodory conditions then F is superpositionally measurable in the sense that for every measurable multifunction $Q : I \to K(E_0)$ the multifunction Φ is measurable.*

Proof. Without loss of generality we can assume that the multimap F is continuous in the second argument for all $t \in I$. Then the multifunction Φ has compact values by Theorem 1.1.7.

Fix $\delta > 0$ and choose a closed set $I_\delta \subset I$ such that $\mu(I \setminus I_\delta) \leq \delta$ and the restrictions of Q to I_δ and F to $I_\delta \times E_0$ are continuous (Theorem 1.3.1 (f) and Theorem 1.3.2). If $i : I_\delta \to I_\delta$ is the identity map then from Theorems 1.2.10 and 1.2.12 it follows that the multifunction $i \times Q : I_\delta \to K(I_\delta \times E_0)$ is continuous. Then the multifunction $\Phi = F \circ (i \times Q)$ is also continuous on I_δ (Theorem 1.2.8). The measurability of the multifunction Φ follows now from Theorem 1.3.1 (f). □

Following the same line of reasoning we can prove the following sufficient condition for superpositional measurability.

Proposition 1.3.1. *If a multimap $F : I \times E_0 \to K(E)$ is u.s.c. or l.s.c. then it is superpositionally measurable.*

It is interesting to note that the property of superpositional measurability disappears if we weaken the Carathéodory conditions to upper ones. Consider the following example [177].

Example 1.3.1. Let $D \subset [0,1]$ be an arbitrary nonmeasurable set. Consider the function $\varphi : [0,1] \to \mathbb{R}$,

$$\varphi(t) = \begin{cases} t, & t \in [0,1] \setminus D, \\ t+1, & t \in D. \end{cases}$$

Let us define the multimap $F : [0,1] \times \mathbb{R} \to K(\mathbb{R})$,

$$F(t,x) = \begin{cases} \{1\}, & (t,x) \in ([0,1] \times \mathbb{R}) \setminus \Gamma_\varphi, \\ [0,1], & (t,x) \in \Gamma_\varphi, \end{cases}$$

where Γ_φ is the graph of φ. It is easy to see that the multimap F satisfies the upper Carathéodory conditions (but not the Carathéodory conditions).

If now $Q = i : [0,1] \to \mathbb{R}$ is the inclusion map then the multimap $\Phi : [0,1] \to K(\mathbb{R})$,

$$\Phi(t) = F(t, Q(t)) = F(t, i(t)) = \begin{cases} [0,1], & t \in [0,1] \setminus D, \\ \{1\}, & t \in D, \end{cases}$$

is not measurable.

This example shows also that the Scorza–Dragoni property (see Theorem 1.3.2) does not hold for multimaps satisfying the upper Carathéodory conditions. In fact it is not difficult to verify that every multimap satisfying the Scorza–Dragoni property is necessarily superpositionally measurable.

Nevertheless, if F satisfies the upper Carathéodory conditions the resulting multifunction Φ possesses some "good" properties. In fact, the following is true.

Theorem 1.3.5. *Let E, E_0 be (not necessarily separable) Banach spaces; and let the multimap $F : I \times E_0 \to K(E)$ be such that*

(i) *for every $x \in E_0$ the multifunction $F(\cdot, x) : I \to K(E)$ has a strongly measurable selection;*

(ii) *for μ-a.e. $t \in I$ the multimap $F(t, \cdot) : E_0 \to K(E)$ is u.s.c.*

Then for every strongly measurable function $q : I \to E_0$ there exists a strongly measurable selection $\varphi : I \to E$ of the multifunction $\Phi : I \to K(E)$,

$$\Phi(t) = F(t, q(t)).$$

Proof. Let $\{q_n\}_{n=1}^\infty$, $q_n : I \to E_0$ be a sequence of step functions converging μ-almost everywhere on I to q. Using condition (i) we can form a sequence of strongly measurable functions $\{\varphi_n\}_{n=1}^\infty$, $\varphi_n : I \to E$,

$$\varphi_n(t) \in F(t, q_n(t)) \quad \text{for } \mu\text{-a.e. } t \in I.$$

Redefining this sequence on a set of a null measure, we can assume that the sequence of functions $\{\varphi_n\}$ takes it values in a separable Banach space $E' \subset E$,

$$E' = \overline{\text{sp}} \bigcup_{n=1}^{\infty} \varphi_n(I).$$

Now for μ-a.e $t \in I$ and $m \geq 1$ we can define

$$\Phi_m(t) = \overline{\bigcup_{k=m}^{\infty} \varphi_k(t)}.$$

From the condition (ii) and Theorem 1.1.7 it follows that the sets $\Phi_m(t)$ are compact and Theorem 1.3.1 (c) implies that $\{\Phi_m\}_{m=1}^{\infty}$, $\Phi_m : I \to K(E')$ is the sequence of measurable multifunctions.

For μ-a.e. $t \in I$ the sets $\Phi_m(t)$, $m \geq 1$ form a decreasing sequence of compact sets and taking into account Corollary 1.3.1 (a) we can conclude that the multifunction $\widetilde{\Phi} : I \to K(E')$,

$$\widetilde{\Phi}(t) = \bigcap_{m=1}^{\infty} \Phi_m(t) \quad \text{for } \mu\text{-a.e. } t \in I$$

is well-defined and measurable. From condition (ii) it follows that

$$\widetilde{\Phi}(t) \subseteq \Phi(t) = F(t, q(t))$$

for μ-a.e. $t \in I$ and the application of Theorem 1.3.1 (c) gives the existence of a function $\varphi \in S_{\widetilde{\Phi}}$ which is strongly measurable since $\widetilde{\Phi}(I) \subset E'$. □

Corollary 1.3.3. *Let the spaces E, E_0 and the multimap F be as in Theorem* 1.3.5. *Then for every strongly measurable multifunction $Q : I \to K(E_0)$ there exists a strongly measurable selection $\varphi : I \to E$ of a multifunction Φ, $\Phi(t) = F(t, Q(t))$.*

Proof. Redefining the multifunction Q on a null set we can assume that its range is contained in a separable subspace of E. Then from Theorem 1.3.1 (c) we obtain the existence of a strongly measurable selection q of a multifunction Q to which we apply the previous theorem. □

Theorem 1.3.5 justifies the following definition.

Definition 1.3.6. Let $F : I \times E_0 \to K(E)$ be a multifunction satisfying the conditions (i) and (ii) of Theorem 1.3.5. The multimap $\mathcal{P}_F$ assigning to every continuous function $q \in C(I, E_0)$ the set of all strongly measurable selections of the multifunction $F(t, q(t))$ is said to be a *superposition multioperator generated by* F.

We will continue the study of the superposition multioperator in Chapter 5.

Bibliographic notes

At present the theory of multivalued maps is a mature branch of modern mathematics with its own clearly delineated subjects and methods. It has found numerous applications in the theory of games, mathematical economics, convex analysis, optimal control, the theory of differential equations and many other fields.

As far as we know, it is in the pioneering work of the famous Polish topologist K. Kuratowski [156] (see also [157]) that the notion of a continuous multimap appears for the first time as an independent mathematical object. In recent years several monographs have been devoted (completely or partially) to the theory of multimaps and its applications. We mention the works of J. Appell, E. De Pascale, H. T. Nguen and P. P. Zabreiko [14], J. P. Aubin [15]–[17], J. P. Aubin and I. Ekeland [19], J. P. Aubin and A. Cellina [18], J. P. Aubin and H. Frankowska [20], C. Berge [27], [28], V. I. Blagodatskih and A. F. Filippov [29], N. A. Bobylev and S. K. Korovin [30], Yu. G. Borisovich, B. D. Gelman, A. D. Myshkis and V. V. Obukhovskii [42], C. Castaing and M. Valadier [56], K. Deimling [72], [73], R. Dragoni, J. W. Macki, P. Nistri and P. Zecca [80], Z. Dzedzej [82], V. V. Fedorchuk and V. V. Filippov [85], O. P. Filatov and M. M. Hapaev [86], K. Goebel and W. A. Kirk [99], L. Górniewicz [102], [103], [104], O. Hadžić [111], S. Hu and N. S. Papageorgiou [118], A. D. Ioffe and V. M. Tihomirov [121], M. Kisielewicz [141], V. B. Kolmanovskii and A. D. Myshkis [143], [144], W. Kryszewski [151], [152], E. Klein and A. Thompson [142], M. A. Krasnoselski and P. P. Zabreiko [148], T. W. Ma [161], V. Makarov and A. M. Rubinov [163], H. Nikaido [171], V. I. Opoitsev [185], A. A. Tolstonogov [210]. The main trends in the development of this theory are also described in a series of surveys of Yu. G. Borisovich, B. D. Gelman, A. D. Myshkis and V. V. Obukhovskii [40], [41], [43], [44] and B. D. Gelman and V. V. Obukhovskii [95]. It is sufficient to point out that the combined references in the surveys [41] and [44] number nearly 2500 items. A detailed bibliographic guide to the general theory of multivalued maps is also contained in the recent book of S. Hu and N. S. Papageorgiou [118]. We refer the reader to the above mentioned sources for details.

The exposition of the first chapter follows, in general, the lines of [40]–[44] and [95]. The investigation of the continuity properties of the intersection of l.s.c. multimaps was developed in the dissertation of V. Obukhovskii (1974) where the notion of a quasi-open multimap was introduced (these results were published in the survey [41]). The approximation result (Theorem 1.2.19) was obtained in the work of A. Lasota and Z. Opial [160] and, for particular cases, in the papers of Yu. G. Borisovich, E. Mukhamadiev, B. D. Gelman and V. V. Obukhovskii [39], A. Cellina [57] and A. Cellina and A. Lasota [58].

For a complete description of the theory of measurable multimaps we refer, besides the monograph of C. Castaing and M. Valadier [56] which is specifically devoted to this subject, to the works of J. Appell, E. De Pascale, H. T. Nguen and P. P. Zabreiko [14], J. P. Aubin and A. Cellina [18], J. P. Aubin and H. Frankowska [20], Yu. G. Borisovich, B. D. Gelman, A. D. Myshkis and V. V. Obukhovskii [41], [42], K. Deimling [73],

S. Hu and N. S. Papageorgiou [118], A. D. Ioffe and V. M. Tihomirov [121], E. Klein and A. Thompson [142], and R. T. Rockafellar [199]. The counter-example 1.3.1 was given in the paper of V. V. Obukhovskii [177] (see also the survey of J. Appell [13] and the monograph [14]). The proof of Theorem 1.3.5, due to C. Castaing, follows [73], Proposition 3.5 (a).

Chapter 2

Measures of noncompactness and condensing multimaps

In this chapter we describe the main results of the theory of measures of noncompactness in Banach spaces. We give also general information on condensing multioperators, i.e., multivalued maps which decrease the measure of noncompactness.

2.1 Measures of noncompactness

We present first the main definitions and we give several examples of measures of noncompactness in certain specific spaces that will be used in the sequel.

2.1.1 Basic notions

We begin with the following general definition.

Definition 2.1.1. Let E be a Banach space and $(A, \geq)$ a partially ordered set. A map $\beta : P(E) \to A$ is called a *measure of noncompactness* (MNC) in E if

$$\beta(\overline{\text{co}}\,\Omega) = \beta(\Omega)$$

for every $\Omega \in P(E)$.

Note that if D is dense in Ω then $\overline{\text{co}}\,\Omega = \overline{\text{co}}\,D$ and hence

$$\beta(\Omega) = \beta(D).$$

A measure of noncompactness β is called:

(i) *monotone* if $\Omega_0, \Omega_1 \in P(E)$, $\Omega_0 \subseteq \Omega_1$ implies $\beta(\Omega_0) \leq \beta(\Omega_1)$;

(ii) *nonsingular* if $\beta(\{a\} \cup \Omega) = \beta(\Omega)$ for every $a \in E$, $\Omega \in P(E)$;

(iii) *invariant with respect to union with compact sets* if $\beta(K \cup \Omega) = \beta(\Omega)$ for every relatively compact set $K \subset E$, and $\Omega \in P(E)$;

(iv) *semiadditive* if $\beta(\Omega_0 \cup \Omega_1) = \max\{\beta(\Omega_0), \beta(\Omega_1)\}$ for every $\Omega_0, \Omega_1 \in P(E)$;

(v) *invariant with respect to reflection* through the origin if $\beta(-\Omega) = \beta(\Omega)$ for every $\Omega \in P(E)$;

(vi) *real* if $A = \overline{\mathbb{R}}_+ = [0, \infty]$ with the natural ordering, and $\beta(\Omega) < +\infty$ for any bounded set $\Omega \in P(E)$;

If A is a cone in a Banach space we will say that the MNC β is

(vii) *algebraically semiadditive* if $\beta(\Omega_0+\Omega_1) \leq \beta(\Omega_0)+\beta(\Omega_1)$ for every $\Omega_0, \Omega_1 \in P(E)$;

(viii) *regular* if $\beta(\Omega) = 0$ is equivalent to the relative compactness of Ω.

Well-known examples of measures of noncompactness satisfying all of the above properties are:
the *Hausdorff MNC*

$$\chi(\Omega) = \inf\{\varepsilon > 0 : \Omega \text{ has a finite } \varepsilon\text{-net}\}$$

and the *Kuratowski MNC*

$$\alpha(\Omega) = \inf\{d > 0 : \Omega \text{ admits a partition into a finite number of sets whose diameters are less than } d\}.$$

It is easy to see that the Hausdorff and Kuratowski MNCs are connected by the following relations:

$$\chi(\Omega) \leq \alpha(\Omega) \leq 2\,\chi(\Omega).$$

We recall some additional properties of the Hausdorff MNC (see [4] for details).

(ix) *The Lipschitz Property*:

$$|\,\chi(\Omega_0) - \chi(\Omega_1)| \leq h(\Omega_0\,, \Omega_1)$$

where h is the Hausdorff metric.

(x) *Semi-homogeneity*:

$$\chi(t\,\Omega) = |\,t\,|\;\chi(\Omega)$$

for every $t \in \mathbb{R}$, $\Omega \in P(E)$.

Note that property (ix) implies the *continuity property* of the MNC χ with respect to the Hausdorff metric.

The Hausdorff MNC can be used to define the following important characteristic of a linear operator in a Banach space.

Definition 2.1.2. Let $L : E \to E$ be a bounded linear operator; $B \subset E$ the unit ball. The number

$$\|\,L\,\|^{(\chi)} \underset{\text{def}}{=} \chi(LB),$$

is said to be *the* (χ)*-norm of the operator* L.

It is easy to verify the following properties:

$$\|L\|^{(\chi)} \leq \|L\|$$
$$\chi(L\Omega) \leq \|L\|^{(\chi)}\chi(\Omega). \tag{2.1.1}$$

The above inequalities imply that

$$\chi(L\Omega) \leq \| L \| \chi(\Omega). \tag{2.1.2}$$

2.1.2 The expression for the Hausdorff MNC in a separable Banach space

Let E be a separable Banach space and $\{E_m\}_{m=1}^{\infty}$ a sequence of finite-dimensional subspaces of E such that $E_m \subseteq E_{m+1}$, $m = 1, 2, \dots$ and $\overline{\bigcup_{m=1}^{\infty} E_m} = E$. Then the Hausdorff MNC of a bounded set $\Omega \subset E$ can be calculated as

$$\chi(\Omega) = \lim_{m\to\infty} d_*(\Omega, E_m) \tag{2.1.3}$$

where $d_*(\Omega, E_m) = \sup_{x\in\Omega} d(x, E_m)$ is the deviation of the set Ω from the subspace E_m.

In fact, for an arbitrary $\varepsilon > 0$, let points $\{y_j\}_{j=1}^{k}$ form the $(\chi(\Omega) + \varepsilon)$-net of the set Ω. Choosing an integer m_0 so large that we can find points $\{x_j\}_{j=1}^{k} \subset E_{m_0}$ with the property that

$$\| x_j - y_j \| < \varepsilon, \quad 1 \leq j \leq k$$

the following inequality holds for $m \geq m_0$,

$$d_*(\Omega, E_m) \leq \chi(\Omega) + 2\varepsilon$$

which implies

$$\chi(\Omega) \geq \lim_{m\to\infty} d_*(\Omega, E_m).$$

Note now that if $\| x \| \leq M$ for all $x \in \Omega$ then

$$d_*(x, E_m) = \inf_{y\in B_{2M}^m(0)} \| x - y \|$$

where $B_{2M}^m(0)$ denotes a ball in E_m of radius $2M$ with center at the origin. Hence for any $\varepsilon > 0$ there exists a compact $(d_*(x, E_m) + \varepsilon)$-net of the set Ω. Therefore we have

$$\chi(\Omega) \leq \lim_{m\to\infty} d_*(\Omega, E_m)$$

obtaining the desired formula.

2.1.3 MNC in spaces of continuous functions

In this section several examples of useful measures of noncompactness in spaces of continuous functions are presented.

Example 2.1.1. It is known (see, for example [4]) that the Hausdorff MNC of a subset Ω of the space $C([a,b];\mathbb{R})$ (of real-valued continuous functions on an interval $[a,b]$) can be evaluated by the formula

$$\chi(\Omega) = \frac{1}{2}\lim_{\delta\to 0}\sup_{x\in\Omega}\max_{|t_1-t_2|\le\delta}|x(t_1)-x(t_2)|. \tag{2.1.4}$$

Example 2.1.2. Consider now the space $C([a,b];E)$ of continuous functions on the interval $[a,b]$ with values in a separable Banach space E equipped with the usual sup-norm. In this space the formula of the *modulus of equicontinuity* of the set of functions $\Omega \subset C([a,b];E)$ has the following form

$$\mathrm{mod}_C(\Omega) = \lim_{\delta\to 0}\sup_{x\in\Omega}\max_{|t_1-t_2|\le\delta}\|x(t_1)-x(t_2)\|. \tag{2.1.5}$$

We show that $\mathrm{mod}_C(\Omega)$ defines a MNC in $C([a,b];E)$. First, we verify its monotonicity. In fact, $\Omega_1 \subseteq \Omega_2$ implies

$$\mathrm{mod}_C(\Omega_1) \le \mathrm{mod}_C(\Omega_2)$$

since

$$\sup_{x\in\Omega_1}\max_{|t_1-t_2|\le\delta}\|x(t_1)-x(t_2)\| \le \sup_{x\in\Omega_2}\max_{|t_1-t_2|\le\delta}\|x(t_1)-x(t_2)\|.$$

Next we show that

$$\mathrm{mod}_C(\overline{\mathrm{co}}\,\Omega) = \mathrm{mod}_C(\Omega).$$

By the monotonocity property it is enough to show that

$$\mathrm{mod}_C(\overline{\mathrm{co}}\,\Omega) \le \mathrm{mod}_C(\Omega).$$

This inequality follows from the following fact.

If $x = \sum_{i=1}^{\infty}\alpha_i y_i$, $\alpha_i \ge 0$, $\sum_{i=1}^{\infty}\alpha_i = 1$, $y_i \in \Omega$, then

$$\begin{aligned}\max_{|t_1-t_2|\le\delta}\|x(t_1)-x(t_2)\| &\le \sup_i\max_{|t_1-t_2|\le\delta}\|y_i(t_1)-y_i(t_2)\|\\ &\le \sup_{x\in\Omega}\max_{|t_1-t_2|\le\delta}\|x(t_1)-x(t_2)\|.\end{aligned}$$

We verify now the semiadditivity property, i.e., we show that

$$\mathrm{mod}_C(\Omega_1\cup\Omega_2) = \max\{\mathrm{mod}_C(\Omega_1),\mathrm{mod}_C(\Omega_2)\}.$$

In fact,

$$\sup_{x\in\Omega_1\cup\Omega_2}\max_{|t_1-t_2|\le\delta}\|x(t_1)-x(t_2)\|$$
$$=\max\{\sup_{x\in\Omega_1}\max_{|t_1-t_2|\le\delta}\|x(t_1)-x(t_2)\|,\ \sup_{x\in\Omega_2}\max_{|t_1-t_2|\le\delta}\|x(t_1)-x(t_2)\|\}.$$

Finally, if the set $K\subset C([a,b];E)$ is relatively compact then the functions of this set form an equicontinuous family and hence

$$\mathrm{mod}_C(K)=0. \tag{2.1.6}$$

From the semiadditivity property it follows that

$$\mathrm{mod}_C(\Omega\cup K)=\max\{\mathrm{mod}_C(\Omega),\mathrm{mod}_C(K)\}=\mathrm{mod}_C(\Omega)$$

proving that $\mathrm{mod}_C(\cdot)$ is invariant with respect to addition of a relatively compact set. In particular, it means that this MNC is nonsingular.

Example 2.1.3. We consider another useful example of MNC in the space of continuous functions $C([a,b];E)$. For $\Omega\subset C([a,b];E)$ define

$$\phi(\Omega)=\sup_{t\in[a,b]}\chi(\Omega(t))$$

where χ is the Hausdorff MNC in E and $\Omega(t)=\{y(t):y\in\Omega\}$. It is easy to verify that this MNC satisfies all the usual properties of measures of noncompactness except regularity.

Nevertheless if the family of functions Ω is equicontinuous then

$$\phi(\Omega)=\chi_C(\Omega)$$

where χ_C is the Hausdorff MNC in the space $C([a,b];E)$.

In fact, the inequality

$$\phi(\Omega)\le\chi_C(\Omega)$$

is obvious since if the collection of functions $\{y_1(\cdot),\ldots,y_n(\cdot)\}$ forms an ε-net for the set Ω, then the points $\{y_1(\cdot),\ldots,y_n(\cdot)\}$ give an ε-net for the set $\Omega(t)$ for every $t\in[a,b]$.

We show now that

$$\phi(\Omega)\ge\chi_C(\Omega).$$

Since the family Ω is equicontinuous, for a given $\varepsilon>0$ we can find $\delta>0$ such that the inequality

$$|t_2-t_1|<\delta$$

implies

$$\|x(t_2)-x(t_1)\|<\frac{\varepsilon}{2}$$

for all $x \in \Omega$. Take n such that $\frac{b-a}{n} < \delta$ and partition the interval $[a, b]$ by points of the form $t_k = a + k\frac{b-a}{n}$, $k = 0, 1, \ldots, n$.

At every point t_k we construct a $\left(\chi(\Omega(t_k)) + \frac{\varepsilon}{2}\right)$-net of the set $\Omega(t_k)$ with the points $\{y_1^k, \ldots, y_{m_k}^k\}$ and consider all possible piece-wise linear functions with vertices in these points. An easy computation shows that this collection of functions forms a $(\phi(\Omega) + \varepsilon)$-net for the set Ω and the desired inequality follows from the arbitrariness of ε.

Example 2.1.4. We consider one more MNC in the space $C([a, b]; E)$. We will use it in the investigation of the properties of the integral multioperator appearing in the Cauchy problem for a differential inclusion in the space E. This MNC is useful when the space E is non separable.

For a bounded $\Omega \subset C([a, b]; E)$ set

$$\nu(\Omega) = \max_{D \in \Delta(\Omega)} \left(\sup_{t \in [a,b]} e^{-Lt} \chi(D(t)), \operatorname{mod}_C(D) \right), \tag{2.1.7}$$

where $\Delta(\Omega)$ is the collection of all denumerable subsets of Ω, L is a constant and $\operatorname{mod}_C(D)$ is given in formula (2.1.5).

The range for the MNC ν is a cone $\mathbb{R}^2_+$, max is taken in the sense of the ordering induced by this cone. We claim that the MNC ν is well defined. In fact, if $\{D'_n\}$ is a sequence of denumerable sets for which

$$\sup_{t \in [a,b]} e^{-Lt} \chi(D'_n(t)) \longrightarrow \sup_{D \in \Delta(\Omega)} \sup_{t \in [a,b]} e^{-Lt} \chi(D(t))$$

and $\{D''_n\}$ is a sequence of denumerable sets for which

$$\operatorname{mod}_C(D''_n) \longrightarrow \sup_{D \in \Delta(\Omega)} \operatorname{mod}_C(D),$$

then for the denumerable set $\widetilde{D} = \left(\bigcup_n D'_n\right) \cup \left(\bigcup_n D''_n\right)$ we have

$$\sup_{t \in [a,b]} e^{-Lt} \chi(\widetilde{D}(t)) = \sup_{D \in \Delta(\Omega)} \sup_{t \in [a,b]} e^{-Lt} \chi(D(t))$$

and

$$\operatorname{mod}_C(\widetilde{D}) = \sup_{D \in \Delta(\Omega)} \operatorname{mod}_C(D).$$

Therefore the maximum on the right-hand side of (2.1.7) is achieved on the set $\widetilde{D}$.

We show now that the MNC ν is monotone. In fact, if $\Omega_1 \subset \Omega_2$ then $\Delta(\Omega_1) \subseteq \Delta(\Omega_2)$ and so the maximum for $\nu(\Omega_2)$ is taken on a larger set than for $\nu(\Omega_1)$, hence

$$\nu(\Omega_1) \leq \nu(\Omega_2).$$

We prove now the equality

$$\nu(\overline{\operatorname{co}}\, \Omega) = \nu(\Omega).$$

By the monotonicity of ν it is sufficient to prove

$$\nu(\overline{\text{co}}\,\Omega) \leq \nu(\Omega).$$

Suppose that $\widehat{D}$ is the set on which the maximum on the left-hand side of the above relation is achieved:

$$\nu(\overline{\text{co}}\,\Omega) = \max_{D\in\Delta(\overline{\text{co}}\,\Omega)} \Big(\sup_{t\in[a,b]} e^{-Lt}\chi(D(t)), \text{mod}_C(D) \Big). \tag{2.1.8}$$

There exists a denumerable set $D' \subset \Omega$ such that $\widehat{D} \subseteq \overline{\text{co}}\,D'$, then by the monotonicity of the Hausdorff MNC and $\text{mod}_C(\Omega)$ we have

$$\begin{aligned} \Big(\sup_{t\in[a,b]} e^{-Lt}\chi(\widehat{D}(t)), \text{mod}_C(\widehat{D}) \Big) &\leq \Big(\sup_{t\in[a,b]} e^{-Lt}\chi((\overline{\text{co}}\,D')(t)), \text{mod}_C(\overline{\text{co}}\,D') \Big) \\ &\leq \Big(\sup_{t\in[a,b]} e^{-Lt}\chi(D'(t)), \text{mod}_C(D') \Big) \leq \nu(\Omega). \end{aligned}$$

We show that ν is invariant with respect to union with a relatively compact set K. In fact, if $\widetilde{D}$ is the set on which the maximum in the right-hand side of the equality

$$\nu(\Omega \cup K) = \max_{D\in\Delta(\Omega\cup K)} \Big(\sup_{t\in[a,b]} e^{-Lt}\chi(D(t)), \text{mod}_C(D) \Big)$$

is achieved then

$$\chi(\widetilde{D}(t)) = \chi((\widetilde{D} \cap \Omega)(t) \cup (\widetilde{D} \cap K)(t)) = \chi((\widetilde{D} \cap \Omega)(t))$$

and

$$\text{mod}_C(\widetilde{D}) = \text{mod}_C((\widetilde{D} \cap \Omega) \cup (\widetilde{D} \cap K)) = \text{mod}_C(\widetilde{D} \cap \Omega).$$

Therefore

$$\nu(\Omega \cup K) = \Big(\sup_{t\in[a,b]} e^{-Lt}\chi((\widetilde{D} \cap \Omega)(t)), \text{mod}_C(\widetilde{D} \cap \Omega) \Big) \leq \nu(\Omega).$$

It remains to prove that the MNC ν is regular. In fact, if the set Ω is relatively compact then also any $D \subseteq \Omega$ is relatively compact, implying $\chi(D(t)) \equiv 0$ and $\text{mod}_C(D) = 0$, so $\nu(\Omega) = 0$. On the other hand, if $\nu(\Omega) = 0$ then for every denumerable subset $D \subseteq \Omega$ we have that

$$\sup_{t\in[a,b]} e^{-Lt}\chi(D(t)) = 0$$

hence $\chi(D(t)) = 0$ for all $t \in [a, b]$. Moreover,

$$\text{mod}_C(D) = 0$$

gives the equicontinuity of the set D. The relative compactness of D, and therefore of Ω follows from the regularity of the Hausdorff MNC χ and the well-known Arzela–Ascoli criterion.

Example 2.1.5. Let us consider one more example of the measure of noncompactness in $C(a, b]; E)$ based on the Hausdorff MNC. This MNC will be used when studying periodic problems for differential inclusions. For convenience we work in $C_T(E)$, the space of continuous T-periodic functions with the values in a Banach space E.

Let $\Omega \subset C_T(E)$ be a bounded subset. Consider the function $\varphi(\Omega)$ given by the formula

$$\varphi(\Omega) = (\chi(\Omega(\cdot)), \mathrm{mod}_C(\Omega)). \tag{2.1.9}$$

The range of φ is the cone $\boldsymbol{K} = \boldsymbol{K}_T^+ \times \mathbb{R}_+$ where $\boldsymbol{K}_T^+$ is the cone of nonnegative T-periodic real-valued functions.

We will show that the formula (2.1.9) defines a monotone, regular MNC, invariant with respect to the union with a relatively compact set.

So, let $\Omega_1 \subseteq \Omega_2$, then obviously $\Omega_1(t) \subseteq \Omega_2(t)$ for any $t \in \mathbb{R}$ and from the monotonicity property of the Hausdorff MNC we have that

$$\chi(\Omega_1(t)) \leq \chi(\Omega_2(t)), \quad t \in \mathbb{R}.$$

The monotonicity of $\mathrm{mod}_C(\Omega)$ was proved earlier.

Let us show that φ satisfies the main property of a MNC:

$$\varphi(\overline{\mathrm{co}}\, \Omega) = \varphi(\Omega).$$

From the monotonicity property of φ it is sufficient to prove the inequality

$$\varphi(\overline{\mathrm{co}}\, \Omega) \leq \varphi(\Omega).$$

We have that $(\overline{\mathrm{co}}\, \Omega)(t) \subseteq \overline{\mathrm{co}}(\Omega(t))$ and so

$$\chi((\overline{\mathrm{co}}\, \Omega)(t)) \leq \chi(\overline{\mathrm{co}}(\Omega(t))) = \chi(\Omega(t)), \quad t \in \mathbb{R},$$

i.e., the first components of $\varphi(\overline{\mathrm{co}}\, \Omega)$ and $\varphi(\Omega)$ coincide. The equality of the second components was proved earlier.

The regularity of the MNC φ follows from the Ascoli–Arzelá criterion. If $\varphi(\Omega) = 0$ then the first component is identically equal to zero, i.e., $\Omega(t)$ is relatively compact for every $t \in \mathbb{R}$, and the vanishing of the second component means the equicontinuity of Ω. The converse follows from the regularity of the Hausdorff MNC and Arzela–Ascoli criterion.

Finally, if $K \subset C_T(E)$ is a relatively compact set then $K(t)$ is relatively compact in E for every $t \in \mathbb{R}$. Therefore

$$\chi((\Omega \cup K)(t)) = \max\{\chi(\Omega(t)), \chi(K(t))\} = \chi(\Omega(t))$$

for every $t \in \mathbb{R}$. The analogous property for the second component was proved earlier.

Remark 2.1.1. Later (see Theorem 4.2.4) we will prove that the function $\chi(\Omega(\cdot))$ is measurable when the Banach space E is separable. In this case it is convenient to consider the cone $\hat{K} = \hat{K}_T^+ \times \mathbb{R}_+$ as the range of φ, where $\hat{K}_T^+$ is the cone of measurable a.e. nonnegative T-periodic real-valued functions. In order to prove the regularity property it is sufficient to note that when the second component is equal to zero, it follows that (see Example 1.1.6 and the Lipschitz property (ix) of χ) the first component is continuous, hence is equal to zero everywhere. All the other properties of φ remain unchanged.

Remark 2.1.2. When the space E has the form

$$E = E_1 \times \cdots \times E_N$$

the MNC φ can be defined by the equality

$$\varphi_N(\Omega) = (\chi_1(\Omega_1(\cdot)), \ldots, \chi_N(\Omega_N(\cdot)), \mathrm{mod}_C(\Omega)) \tag{2.1.10}$$

where χ_i is the Hausdorff MNC in the space E_i and Ω_i is the projection of the set Ω on E_i, $i = 1, \ldots, N$.

The MNC φ_N has the same properties as the MNC φ. To verify this fact, it is enough to note that if $\Omega' \subseteq \Omega''$ then

$$\mathrm{Pr}_i(\Omega') \subseteq \mathrm{Pr}_i(\Omega'')$$

and

$$\overline{\mathrm{co}}\,\mathrm{Pr}_i(\Omega) = \mathrm{Pr}_i(\overline{\mathrm{co}}\,\Omega),$$

$i = 1, \ldots, N$.

2.2 Condensing multimaps

This section is principal in our treatment of the theory of condensing multivalued maps. We will show that every multimap condensing with respect to a "good" measure of noncompactness is restrictible to a fundamental compact set. In the next chapter these results will be the basis for the construction of a topological degree theory for condensing multimaps.

Everywhere in this section the symbol E denotes a real Banach space.

Given $X \subseteq E$, every multimap $F : X \to P(E)$ defines a multimap $\Phi : X \to P(E)$, determined by

$$\Phi(x) = x - F(x)$$

which is called the *multivalued vector field* (multifield) corresponding to the multimap F. Denoting by $i : X \to E$ the inclusion map, we can write

$$\Phi = i - F.$$

If Λ is a space of parameters and $G : \Lambda \times X \to P(E)$ is a family of multimaps, then $\Psi : \Lambda \times X \to P(E)$, given by

$$\Psi(\lambda, x) = x - G(\lambda, x),$$

is said to be a *family of multifields*.

A point $x \in X$ such that

$$x \in F(x)$$

is called a *fixed point* of a multimap F. The set of all fixed points of F will be denoted by the symbol Fix F. It is clear that every fixed point x of a multimap is a *singular point* for the corresponding multifield Φ, i.e.,

$$0 \in \Phi(x),$$

and vice versa.

If Fix $F = \emptyset$ the multifield $\Phi = i - F$ is called *nonsingular*, i.e., $\Phi : X \to P(E \setminus 0)$. A family $\Psi : \Lambda \times X \to P(E \setminus 0)$ is also called nonsingular.

An u.s.c. and compact multimap F is called *completely continuous*. The corresponding multifield $\varphi = i - F$ is also said to be completely continuous.

If Λ is a topological space, then there is a natural definition (replacing X by $\Lambda \times X$) for a completely continuous family of multimaps $G : \Lambda \times X \to P(E)$ and the corresponding family of multifields $\Psi = i - G$.

We describe now the principal classes of noncompact multimaps which we will consider in the sequel. Let $X \subset E$ be a closed set. We begin with the following important definition.

Definition 2.2.1. A closed convex set $T \subseteq E$ is called *fundamental* for a multimap $F : X \to K(E)$ (or for its corresponding multifield $\Phi = i - F$) if

(1) $F(X \cap T) \subseteq T$;

(2) $x_0 \in \overline{co}(F(x_0) \cup T)$ implies $x_0 \in T$.

We emphasize that this definition does not exclude the case $T = \emptyset$ or $X \cap T = \emptyset$, which necessarily implies that Fix $F = \emptyset$.

Definition 2.2.2. Let Λ be a compact topological space. A closed convex set T is called fundamental for a family of multimaps $G : \Lambda \times X \to K(E)$ if it is fundamental for every multimap $G(\lambda, \cdot)$, $\lambda \in \Lambda$.

The entire space E and $\overline{co}F(X)$ are natural examples of fundamental sets.

The following properties of fundamental sets can be easily verified.

Lemma 2.2.1. *The set* Fix F *is included in any fundamental set for* F.

Lemma 2.2.2. *If T is a fundamental set for the family of multimaps $G : \Lambda \times X \to K(E)$, and $P \subset T$, then the set $\widetilde{T} = \overline{\text{co}}(G(\Lambda \times (X \cap T)) \cup P)$ is also fundamental.*

Lemma 2.2.3. *If $\{T_\alpha\}$ is a system of fundamental sets of a multimap $F : X \to K(E)$ then the set $\bigcap_\alpha T_\alpha$ is also fundamental.*

We can now describe the main classes of noncompact multimaps which we will consider in the sequel.

Definition 2.2.3. A multimap $F : X \to K(E)$ or a family of multimaps $G : \Lambda \times X \to K(E)$ is called *fundamentally restrictible* if there exists a fundamental set T of F (or G) such that the restriction of F (or G) to $X \cap T$ (or, respectively, to $\Lambda \times (X \cap T)$) is compact.

Note that the multimap F or the family G is fundamentally restrictible if admits an empty fundamental set T or if $X \cap T = \emptyset$. From the definition it follows easily that the restriction of a fundamentally restrictible multimap F or a family G to a closed subset $X_1 \subset X$ (respectively to $\Lambda \times X_1$) is fundamentally restrictible.

To select a narrow class of multimaps we will need the following.

Definition 2.2.4. A fundamental set T of a multimap $F : X \to K(E)$ or a family of multimaps $G : \Lambda \times X \to K(E)$ is said to be *essential* if $X \cap T \neq \emptyset$ and the restriction of F to $X \cap T$ or, respectively G to $\Lambda \times (X \cap T)$ is compact.

Definition 2.2.5. If a multimap $F : X \to K(E)$ or a family of multimaps $G : \Lambda \times X \to K(E)$ has an essential fundamental set T then F (or G) is called *completely fundamentally restrictible* (on T). The corresponding multifield $\Phi = i - F$ (or family $\Psi = i - G$) is also called completely fundamentally restrictible.

Every compact multimap $F : X \to K(E)$ is completely fundamentally restrictible on E or on $\overline{\text{co}}(F(X) \cup \{a\})$, $a \in X$ arbitrary.

We will consider condensing multimaps as a principal class of fundamentally restrictible multimaps. Fundamental restrictibility is a very convenient tool, allowing us to treat different classes of multimaps from a unified point of view.

We give the necessary definitions.

Let $\beta : P(E) \to (A, \geq)$ be a MNC in E.

Definition 2.2.6. A multimap $F : X \to K(E)$ or a family of multimaps $G : \Lambda \times X \to K(E)$ is called *condensing relative to a MNC β* (or *β-condensing*) if for every $\Omega \subseteq X$ that is not relatively compact we have, respectively

$$\beta(F(\Omega)) \not\geq \beta(\Omega) \quad \text{or} \quad \beta(G(\Lambda \times \Omega)) \not\geq \beta(\Omega).$$

We introduce the following important class of condensing multimaps.

Definition 2.2.7. Let β be a real measure of noncompactness and $0 \le k < 1$. A multimap $F : X \to K(E)$ or a family of multimaps $G : \Lambda \times X \to K(E)$ is said to be (k, β)-*condensing* if, respectively,

$$\beta(F(\Omega)) \le k\,\beta(\Omega) \quad \text{or} \quad \beta(G(\Lambda \times \Omega)) \le k\,\beta(\Omega) \quad \text{for every } \Omega \subseteq X.$$

The corresponding multifields or families of multifields are also called β- or (k, β)-condensing.

We consider the following examples of condensing families of multimaps.

Proposition 2.2.1.

(i) *Let β be a monotone, semiadditive MNC, and $F_0, F_1 : X \to K(E)$ be β-condensing multimaps;*

or,

(ii) *Let β be a monotone MNC, invariant with respect to union with compact sets, $F_0 : X \to K(E)$ be β-condensing multimap and $F_1 : X \to K(E)$ a compact multimap.*

Then the family of multimaps $G : [0, 1] \times X \to K(E)$,

$$G(\lambda, x) = \lambda\, F_1(x) + (1 - \lambda)\, F_0(x)$$

is β-condensing.

Proof. In both cases, let $\beta(G(\Lambda \times \Omega)) \ge \beta(\Omega)$ for some $\Omega \subseteq X$. Then we have

$$\beta(G(\Lambda \times \Omega)) \le \beta(\overline{\mathrm{co}}(\,F_0(\Omega) \cup F_1(\Omega))) = \beta\,(\,F_0(\Omega) \cup F_1(\Omega)).$$

In case (i) we obtain

$$\beta\,(\,F_0(\Omega) \cup F_1(\Omega)) \le \max\,\{\beta\,(\,F_0(\Omega))\,,\ \beta\,(\,F_1(\Omega))\}$$

therefore $\max\,\{\beta\,(\,F_0(\Omega))\,,\ \beta\,(\,F_1(\Omega))\} \ge \beta(\Omega)$, hence Ω is relatively compact.

In case (ii) we have $\beta(F_0(\Omega) \cup F_1(\Omega)) = \beta(F_0(\Omega))$ and hence $\beta(F_0(\Omega)) \ge \beta(\Omega)$, implying the same conclusion. □

We give now conditions for the fundamental restrictibility and complete fundamental restrictibility of condensing multimaps.

Theorem 2.2.1. *Let $\mathcal{K} \subseteq E$ be a nonempty convex closed subset of E and X be a closed subset of $\mathcal{K}$ (in the relative topology).*

(a) *A family of multimaps $G : \Lambda \times X \to K(\mathcal{K})$ is fundamentally restrictible in each of the following cases:*

(i) *G is u.s.c. and β-condensing with respect to a monotone MNC β defined on subsets of $\mathcal{K}$;*

(ii) *X is bounded and G is (k, β)-condensing with respect to a real monotone and regular MNC β defined on subsets of $\mathcal{K}$.*

(b) *In both cases, if the MNC β is nonsingular then the family G is completely fundamentally restrictible. Moreover, given any finite set $P \subset X$, G admits a compact essential fundamental set containing P.*

Proof. (a) In both cases (i) and (ii), let $\{T_\alpha\}$ be the collection of all fundamental sets of G. (Note that the set $\mathcal{K}$ belongs to this collection).

Consider the set $T = \bigcap_\alpha T_\alpha$. From Lemmas 2.2.3 and 2.2.2 it follows that

$$T = \overline{\text{co}}(G(\Lambda \times (X \cap T))).$$

Then we have

$$\beta(X \cap T) \leq \beta(T) = \beta(\overline{\text{co}}(G(\Lambda \times (X \cap T))) = \beta(G(\Lambda \times (X \cap T))).$$

Now in case (i) the set $X \cap T$ is compact and the conclusion follows from Theorem 1.1.7.

In case (ii) we obtain $\beta(X \cap T) = \beta(G(\Lambda \times (X \cap T))) = 0$ and the assertion follows from the regularity of β.

(b) Consider the collection $\{T_\alpha\}$ of all fundamental sets of G containing the set P. Again we have $\mathcal{K} \in \{T_\alpha\}$.

Then for $T = \bigcap_\alpha T_\alpha$ we have $\emptyset \neq P \subset X \cap T$ and

$$T = \overline{\text{co}}(G(\Lambda \times (X \cap T)) \cup P) \tag{2.2.1}$$

therefore,

$$\begin{aligned}\beta(X \cap T) &\leq \beta(T) = \beta(\overline{\text{co}}(G(\Lambda \times (X \cap T)) \cup P)) \\ &= \beta(G(\Lambda \times (X \cap T) \cup P)) = \beta(G(\Lambda \times (X \cap T))).\end{aligned}$$

The same argument as above shows that T is an essential fundamental set. Compactness follows from the equality (2.2.1) and Mazur's theorem. □

Remark 2.2.1. If X is convex and $G(\Lambda \times X) \subset X$ we can assume, without loss of generality, that $X \cap T = T$ for any fundamental set T of G. Hence in both cases the condition of monotonicity of β is superfluous. In case (i) the condition of upper semicontinuity of G can be omitted.

Remark 2.2.2. From the proof can be noted that in case (ii), if β is real, monotone, nonsingular, but in general non-regular, we can guarantee only the existence of a fundamental set T of G such that $X \cap T \neq \emptyset$ and

$$\beta(X \cap T) = 0$$

or

$$\beta(T) = 0$$

if X is convex and $G(\Lambda \times X) \subseteq X$.

We will use the following important concepts.

Definition 2.2.8. A completely fundamentally restrictible multimap F is said to be 1- or 2-*completely fundamentally restrictible* if it possesses an essential fundamental set containing any prescribed point (respectively, any prescribed pair of points) from its domain. The corresponding multifield $\Phi = i - F$ will be called 1- or 2-completely fundamentally restrictible also.

Note that the condensing multimaps in Theorem 2.2.1 (b) are 1- and 2-completely fundamentally restrictible.

The following is an extension of Definition 2.2.7.

Definition 2.2.9. Let E_0, E_1 be Banach spaces and β_0, β_1 real measures of noncompactness in E_0 and E_1 respectively. A multimap $F : X \subseteq E_0 \to K(E_1)$ is said to be (k, β_0, β_1)*-bounded* if $\beta_1(F(\Omega)) \leq k\,\beta_0(\Omega)$ for all $\Omega \subseteq X$ and some $k \geq 0$.

It is obvious that in case $E_0 = E_1 = E$, $\beta_0 = \beta_1 = \beta$, $k < 1$ we return to the definition of a (k, β)-condensing multimap.

The following statement can be used as a test to verify the (k, χ_0, χ_1)-boundedness (or (k, χ)-condensivity) of a multimap.

Proposition 2.2.2. *Let E_0, E_1 be Banach spaces, χ_0, χ_1 Hausdorff measures of noncompactness in E_0 and E_1 respectively, $X \subseteq E_0$ and $\mathcal{M}$ a collection of bounded subsets of X.*

Suppose that the multimap $B : X \times E_0 \to K(E_1)$ satisfies the following conditions:

(i) *for any $x \in X$ the multimap $B(x, \cdot) : E_0 \to K(E_1)$ is k-Lipschitz with respect to the Hausdorff metric h on $K(E_1)$, i.e.,*

$$h(B(x, y_0), B(x, y_1)) \leq k\| y_0 - y_1 \|$$

for any $y_0, y_1 \in E_0$ where $k \in \mathbb{R}_+$ does not depend on x;

(ii) *the set $B(\Omega \times \{y\})$ is relatively compact in E_1 for any $\Omega \in \mathcal{M}$ and $y \in E_0$.*

Then the multimap $A : X \to K(E_1)$ defined as $A(x) = B(x, x)$ is (k, χ_0, χ_1)-bounded on $\mathcal{M}$, i.e.,

$$\chi_1(A(\Omega)) \leq k\,\chi_0(\Omega) \quad \textit{for any } \Omega \in \mathcal{M}.$$

Proof. Let $\Omega \in \mathcal{M}$, $\varepsilon > 0$ and S be a finite $(\chi_0(\Omega)+\varepsilon)$-net of Ω in E_0. We shall show that the relatively compact set $B(\Omega \times S)$ is a $k(\chi_0(\Omega)+\varepsilon)$-net for the set $A(\Omega) \subset E_1$.

Let $z \in A(\Omega)$, hence $z \in B(x, x)$ where $x \in X$. Take $y \in S$ such that $\| x - y \| \leq \chi_0(\Omega) + \varepsilon$.

From condition (i) it follows that there exists an element $z_1 \in B(x, y) \subset B(\Omega \times S)$ such that $\| z - z_1 \| \leq k\| x - y \| \leq k(\chi_0(\Omega) + \varepsilon)$.

Thus $\chi_1(A(\Omega)) \leq k(\chi_0(\Omega) + \varepsilon)$, and the assertion follows from the arbitrariness of ε. □

Corollary 2.2.1. *Let E be a Banach space; $X \subset E$ a bounded closed subset. If the multimap $F_0 : X \to K(E)$ is k-Lipschitz, $0 \leq k < 1$, with respect to the Hausdorff metric and the multimap $F_1 : X \to K(E)$ is compact then their sum $F_0 + F_1 : X \to K(E)$ is (k, χ)-condensing.*

Bibliographic notes

The definition of the MNC α and some applications of this concept to topology are again due to K. Kuratowski (see [155] and [157]). The MNC χ was studied in the work of L. S. Goldenshtein, I. Gohberg and A. S. Markus [101]. The further development of the concept of a MNC is connected with the names of A. Ambrosetti, Yu. G. Borisovich, J. Daneš, G. Darbo, M. Furi, R. D. Nussbaum, W. V. Petryshyn, B. N. Sadovskii, Yu. I. Sapronov, A. Vignoli and others (we refer the reader to the monographs of R. R. Akhmerov, M. I. Kamenskii, A. S. Potapov, A. E. Rodkina and B. N. Sadovskii [4], J. Banas and K. Goebel [25] and to the surveys [203] and [5].) The abstract definition of a MNC was given by B. N. Sadovskii (see [202], [203]).

Concrete formulas for the evaluation of different MNC's in separable Banach spaces were suggested by many authors (see, for example [25], [72], [101] and others). Measures of noncompactness in the space of continuous real-valued functions were studied in the works [101], [4] and others. Let us mention that the modulus of equicontinuity as MNC (see Example 2.1.2) was studied in the paper of J. F. Couchouron and M. I. Kamenskii [67] and the MNC given by the formula (2.1.10) was described in the work of M. I. Kamenskii [127].

Definitions of condensing maps in the single-valued case are due to the works of G. Darbo [70], L. S. Goldenshtein, I. Gohberg and A. S. Markus [101], B. N. Sadovskii [201], Yu. G. Borisovich and Yu. I. Sapronov [47], R. D. Nussbaum [174]. The general definition of a condensing operator was given by B. N. Sadovskii [201], [203].

The first treatments of condensing multivalued maps appeared in the works of J. Danes [69], C. J. Himmelberg, J. R. Porter and F. S. Van Vleck [117] and V. V. Obukhovskii [175]. In this exposition we consider the condensing multimaps as fundamentally restrictible ones. The idea of the restriction of a condensing map to a fundamental compact set is due to Yu. I. Sapronov [200] and V. V. Obukhovskii [175]. This fact permits to consider different classes of condensing (and even other noncompact) maps from a unified point of view. This is very convenient for the construction of topological degree theory and was exploited in many works (see [35]–[38], [40], [41], [43], [92], [172], [179], [181] and others). In conclusion we mention that Proposition 2.2.2, giving a useful test for verifying the condensivity of a multimap, is a generalization of Theorem 1.5.7 from [4] and was proved in the paper [94].

Chapter 3

Topological degree theory for condensing multifields

3.1 The relative topological degree and fixed points of compact multimaps

Our principal aim in this chapter is to define and investigate the topological degree for condensing multifields in a Banach space. In particular we focus our attention on the theory of topological degree for completely continuous convex-valued multifields. In this section we recall briefly the idea of its construction and its main properties.

We start with the following notion (details can be found in [34]).

Let $U \subset E$ be an open set and $\boldsymbol{K} \subseteq E$ be a nonempty convex closed set. The closure and the boundary of the set $U_{\boldsymbol{K}} \underset{\text{def}}{=} U \cap \boldsymbol{K}$ in the relative topology of the space $\boldsymbol{K}$ will be denoted by $\overline{U}_{\boldsymbol{K}}$ and $\partial U_{\boldsymbol{K}}$ respectively. We suppose, at first, that $\overline{U}_{\boldsymbol{K}}$ and $\partial U_{\boldsymbol{K}}$ are nonempty.

To every completely continuous (i.e., compact and continuous) map $f : \partial U_{\boldsymbol{K}} \to \boldsymbol{K}$ such that $x \neq f(x)$ for all $x \in \partial U_{\boldsymbol{K}}$, one can associate an integer defined as the *relative topological degree* of the corresponding vector field $\varphi = i - f$, $\deg_{\boldsymbol{K}}(\varphi, \partial U_{\boldsymbol{K}})$ with the following principal properties:

(i) *Normalization property.* If $f(x) \equiv x_0$ for all $x \in \partial U_{\boldsymbol{K}}$ then

$$\deg_{\boldsymbol{K}}(\varphi, \partial U_{\boldsymbol{K}}) = \begin{cases} 1, & \text{if } x_0 \in U_{\boldsymbol{K}}, \\ 0, & \text{if } x_0 \notin U_{\boldsymbol{K}}. \end{cases}$$

(ii) *Homotopy invariance.* Assume that the vector fields $\varphi_0 = i - f_0$ and $\varphi_1 = i - f_1$ corresponding to completely continuous maps $f_0, f_1 : \partial U_{\boldsymbol{K}} \to \boldsymbol{K}$ are homotopic ($\varphi_0 \sim \varphi_1$), i.e., there exists a completely continuous map $g : [0, 1] \times \partial U_{\boldsymbol{K}} \to \boldsymbol{K}$ such that

(a) $x \neq g(\lambda, x)$ for all $\lambda \in [0, 1]$; $x \in \partial U_{\boldsymbol{K}}$;

(b) $g(0, \cdot) = f_0$, $g(1, \cdot) = f_1$.

Then

$$\deg_{\boldsymbol{K}}(\varphi_0, \partial U_{\boldsymbol{K}}) = \deg_{\boldsymbol{K}}(\varphi_1, \partial U_{\boldsymbol{K}}).$$

(iii) *Additive dependence on the domain.* Let $\{U_{jK}\} j \in J$ be a family of disjoint subsets of U_K open in K, and $f : \overline{U}_K \to K$ be a completely continuous map, having no fixed points on $\overline{U}_K \setminus \bigcup_{j \in J} U_{jK}$. Then the degrees $\deg_K(\varphi, \partial U_{jK})$ are nonzero for only finitely many indices $j \in J$ and

$$\deg_K(\varphi, \partial U_K) = \sum_{j \in J} \deg_K(\varphi, \partial U_{jK}).$$

(iv) *The principle of map restriction.* Let K_1 be a closed, convex subset of E, $K_1 \subset K$. If $f : \partial U_K \to K$ is a completely continuous map, $x \neq f(x)$ for all $x \in \partial U_K$ and $f(\partial U_K) \subseteq K_1$ then

$$\deg_K(\varphi, \partial U_K) = \deg_{K_1}(\varphi, \partial U_{K_1}).$$

(v) *The odd field property.* Let K and U be symmetric with respect to the origin and $0 \in U_K$. Let $f : \partial U_K \to K$ be a completely continuous map such that for all $x \in \partial U_K$ we have $x \neq f(x)$ and $f(-x) = -f(x)$. Then the degree $\deg_K(\varphi, \partial U_K)$ is an odd number.

(vi) *The fixed point principle.* If $f : \overline{U}_K \to K$ is a completely continuous map such that $\deg_K(\varphi, \partial U_K) \neq 0$, then f admits a fixed point, that is there exists $x \in U_K$ such that $x = f(x)$.

We start now the construction of the relative topological degree for a completely continuous convex-valued multifield. To this aim we introduce the following concepts.

Definition 3.1.1. Two completely continuous multimaps $F_0, F_1 : \partial U_K \to Kv(K)$ and the corresponding multifields $\Phi_0 = i - F_0$ and $\Phi_1 = i - F_1$ are said to be *homotopic*

$$\Phi_0 \underset{K}{\sim} \Phi_1$$

if there exists a completely continuous family $G : [0, 1] \times \partial U_K \to Kv(K)$ such that $\operatorname{Fix} G(\lambda, \cdot) \cap \partial U_K = \emptyset$ for all $\lambda \in [0, 1]$ and $G(0, \cdot) = F_0$, $G(1, \cdot) = F_1$.

Definition 3.1.2. A completely continuous vector field $\varphi = i - f$, $f : \partial U_K \to K$ is said to be a *single-valued homotopic approximation* of the completely continuous multifield $\Phi = i - F$, $F : \partial U_K \to Kv(K)$, $\operatorname{Fix} F \cap \partial U_K = \emptyset$, if $\varphi \underset{K}{\sim} \Phi$.

To prove the existence of single-valued homotopic approximations we use the following result.

Lemma 3.1.1. *Let Λ be a compact topological space and $X \subseteq E$ be a closed set. Let $G : \Lambda \times X \to K(E)$ be a completely continuous family of multimaps, and $\Psi = i - G$ the corresponding family of multifields. Then the set $\Psi(\Lambda \times X)$ is closed.*

Proof. Let $\{z_n\} \subset \Psi(\Lambda \times X)$, $z_n \to z_0$. For every z_n we have the representation $z_n = x_n - y_n$ where $x_n \in X$, $y_n \in G(\lambda_n, x_n)$. Since the sequences $\{\lambda_n\}, \{y_n\}$ are contained in compact sets, we can assume without loss of generality that $\lambda_n \to \lambda_0$, $y_n \to y_0$. Then $x_n = y_n + z_n \to x_0 = y_0 + z_0 \in X$. As the multimap G is closed (see Theorem 1.1.4), it follows that $y_0 \in G(\lambda_0, x_0)$, hence $z_0 = x_0 - y_0 \in x_0 - G(\lambda_0, x_0) \subset \Psi(\Lambda \times X)$. □

Theorem 3.1.1. *Every completely continuous multifield $\Phi = i - F$, $F : \partial U_K \to Kv(\boldsymbol{K})$,* Fix $F \cap \partial U_K \neq \emptyset$ *admits a single-valued homotopic approximation.*

Proof. From the previous lemma it follows that the distance δ from the origin to the set $\Phi(\partial U_K)$ is positive. Then as a single valued homotopic approximation we can take the field $\varphi = i - f$ where $f \colon \partial U_K \to \boldsymbol{K}$ is an arbitrary ε-approximation $(0 < \varepsilon < \delta/2)$ such that $f(\partial U_K) \subset \operatorname{co} F(\partial U_K)$ (see Theorem 1.2.19). In fact it follows from Mazur's theorem that the set $\operatorname{co} F(\partial U_K)$ is relatively compact, and the homotopy between φ and Φ is realized by the multimap $G : [0,1] \times \partial U_K \to Kv(E)$, defined as $G(\lambda, x) = \lambda f(x) + (1-\lambda) F(x)$. □

We can give the following definition.

Definition 3.1.3. The *relative topological degree* $\deg_{\boldsymbol{K}}(\Phi, \partial U_K)$ of a completely continuous multifield $\Phi = i - F$, $F : \partial U_K \to Kv(\boldsymbol{K})$, Fix $F \cap \partial U_K = \emptyset$, is the relative topological degree $\deg_{\boldsymbol{K}}(\varphi, \partial U_K)$ of an arbitrary single-valued homotopic approximation of Φ.

The notion is well defined. In fact, we can prove the following statement.

Theorem 3.1.2. *Let $\varphi_0 = i - f_0$, $\varphi_1 = i - f_1$ be single -valued homotopic approximations of a completely continuous multifield $\Phi = i - F$, $F : \partial U_K \to Kv(\boldsymbol{K})$,* Fix $F \cap \partial U_K = \emptyset$. *Then φ_0 and φ_1 are homotopic in the class of single-valued completely continuous vector fields.*

Proof. It follows easily from the definition of single valued homotopic approximation that φ_0 and φ_1 are homotopic in the class of completely continuous multifields. Let $G : [0,1] \times \partial U_K \to Kv(E)$ be a completely continuous multimap realizing this homotopy, i.e., $G(0,\cdot) = f_0$, $G(1,\cdot) = f_1$. If $\Psi = i - G$ is the corresponding family of multifields then from Lemma 3.1.1 it follows that the set $\Psi([0,1] \times \partial U_K)$ is closed and hence its distance δ from 0 is positive. Now, let $g : [0,1] \times \partial U_K \to \boldsymbol{K}$ be an arbitrary single-valued ε-approximation of the multimap G, $(0 < \varepsilon < \delta/2)$ such that $g([0,1] \times \partial U_K) \subset \operatorname{co} G([0,1] \times \partial U_K)$. Then the required homotopy connecting φ_0 and φ_1 is given by the completely continuous map $h : [0,1] \times \partial U_K \to \boldsymbol{K}$, defined as

$$h(\lambda, x) = \begin{cases} 3\lambda\, g(0,x) + (1 - 3\lambda)\, f_0(x), & 0 \le \lambda \le 1/3, \\ g(3\lambda - 1, x), & 1/3 \le \lambda \le 2/3, \\ (3 - 3\lambda)\, g(1,x) + (3\lambda - 2)\, f_1(x), & 2/3 \le \lambda \le 1. \end{cases}$$ □

From this statement and the homotopy invariance property of the degree it follows that the topological degrees of all single-valued homotopic approximations of $\Phi = i - F$ relative to $\boldsymbol{K}$ are equal, justifying Definition 3.1.3.

Let us consider the main properties of this degree.

(a) *The normalization property.*

If $F(x) \equiv A$ for all $x \in \partial U_{\boldsymbol{K}}$ then

$$\deg_{\boldsymbol{K}}(i - F, \partial U_{\boldsymbol{K}}) = \begin{cases} 1, & \text{if } A \cap U_{\boldsymbol{K}} \neq \emptyset, \\ 0, & \text{if } A \cap \overline{U}_{\boldsymbol{K}} = \emptyset. \end{cases}$$

Proof. It is sufficient to take as a single-valued homotopic approximation $\phi = i - f$ where $f(x) \equiv x_0 \in A \cap U_{\boldsymbol{K}}$ in the first case and $f(x) \equiv x_0 \in A$ in the second case. □

(b) *Homotopy invariance.*

If $\Phi_0 \underset{\boldsymbol{K}}{\sim} \Phi_1$ then $\deg_{\boldsymbol{K}}(\Phi_0, \partial U_{\boldsymbol{K}}) = \deg_{\boldsymbol{K}}(\Phi_1, \partial U_{\boldsymbol{K}})$.

Proof. It follows immediately from Definitions 3.1.1–3.1.3. □

The following two properties can be proved by direct transfer to single-valued approximations.

(c) *Additive dependence on the domain.*

Let $\{U_{j\boldsymbol{K}}\}_{j \in J}$ be a family of open in $\boldsymbol{K}$ disjoint subsets of $U_{\boldsymbol{K}}$ and $F : \overline{U}_{\boldsymbol{K}} \to Kv(\boldsymbol{K})$ be a completely continuous multimap having no fixed points on $\overline{U}_{\boldsymbol{K}} \setminus \bigcup_{j \in J} U_{j\boldsymbol{K}}$. Then the degrees $\deg_{\boldsymbol{K}}(\Phi, \partial U_{j\boldsymbol{K}})$ are nonzero for only finitely many indices $j \in J$ and

$$\deg_{\boldsymbol{K}}(\Phi, \partial U_{\boldsymbol{K}}) = \sum_{j \in J} \deg_{\boldsymbol{K}}(\Phi, \partial U_{j\boldsymbol{K}}).$$

(d) *The principle of map restriction.*

Let $\boldsymbol{K}_1$ be a convex closed subset of E, $\boldsymbol{K}_1 \subset \boldsymbol{K}$. If a multimap $F : \partial U_{\boldsymbol{K}} \to Kv(\boldsymbol{K})$ is completely continuous, $\operatorname{Fix} F \cap \partial U_{\boldsymbol{K}} = \emptyset$ and $F(\partial U_{\boldsymbol{K}}) \subseteq \boldsymbol{K}_1$ then

$$\deg_{\boldsymbol{K}}(\Phi, \partial U_{\boldsymbol{K}}) = \deg_{\boldsymbol{K}_1}(\Phi, \partial U_{\boldsymbol{K}_1}).$$

We can now formulate the basic fixed point principle.

Theorem 3.1.3. *Let $\Phi = i - F$ be the multifield corresponding to the completely continuous multimap $F : \overline{U}_{\boldsymbol{K}} \to Kv(\boldsymbol{K})$ with* $\operatorname{Fix} F \cap \partial U_{\boldsymbol{K}} = \emptyset$. *If* $\deg_{\boldsymbol{K}}(\Phi, \partial U_{\boldsymbol{K}}) \neq 0$ *then* $\emptyset \neq \operatorname{Fix} F \subset U_{\boldsymbol{K}}$.

Proof. In fact, assume that F has no fixed points in $\overline{U}_{\boldsymbol{K}}$. According to Lemma 3.1.1 the set $\Phi(\overline{U}_{\boldsymbol{K}})$ is closed and hence the distance δ of this set from zero is positive. Then for each $0 < \varepsilon < \delta/2$, any ε-approximation $f : \overline{U}_{\boldsymbol{K}} \to \boldsymbol{K}$ of the multimap F has no fixed points, hence we have $\deg_{\boldsymbol{K}}(\varphi, \partial U_{\boldsymbol{K}}) = 0$ implying $\deg_{\boldsymbol{K}}(\Phi, \partial U_{\boldsymbol{K}}) = 0$. □

This principle allows a simple and geometrically clear proof of many fixed point theorems used in applications.

We give some examples.

Theorem 3.1.4. *Let $U_{\boldsymbol{K}}$ be a nonempty convex set and $F : \overline{U}_{\boldsymbol{K}} \to Kv(\boldsymbol{K})$ be a completely continuous multimap such that $F(x) \cap \overline{U}_{\boldsymbol{K}} \neq \emptyset$ for all $x \in \partial U_{\boldsymbol{K}}$.*

Then Fix $F \neq \emptyset$. *If in addition* Fix $F \cap \partial U_{\boldsymbol{K}} = \emptyset$ *then* $\deg_{\boldsymbol{K}}(\Phi, \partial U_{\boldsymbol{K}}) = 1$.

Proof. Let Fix $F \cap \partial U_{\boldsymbol{K}} = \emptyset$. From Corollary 1.2.1 it follows that the multimap $F_1 : \partial U_{\boldsymbol{K}} \to Kv(\boldsymbol{K})$, $F_1(x) = F(x) \cap \overline{U}_{\boldsymbol{K}}$ is completely continuous. Let $x_0 \in U_{\boldsymbol{K}}$ be an arbitrary point. Then it is easy to see that the multimap $G : [0, 1] \times \partial U_{\boldsymbol{K}} \to Kv(E)$,

$$G(\lambda, x) = \begin{cases} (1 - 2\lambda)F(x) + 2\lambda\, F_1(x), & 0 \leq \lambda \leq 1/2, \\ (2 - 2\lambda)F_1(x) + (2\lambda - 1)\, x_0, & 1/2 \leq \lambda \leq 1, \end{cases}$$

generates a homotopy between the multifield $\Phi = i - F$ and the multifield $\Phi_0 = i - F_0$, where $F_0(x) \equiv x_0$. The assertion follows from the properties of homotopy invariance and normalization and Theorem 3.1.3. □

The well-known Bohnenblust–Karlin fixed point theorem ([31]) is a corollary of Theorem 3.1.4. To prove this, let us recall that from the Tietze–Dugundji extension theorem it follows that every convex closed subset of E is a retract of E.

Theorem 3.1.5. *If M is a closed convex subset of E and $F : M \to Kv(M)$ is a completely continuous multimap, then* Fix $F \neq \emptyset$.

Proof. Let us take $\boldsymbol{K} = E$ and U an arbitrary open bounded set containing the compact set $M_1 = \overline{\text{co}}\, F(M)$. Consider now the retraction $\rho : \overline{U} \to M_1$ and the multimap $F_1 : \overline{U} \to Kv(M)$ defined as $F_1 = F \circ \rho$. Such a map is obviously completely continuous (cf. Theorem 1.2.8) and satisfies the conditions of Theorem 3.1.4, therefore Fix $F_1 \neq \emptyset$. But if $x \in$ Fix F_1 then $x \in M_1$ and hence $\rho(x) = x$. Thus $x \in F_1(x) = F(\rho(x)) = F(x)$, i.e., $x \in$ Fix F. □

Remark 3.1.1. We can now eliminate the assumption that $\overline{U}_{\boldsymbol{K}}$ and $\partial U_{\boldsymbol{K}}$ are nonempty sets. In case $\overline{U}_{\boldsymbol{K}} = \emptyset$ we set $\deg_{\boldsymbol{K}}(\Phi, \partial U_{\boldsymbol{K}}) = 0$ and in case $\boldsymbol{K} \subset U$ it is natural to assume that $\deg_{\boldsymbol{K}}(\Phi, \partial U_{\boldsymbol{K}}) = 1$.

We can now formulate the following fixed point theorem.

Theorem 3.1.6. *Let $\partial U_{\boldsymbol{K}} \neq \emptyset$ and $F : \overline{U}_{\boldsymbol{K}} \to Kv(\boldsymbol{K})$ be a completely continuous multimap, $\Phi = i - F$ be the corresponding multifield. Suppose that for every $x \in \partial U_{\boldsymbol{K}}$ we have*

$$\Phi(x) \cap L_x^a = \emptyset$$

where

$$L_x^a = \{z : z \in E,\ z = \mu(x - a),\ \mu \leq 0\} \ \text{for a certain point } a \in U_{\boldsymbol{K}}.$$

Then $\deg_{\boldsymbol{K}}(\Phi, \partial U_{\boldsymbol{K}}) = 1$ *and hence* Fix $F \neq \emptyset$.

Proof. The multifields Φ and $\Phi_0 = i - F_0$, where $F_0(x) \equiv a$, are connected on $\partial U_{\boldsymbol{K}}$ by the homotopy realized by the multimap $G : [0, 1] \times \partial U_{\boldsymbol{K}} \to Kv(E)$ defined as

$$G(\lambda, x) = \lambda\, F_0(x) + (1 - \lambda)\, F(x).$$

Consequently, $\deg_{\boldsymbol{K}}(\Phi, \partial U_{\boldsymbol{K}}) = \deg_{\boldsymbol{K}}(\Phi_0, \partial U_{\boldsymbol{K}})$ and the assertion follows from the normalization property. □

As a corollary we can obtain the following fixed point principle for multimaps satisfying the so called *Leray–Schauder boundary condition.*

Corollary 3.1.1. *Let $0 \in U_{\boldsymbol{K}}$, $\partial U_{\boldsymbol{K}} \neq \emptyset$ and $F : \overline{U}_{\boldsymbol{K}} \to Kv(\boldsymbol{K})$ a completely continuous multimap such that*

$$F(x) \cap \gamma x = \emptyset, \quad \forall \gamma \geq 1,\ \forall x \in \partial U_{\boldsymbol{K}}.$$

Then $\deg_{\boldsymbol{K}}(\Phi, \partial U_{\boldsymbol{K}}) = 1$ *and hence* Fix $F \neq \emptyset$.

Another interesting corollary is the fixed point principle for multimaps "*directed inward*" on $\partial U_{\boldsymbol{K}}$.

Corollary 3.1.2. *Suppose that $U_{\boldsymbol{K}}$ is a convex set, $0 \in U_{\boldsymbol{K}}$, $\partial U_{\boldsymbol{K}} \neq \emptyset$. Let $F : \overline{U}_{\boldsymbol{K}} \to Kv(\boldsymbol{K})$ be a completely continuous multimap having no fixed point on $\partial U_{\boldsymbol{K}}$ and such that*

$$F(x) \subset I_{\overline{U}}(x) = \{y : y \in E,\ x + \gamma(y - x) \in \overline{U}_{\boldsymbol{K}} \text{ for some } \gamma > 0\}, \quad \forall x \in \partial U_{\boldsymbol{K}}.$$

Then $\deg_{\boldsymbol{K}}(\Phi, \partial U_{\boldsymbol{K}}) = 1$ *and hence* Fix $F \neq \emptyset$.

We can prove now the *odd field theorem.*

Theorem 3.1.7. *Suppose that $\boldsymbol{K}$ and $U_{\boldsymbol{K}}$ are symmetric with respect to the origin, $0 \in U_{\boldsymbol{K}}$ and $\partial U_{\boldsymbol{K}} \neq \emptyset$. Let $F : \overline{U}_{\boldsymbol{K}} \to Kv(\boldsymbol{K})$ be a completely continuous multimap such that the corresponding multifield $\Phi = i - F$ satisfies the condition*

$$\Phi(x) \cap \mu\Phi(-x) = \emptyset, \quad \forall x \in \partial U_{\boldsymbol{K}},\ 0 \leq \mu \leq 1.$$

Then $\deg_{\boldsymbol{K}}(\Phi, \partial U_{\boldsymbol{K}})$ *is an odd number and hence* Fix $F \neq \emptyset$.

Proof. We start showing that there exists $\delta > 0$ such that for any ε-approximation f of F on ∂U_K, $0 < \varepsilon < \delta$, the associated field $\varphi = i - f$ satisfies the same boundary condition

$$\varphi(x) \neq \mu\varphi(-x)$$

for all $x \in \partial U_K$ and $0 \leq \mu \leq 1$. In fact, supposing the contrary, we will have a sequence of ε_n-approximations f_n of F with $\varepsilon_n \to 0$ such that $\varphi_n(x_n) = \mu_n\varphi_n(-x_n)$ for some $x_n \in \partial U_K$, $\mu_n \in [0, 1]$ where $\varphi_n = i - f_n$. It means that

$$x_n = \frac{1}{(1+\mu_n)}\left[f_n(x_n) - \mu_n f_n(-x_n)\right].$$

Since the sequences $\{f_n(x_n)\}$,$\{f_n(-x_n)\}$ are contained in a relatively compact set $\operatorname{co} F(\partial U_K)$ we can assume without loss of generality that $f_n(x_n) \to y_0$, $f_n(-x_n) \to z_0$ and also $\mu_n \to \mu_0 \in [0, 1]$. Therefore $x_n \to x_0 = \frac{1}{(1+\mu_0)}(y_0 - \mu_0 z_0)$ and so $x_0 - y_0 = \mu_0(-x_0 - z_0)$. But $(x_n, f_n(x_n)), (-x_n, f_n(-x_n)) \in \Gamma_{f_n} \subset W_{\varepsilon_n}(\Gamma_F)$ and since the set Γ_F is closed (Theorem 1.1.4) we have $(x_0, y_0), (-x_0, z_0) \in \Gamma_F$ and hence $\Phi(x_0) \bigcap \mu_0\Phi(-x_0)) \neq \emptyset$ contrary to the assumption.

The homotopy generated by the family of fields $\psi : [0, 1] \times \partial U_K \to E \setminus 0$, defined as $\psi(\lambda, x) = 1/(1+\lambda)\,[\varphi(x) - \lambda\varphi(-x)]$, connects the field $\varphi(\cdot)$ with the field $\psi_1(\cdot) = \psi(1, \cdot)$ which is odd, i.e., $\psi_1(-x) = -\psi_1(x)$ for all $x \in \partial U_K$. Considering the field φ as a single-valued homotopic approximation of the multifield Φ, we obtain that $\deg_K(\Phi, \partial U_K) = \deg_K(\varphi, \partial U_K) = \deg_K(\psi_1, \partial U_K)$ is an odd number, proving the theorem. □

3.2 Topological degree for condensing multifields

3.2.1 The definition of the degree

We approach now the problem of defining the relative topological degree for condensing convex-valued multifields. Let U be an open set in a Banach space E, $K \subseteq E$ a convex, closed set such that $U_K \neq \emptyset$. Let $F : \overline{U}_K \to Kv(K)$ be a closed fundamentally restrictible multimap, $\operatorname{Fix} F \cap \partial U_K = \emptyset$.

Definition 3.2.1. The topological degree of the multifield $\Phi = i - F$ is defined as

$$\deg_K(\Phi, \overline{U}_K) \underset{\text{def}}{=} 0$$

if F is not completely fundamentally restrictible. If F is completely fundamentally restrictible, then

$$\deg_K(\Phi, \overline{U}_K) \underset{\text{def}}{=} \deg_T(\Phi, \partial U_T)$$

where $T \subseteq K$ is an arbitrary essential fundamental set of F.

Theorem 3.2.1. *Definition* 3.2.1 *is consistent, that is the topological degree does not depend on the choice of the essential fundamental set* $T \subseteq \boldsymbol{K}$ *when* F *is completely fundamentally restrictible.*

Proof. Note first that by Theorem 1.1.5, the restriction of F to ∂U_T is u.s.c.

Let T_0, T_1 be two essential fundamental sets of F. Then by Lemma 2.2.3 the set $T = T_0 \cap T_1$ is also fundamental. If $\overline{U}_T = \emptyset$ then from Lemma 2.2.1 it follows that $\operatorname{Fix} F = \emptyset$ and hence

$$\deg_{T_0}(\Phi, \partial U_{T_0}) = \deg_{T_1}(\Phi, \partial U_{T_1}) = 0.$$

In case $\overline{U}_T \neq \emptyset$ we prove that

$$\deg_{T_i}(\Phi, \partial U_{T_i}) = \deg_T(\Phi, \partial U_T), \quad i = 0, 1.$$

Consider the retraction $\rho : E \to \overline{\operatorname{co}}(F(\overline{U}_T))$ and the completely continuous multimap $\widehat{F} : \overline{U}_{T_0} \to Kv(T)$, $\widehat{F}(x) = \overline{\operatorname{co}}(\rho \circ F)(x)$. The family $G : [0, 1] \times \overline{U}_{T_0} \to Kv(T_0)$ defined as

$$G(\lambda, x) = (1 - \lambda) F(x) + \lambda \widehat{F}(x)$$

creates a homotopy $\Phi \underset{T_0}{\sim} \widehat{\Phi}$ on ∂U_{T_0} where $\widehat{\Phi} = i - \widehat{F}$.

In fact, it easy to see that the family G is completely continuous. Now let $x_0 \in G(\lambda_0, x_0)$ for some $\lambda_0 \in [0, 1], x_0 \in \partial U_{T_0}$. Clearly $x_0 \in \overline{\operatorname{co}}(F(x_0) \cup T)$ hence $x_0 \in T$. Therefore $G(\lambda_0, x_0) = F(x_0)$ and $x_0 \in \operatorname{Fix} F$, contradicting the assumption that $\operatorname{Fix} F \cap \partial U_{\boldsymbol{K}} = \emptyset$.

From the homotopy property of the degree it follows that

$$\deg_{T_0}(\Phi, \partial U_{T_0}) = \deg_{T_0}(\widehat{\Phi}, \partial U_{T_0}),$$

but the map restriction property implies that

$$\deg_{T_0}(\widehat{\Phi}, \partial U_{T_0}) = \deg_T(\widehat{\Phi}, \partial U_T) = \deg_T(\Phi, \partial U_T).$$

Therefore $\deg_{T_0}(\Phi, \partial U_{T_0}) = \deg_T(\Phi, \partial U_T)$.

The equality $\deg_{T_1}(\Phi, \partial U_{T_1}) = \deg_T(\Phi, \partial U_T)$ can be proved analogously. □

Our aim now is to describe the main properties of the topological degree for fundamentally restrictible multifields. We start with the following notion.

Definition 3.2.2. Two closed (completely) fundamentally restrictible multimaps F_0, $F_1 : \overline{U}_{\boldsymbol{K}} \to Kv(\boldsymbol{K})$ and the corresponding multifields $\Phi_0 = i - F_0$, $\Phi_1 = i - F_1$ are said to be homotopic

$$\Phi_0 \underset{\boldsymbol{K}}{\sim} \Phi_1$$

if there exists a closed (completely) fundamentally restrictible family $G : [0, 1] \times \overline{U}_{\boldsymbol{K}} \to Kv(\boldsymbol{K})$ such that $\operatorname{Fix} G(\lambda, \cdot) \cap \partial U_{\boldsymbol{K}} = \emptyset$ for all $\lambda \in [0, 1]$, and $G(0, \cdot) = F_0(\cdot)$, $G(1, \cdot) = F_1(\cdot)$.

As a direct consequence of Definitions 3.2.1 and 3.2.2 we obtain from (b) of Section 3.1 the following invariance property.

Property 3.2.1. *Homotopy invariance.* If the closed fundamentally restrictible multifields Φ_0 and Φ_1 are homotopic, $\Phi_0 \underset{K}{\sim} \Phi_1$, then

$$\deg_{\boldsymbol{K}}(\Phi_0, \overline{U}_{\boldsymbol{K}}) = \deg_{\boldsymbol{K}}(\Phi_1, \overline{U}_{\boldsymbol{K}}).$$

We can also deduce the following extensions of the corresponding properties of the topological degree for the case of fundamentally restrictible multifields.

Property 3.2.2. *Additive dependence on the domain.* Let $\{U_{j\boldsymbol{K}}\}_{j\in J}$ be a family of open in $\boldsymbol{K}$ disjoint subsets of $U_{\boldsymbol{K}}$ and $F : \overline{U}_{\boldsymbol{K}} \to Kv(\boldsymbol{K})$ be a fundamentally restrictible multimap having no fixed points on $\overline{U}_{\boldsymbol{K}} \setminus \cup U_{j\boldsymbol{K}}$. Then the degrees $\deg_{\boldsymbol{K}}(\Phi, \overline{U}_{j\boldsymbol{K}})$ are nonzero for only finitely many indices $j \in J$ and

$$\deg_{\boldsymbol{K}}(\Phi, \overline{U}_{\boldsymbol{K}}) = \sum_{j\in J} \deg_{\boldsymbol{K}}(\Phi, \overline{U}_{j\boldsymbol{K}}).$$

Property 3.2.3. *The principle of map restriction.* Let $\boldsymbol{K}_1$ be a convex closed subset of E, $\boldsymbol{K}_1 \subset \boldsymbol{K}$. If the multimap $F : \overline{U}_{\boldsymbol{K}} \to Kv(\boldsymbol{K})$ is fundamentally restrictible, $\operatorname{Fix} F \cap \partial U_{\boldsymbol{K}} = \emptyset$ and $F(\overline{U}_{\boldsymbol{K}}) \subseteq \boldsymbol{K}_1$ then

$$\deg_{\boldsymbol{K}}(\Phi, \overline{U}_{\boldsymbol{K}}) = \deg_{\boldsymbol{K}_1}(\Phi, \overline{U}_{\boldsymbol{K}_1}).$$

3.2.2 The relative topological degree for fundamentally restrictible multifields defined on the boundary

We need to pay some attention to the following situation. In accordance with Definition 3.2.1 the topological degree $\deg_{\boldsymbol{K}}(\Phi, \overline{U}_{\boldsymbol{K}})$ is defined by the values of the multifield Φ on the relative boundary ∂U_T. On the other hand we take an essential fundamental set T of the multimap F given on the whole domain $\overline{U}_{\boldsymbol{K}}$, therefore, in the general case, the degree of a fundamentally restrictible multifield $\deg_{\boldsymbol{K}}(\Phi, \overline{U}_{\boldsymbol{K}})$ is really defined on $\overline{U}_{\boldsymbol{K}}$. In this connection it would be interesting to learn for which classes of noncompact multifields the topological degree can be correctly defined (as for completely continuous ones) on the relative boundary $\partial U_{\boldsymbol{K}}$ and consequently would not depend on values taken by a multifield "inside" the domain, i.e., on $U_{\boldsymbol{K}}$. To clear up this problem let us recall (see Definition 2.2.8) that a completely fundamentally restrictible multimap is said to be 2-completely fundamentally restrictible if there exists an essential fundamental set containing any two prescribed points from its domain.

The collection of closed 2-completely fundamentally restrictible multifields $\Phi = i - F : \partial U_{\boldsymbol{K}} \to Kv(\boldsymbol{K})$, $\operatorname{Fix} F \cap \partial U_{\boldsymbol{K}} = \emptyset$ is just the class of multifields for which the topological degree $\deg_{\boldsymbol{K}}(\Phi, \partial U_{\boldsymbol{K}})$ can be correctly defined on $\partial U_{\boldsymbol{K}}$ as $\deg_T(\Phi, \partial U_T)$ where T is an arbitrary essential fundamental set of Φ.

In fact, if T_0, T_1 are two essential fundamental sets of F then, due to the fact that Φ is 2-completely fundamentally restrictible, there exists an essential fundamental set T' such that $T' \cap T_0 \cap \partial U_K \neq \emptyset$ and $T' \cap T_1 \cap \partial U_K \neq \emptyset$. Then, repeating the reasoning made during the proof of Theorem 3.2.1 we can verify that

$$\deg_{T_i}(\Phi, \partial U_{T_i}) = \deg_{T'}(\Phi, \partial U_{T'}), \quad i = 0, 1,$$

and therefore

$$\deg_{T_0}(\Phi, \partial U_{T_0}) = \deg_{T_1}(\Phi, \partial U_{T_1}),$$

justifying the correctness.

3.2.3 Topological degree for condensing multifields via compact homotopy approximations

Let β be a MNC in E which is monotone and invariant with respect to the union with compact sets and $F_0, F_1 : \partial U_K \multimap Kv(K)$, $\operatorname{Fix} F_i = \emptyset$, $i = 0, 1$ be u.s.c. β-condensing multimaps. We give the following refinement of Definition 3.2.2.

Definition 3.2.3. The multifields $\Phi_0 = i - F_0$, $\Phi_1 = i - F_1$ are said to be β-*homotopic* if there exists an u.s.c. β-condensing family of multimaps $G : [0, 1] \times \partial U_K \multimap Kv(K)$ such that $\operatorname{Fix} G(\lambda, \cdot) = \emptyset$ and $G(0, \cdot) = F_0$, $G(1, \cdot) = F_1$.

The following fact is crucial for our program.

Lemma 3.2.1. *The class of multifields β-homotopic to a given u.s.c. β-condensing multifield $\Phi = i - F : \partial U_K \multimap Kv(K)$, $\operatorname{Fix} F = \emptyset$ contains a completely continuous representative $\widehat{\Phi}$ and moreover, the β-homotopy $\Phi \underset{\beta}{\sim} \widehat{\Phi}$ is realized by a family $G : [0, 1] \times \partial U_K \multimap Kv(K)$ of the form*

$$G(\lambda, x) = \lambda \widehat{F}(x) + (1 - \lambda) F(x).$$

Proof. In fact, if T is an essential fundamental set of F and $\rho : E \to \overline{\text{co}}\, F(\partial U_K \cap T)$ is a retraction then $\widehat{\Phi}$ can be given by the formula $\widehat{\Phi} = i - \widehat{F}$,

$$\widehat{F}(x) = \overline{\text{co}} \rho \circ F(x).$$

Indeed, then the family $G(\lambda, x) = \lambda \widehat{F}(x) + (1 - \lambda) F(x)$ is β-condensing due to Proposition 2.2.1 (ii) and fixed point free since $x \in G(\lambda, x)$, $x \in \partial U_K$ easily implies that (see the proof of Theorem 3.2.1) $x \in \operatorname{Fix} F$, which is the empty set. □

The last observation justifies the introduction of the following concept.

Definition 3.2.4. A completely continuous multifield $\widehat{\Phi} = i - F : \partial U_{\boldsymbol{K}} \to Kv(\boldsymbol{K})$, $\operatorname{Fix} \widehat{F} = \emptyset$ is said to be a *compact homotopy approximation* of an u.s.c. β-condensing multifield $\Phi = i - F : \partial U_{\boldsymbol{K}} \to Kv(\boldsymbol{K})$, $\operatorname{Fix} F = \emptyset$ provided

$$\Phi \underset{\beta}{\sim} \widehat{\Phi}.$$

We can now give the following definition for the topological degree of a β-condensing multifield, often useful in applications.

Definition 3.2.5. Let $\Phi = i - F : \partial U_{\boldsymbol{K}} \to Kv(\boldsymbol{K})$, $\operatorname{Fix} F = \emptyset$ be an u.s.c. β-condensing multifield. Then

$$\deg_{\boldsymbol{K}}(\Phi, \partial U_{\boldsymbol{K}}) = \deg_{\boldsymbol{K}}(\widehat{\Phi}, \partial U_{\boldsymbol{K}})$$

where $\widehat{\Phi}$ is an arbitrary compact homotopy approximation of Φ.

Lemma 3.2.2. *Definition* 3.2.5 *is correct and agrees with Definition* 3.2.1.

Proof. Let $\widehat{\Phi}_0 = i - \widehat{F}_0$, $\widehat{\Phi}_1 = i - \widehat{F}_1$ be two arbitrary compact homotopy approximations of a multifield Φ and $G : [0,1] \times \partial U_{\boldsymbol{K}} \to Kv(\boldsymbol{K})$, $G(0,\cdot) = \widehat{F}_0$, $G(1,\cdot) = \widehat{F}_1$ be an u.s.c. β-condensing family of multimaps realizing the homotopy $\widehat{\Phi}_0 \underset{\beta}{\sim} \widehat{\Phi}_1$.

Now let T be an arbitrary essential fundamental set of the family G and $\rho : E \to \overline{\operatorname{co}}\, G([0,1] \times \partial U_{\boldsymbol{K}})$ be an arbitrary retraction. Define a completely continuous family $G' : [0,1] \times \partial U_{\boldsymbol{K}} \to Kv(\boldsymbol{K})$,

$$G'(\lambda, x) = \overline{\operatorname{co}} \rho \circ G(\lambda, x).$$

Then the completely continuous family $\widetilde{G} : [0,1] \times \partial U_{\boldsymbol{K}} \to Kv(\boldsymbol{K})$,

$$\widetilde{G}(x,\lambda) = \begin{cases} 3\lambda G'(0,x) + (1-3\lambda)\widehat{F}_0(x), & 0 \le \lambda \le 1/3; \\ G'(3\lambda - 1, x), & 1/3 \le \lambda \le 2/3; \\ (3-3\lambda)G'(1,x) + (3\lambda - 2)\widehat{F}_1(x), & 2/3 \le \lambda \le 1 \end{cases}$$

realizes the completely continuous homotopy $\widehat{\Phi}_0 \sim \widehat{\Phi}_1$ and due to the Homotopy Invariance Property for completely continuous multifields (see (b) of Section 3.1)

$$\deg_{\boldsymbol{K}}(\widehat{\Phi}_0, \partial U_{\boldsymbol{K}}) = \deg_{\boldsymbol{K}}(\widehat{\Phi}_1, \partial U_{\boldsymbol{K}}).$$

As we have seen above we can find a compact homotopy approximation $\widehat{\Phi} = i - \widehat{F}$ of a multifield $\Phi = i - F$ such that $\widehat{F}(\partial U_{\boldsymbol{K}}) \subseteq T$ and $\widehat{F}|_{\partial U_{\boldsymbol{K}} \cap T} = F|_{\partial U_{\boldsymbol{K}} \cap T}$ where T is an arbitrary essential fundamental set of F. Then from the principle of map restriction (d) of Section 3.1 it follows that

$$\deg_{\boldsymbol{K}}(\widehat{\Phi}, \partial U_{\boldsymbol{K}}) = \deg_T(\widehat{\Phi}, \partial U_T) = \deg_T(\Phi, \partial U_T). \qquad \square$$

3.3 The evaluation of the degree

3.3.1 Some fixed point theorems

In this section we will evaluate the topological degree for multifields in different situations. We use the following general principle, which is a direct consequence of Definition 3.2.1 and Theorem 3.1.3 to obtain, as corollaries, the corresponding fixed point theorems.

Theorem 3.3.1. *Let $\Phi = i - F$ be a multifield corresponding to a completely fundamentally restrictible multimap $F : \overline{U}_{\boldsymbol{K}} \to Kv(\boldsymbol{K})$ such that* Fix $F \cap \partial U_{\boldsymbol{K}} = \emptyset$. *If* $\deg_{\boldsymbol{K}}(\Phi, \overline{U}_{\boldsymbol{K}}) \neq 0$ *then* $\emptyset \neq$ Fix $F \subset U_{\boldsymbol{K}}$.

Let $F : \overline{U}_{\boldsymbol{K}} \to Kv(\boldsymbol{K})$ be a completely fundamentally restrictible multimap.

Suppose that N is an isolated component of the fixed point set Fix F, i.e., there exists $\varepsilon > 0$ such that

$$W_\varepsilon(N) \cap \operatorname{Fix} F = N$$

where $W_\varepsilon(N)$ denotes the ε-neighborhood of the set N in the relative topology of the set $\boldsymbol{K}$. From Property 3.2.2, the additive dependence on the domain, it follows that the topological degree

$$\deg_{\boldsymbol{K}}(i - F, \overline{W_{\varepsilon_1}(N)})$$

does not depend on the choice of $\varepsilon_1, 0 < \varepsilon_1 < \varepsilon$. This generic value of the topological degree will be called the *index of the set* N. It will be denoted as

$$\operatorname{ind}(i - F, N).$$

In particular, we denote by $\operatorname{ind}(i - F, x_0)$ the index of an isolated fixed point x_0.

It is easy to see that when a completely fundamentally restrictible multimap F has only a finite number of fixed points $x_1, \ldots, x_n$ in $U_{\boldsymbol{K}}$ we have

$$\deg_{\boldsymbol{K}}(i - F, \overline{U}_{\boldsymbol{K}}) = \sum_{i=1}^{n} \operatorname{ind}(i - F, x_i).$$

Using Theorem 3.3.1 we can formulate the “noncompact” analog of the fixed point principle expressed in Theorem 3.1.4.

Theorem 3.3.2. *Let $U_{\boldsymbol{K}}$ be a convex set and $F : \overline{U}_{\boldsymbol{K}} \to Kv(\boldsymbol{K})$ a closed completely fundamentally restrictible multimap such that*

$$F(x) \cap \overline{U}_{\boldsymbol{K}} \neq \emptyset$$

for all $x \in \partial U_{\boldsymbol{K}}$. Then Fix $F \neq \emptyset$. *If, in addition,* Fix $F \cap \partial U_{\boldsymbol{K}} = \emptyset$ *then*

$$\deg_{\boldsymbol{K}}(\Phi, \overline{U}_{\boldsymbol{K}}) = 1.$$

Proof. Let T be an essential fundamental set of F. If $\partial U_T = \emptyset$ then the assertion follows from Theorem 3.1.5 and Remark 3.1.1. If $\partial U_T \neq \emptyset$ we can apply Theorem 3.1.4. □

The following fixed point theorem generalizes the Bohnenblust–Karlin Theorem 3.1.5 and some fixed point principles for noncompact multimaps.

Theorem 3.3.3. *If M is a closed convex subset of E and $F : M \to Kv(M)$ is a closed completely fundamentally restrictible multimap, then* $\operatorname{Fix} F \neq \emptyset$.

Proof. If $T \subseteq M$ is an essential fundamental set of F we can apply Theorem 3.1.5 to the restriction of F to T. □

Keeping in mind Theorem 2.2.1 and Remark 2.2.1 we can state the following fixed point principles for condensing multimaps.

Corollary 3.3.1. *If M is a convex closed subset of E, and $F : M \to Kv(M)$ is a closed β-condensing multimap, where β is a nonsingular MNC defined on subsets of M, then* $\operatorname{Fix} F \neq \emptyset$.

Corollary 3.3.2. *If M is a bounded convex closed subset of E, and $F : M \to Kv(M)$ is a closed (k, β)-condensing multimap, where β is a real nonsingular and regular MNC defined on subsets of M, then* $\operatorname{Fix} F \neq \emptyset$.

The following fixed point principles are valid for a class of 1-completely fundamentally restrictible multimaps.

Theorem 3.3.4. *Let $\partial U_{\boldsymbol{K}} \neq \emptyset$ and $F : \overline{U}_{\boldsymbol{K}} \to Kv(\boldsymbol{K})$ be a closed 1-completely fundamentally restrictible multimap. Suppose there exists $a \in U_{\boldsymbol{K}}$ such that*

$$\Phi(x) \cap L_x^a = \emptyset$$

for the corresponding multifield $\Phi = i - F$ at every $x \in \partial U_{\boldsymbol{K}}$, and

$$L_x^a = \{z : z \in E,\ z = \mu(x - a),\ \mu \leq 0\}.$$

Then $\deg_{\boldsymbol{K}}(\Phi, \partial U_{\boldsymbol{K}}) = 1$ *and hence* $\operatorname{Fix} F \neq \emptyset$.

Proof. Take an essential fundamental set T of F containing a point $\{a\}$. In case $\partial U_T \neq \emptyset$ apply Theorem 3.1.6 and use Remark 3.1.1 if $\partial U_T = \emptyset$. □

Corollary 3.3.3. *Let $0 \in U_{\boldsymbol{K}}$, $\partial U_{\boldsymbol{K}} \neq \emptyset$ and $F : \overline{U}_{\boldsymbol{K}} \to Kv(\boldsymbol{K})$ a closed 1-completely fundamentally restrictible multimap such that*

$$F(x) \cap \gamma x = \emptyset$$

for all $\gamma \geq 1$ at each point $x \in \partial U_{\boldsymbol{K}}$. Then $\deg_{\boldsymbol{K}}(\Phi, \overline{U}_{\boldsymbol{K}}) = 1$ *and hence* $\operatorname{Fix} F \neq \emptyset$.

Corollary 3.3.4. *Suppose that U_K is convex, $0 \in U_K$ and $\partial U_K \neq \emptyset$. Let $F : \overline{U_K} \to Kv(K)$ be a closed 1-completely fundamentally restrictible multimap having no fixed points on ∂U_K and such that*

$$F(x) \subset I_{\overline{U}} = \{y : y \in E,\ x + \gamma(y - x) \in \overline{U}_K \text{ for some } \gamma > 0\}$$

at every $x \in \partial U_K$. Then $\deg_K(\Phi, \overline{U}_K) = 1$ and hence $\operatorname{Fix} F \neq \emptyset$.

Theorem 3.3.5. *Suppose that K and U_K are symmetric with respect to the origin, $0 \in U_K$ and $\partial U_K \neq \emptyset$. Let β be a monotone, semiadditive MNC invariant with respect to reflection at the origin. Assume also that a multimap $F : \overline{U}_K \to Kv(K)$ satisfies the one of the following conditions:*

(i) *F is u.s.c. and β-condensing;*

or

(ii) *the MNC β is real and regular and F is closed and (k, β)-condensing.*

Let the corresponding multifield $\Phi = i - F$ satisfy the boundary condition

$$\Phi(x) \cap \mu\Phi(-x) = \emptyset$$

for all $x \in \partial U_K$ and $0 \le \mu \le 1$. Then $\deg_K(\Phi, \overline{U}_K)$ is an odd number and hence $\operatorname{Fix} F \neq \emptyset$.

Proof. Consider the collection $\{T_j\}$ of all fundamental sets of the multimap F which are symmetric with respect to the origin. This collection is nonempty since it contains K. Then

$$T = \bigcap_{j \in J} T_j$$

is an essential fundamental set of F which is symmetric with respect to the origin. In fact, T obviously contains the origin, therefore $\overline{U}_T \neq \emptyset$. Further, from Lemma 2.2.2 it follows that the set

$$\overline{\text{co}}(F(\overline{U}_T) \cup (-F(\overline{U}_T)))$$

belongs to the collection $\{T_j\}_{j \in J}$ and hence

$$T = \overline{\text{co}}(F(\overline{U}_T) \cup (-F(\overline{U}_T))).$$

Therefore

$$\begin{aligned}\beta(\overline{U}_T) \le \beta(T) &= \beta(\overline{\text{co}}(F(\overline{U}_T) \cup (-F(\overline{U}_T))))\\ &= \max\{\beta(F(\overline{U}_T)), \beta(-F(\overline{U}_T))\} = \beta(F(\overline{U}_T))\end{aligned}$$

and we apply the same argument as in the proof of Theorem 2.2.1 to conclude that the restriction F on $\overline{U}_T$ is compact.

If $\partial U_T = \emptyset$ we have $\deg_T(\Phi, \partial U_T) = \deg_K(\Phi, \overline{U}_K) = 1$ (see Remark 3.1.1). If $\partial U_T \neq \emptyset$ we can use the Odd Field Theorem 3.1.7. □

3.3.2 Topological degree of equivariant multifields

We start with the following definition.

Definition 3.3.1. A linear operator $d : E \to E$ is said to be *periodic of period* $p \geq 2$ provided $d^p = \mathrm{id}_E$ and $d^k(x) \neq x$ for all $k \not\equiv 0 \pmod p$ and $x \in E \setminus L$ where $L = \operatorname{Fix} d$ is a closed subspace of E.

Without loss of generality we will assume that the periodic operator d is isometric, i.e., $\| d \| = 1$ since otherwise we can pass to the equivalent norm

$$\|x'\| = \sum_{k=1}^{p-1} \|d^k x\|.$$

Let $B \subset E$ be the unit ball with the center at the origin, $S = \partial B$.

Definition 3.3.2. A multifield $\Phi = i - F : S \to Kv(E)$ is said to be *equivariant* (*with respect to a periodic operator* d) provided

$$\Phi(dx) = d(\Phi(x))$$

for all $x \in S$.

Our aim is to evaluate the topological degree for equivariant multifields. We begin with the case of a completely continuous multifield.

Theorem 3.3.6. *Let* $\Phi = i - F : S \to Kv(E)$, $\operatorname{Fix} F \cap S = \emptyset$ *be a completely continuous multifield equivariant with respect to a periodic operator d of period* $p \geq 2$ *and such that* $\operatorname{Fix} d = \{0\}$. *Then*

$$\deg(\Phi, S) \equiv 1 \pmod p.$$

Proof. We follow the method used in the proof of Theorem 1.2.19 and apply a special form of the construction given there in order to obtain an equivariant single-valued approximation of the multifield Φ.

Given $\varepsilon > 0$, for each $x \in S$ let the number $\delta'(x) > 0$ be such that $F(B_{\delta'(x)}(x)) \subset W_\varepsilon(F(x))$ and take

$$\delta(x) = \min_{0 \leq k \leq p-1} \delta'(d^k x).$$

Now consider the cover of S of the form $\{B_{\eta(x)}(x)\}$ where $\eta(x) = (1/4)\delta(x)$ and a subordinated locally finite cover $\{V_j\}_{j \in J}$ of S with the corresponding partition of unity $\{q_j\}_{j \in J}$. We add to the cover $\{V_j\}_{j \in J}$ the collection of sets $\{d^k V_j\}_{k=1}^{p-1}{}_{j \in J}$ combined with the functions $\{q_j(d^{-k}x)\}_{k=1}^{p-1}{}_{j \in J}$.

It is easy to see that the resulting system $\{d^k V_j\}_{k=0}^{p-1}{}_{j \in J}$ forms a locally finite cover of S and $\{(1/p)q_j(d^{-k}x)\}_{k=0}^{p-1}{}_{j \in J}$ is the corresponding partition of unity. Moreover, this covering is subordinated to the covering $\{B_{\eta(x)}(x)\}$. In fact, take a set $d^{k_0}V_{j_0}$, $1 \le k_0 \le p-1$, $j_0 \in J$ and let $V_{j_0} \subset B_{\eta(x_0)}(x_0)$. Then

$$d^{k_0}V_{j_0} \subset d^{k_0}B_{\eta(x_0)}(x_0) = B_{\eta(d^{k_0}x_0)}(d^{k_0}x_0)$$

due to the isometry of the operator d.

To construct the ε-approximation take arbitrary points $y_j \in F(V_j)$, $j \in J$ and note that the points $d^k y_j$, $1 \le k \le p-1$, $j \in J$ belong to the sets $F(d^k V_j)$ due to the equivariance of the multimap F:

$$d^k y_j \in d^k F(V_j) = F(d^k V_j).$$

Then the ε-approximation corresponding to the choice of the points $d^k y_j$, $0 \le k \le p-1$, $j \in J$ has the form

$$f_\varepsilon(x) = \frac{1}{p}\sum_{k=0}^{p-1}\sum_{j\in J} q_j(d^{-k}x)\cdot d^k y_j.$$

To verify that the map f_ε is equivariant we evaluate $f_\varepsilon(dx)$ for arbitrary $x \in S$. We have

$$\begin{aligned} f_\varepsilon(dx) &= \frac{1}{p}\Big[\sum_{j\in J} q_j(dx)\cdot y_j + \sum_{j\in J} q_j(x)\cdot dy_j + \sum_{j\in J} q_j(d^{-1}x)\cdot d^2 y_j \\ &\qquad + \cdots + \sum_{j\in J} q_j(d^{-p+2}x)\cdot d^{p-1}y_j\Big] \\ &= \frac{1}{p}\Big[\sum_{j\in J} q_j(x)\cdot dy_j + \sum_{j\in J} q_j(d^{-1}x)\cdot d(dy_j) \\ &\qquad + \cdots + \sum_{j\in J} q_j(d^{-p+2}x)\cdot d(d^{p-2}y_j) \\ &\qquad + \sum_{j\in J} q_j(d^{-p+1}x)\cdot d(d^{p-1}y_j)\Big] \end{aligned}$$

since obviously

$$q_j(dx)\cdot y_j = q_j(d^{-p+1}x)\cdot d(d^{p-1}y_j)$$

and therefore

$$f_\varepsilon(dx) = d\Big(\frac{1}{p}\sum_{k=0}^{p-1}\sum_{j\in J} q_j(d^{-k}x)\cdot d^k y_j\Big) = df_\varepsilon(x).$$

Taking now $\varepsilon > 0$ small enough we have

$$\deg(\Phi, S) = \deg(i - f_\varepsilon, S) \equiv 1 \pmod p$$

in accordance with the well-known formula for the topological degree of equivariant single-valued vector fields (see [148]). □

We consider now the more general situation when $\Phi = i - F$ is a equivariant fundamentally restrictible multifield.

Theorem 3.3.7. *Let* $\Phi = i - F : S \to Kv(E)$, $\operatorname{Fix} F \cap S = \emptyset$ *be an u.s.c. 2-fundamentally restrictible multifield equivariant on* S *with respect to a periodic operator* d *of a period* $p \geq 2$. *Assume that* Φ *has an essential fundamental set* T *such that* $dT \subseteq T$ *and that* $\operatorname{Fix} d = \{0\}$. *Then*

$$\deg(\Phi, S) \equiv 1 \ (\mathrm{mod}\ p).$$

Proof. Let T be any essential d-invariant fundamental set of F and $\rho : E \to \overline{\mathrm{co}}\, F(S_T)$ any retraction. From the equivariance condition it follows that the set $\overline{\mathrm{co}}\, F(S_T)$ is also d-invariant and therefore we can see that the map $\widetilde{\rho} : E \to \overline{\mathrm{co}}\, F(S_T)$,

$$\widetilde{\rho}(x) = \frac{1}{p} \sum_{j=0}^{p-1} d^{p-j} \rho(d^j x)$$

is a equivariant retraction: $\widetilde{\rho} \circ d = d \circ \widetilde{\rho}$.

Then $\widetilde{F} : S \to Kv(E)$,

$$\widetilde{F}(x) = \overline{\mathrm{co}}\, \widetilde{\rho} \circ F(x)$$

is a completely continuous equivariant extension of the multimap F:

$$\widetilde{F}|_{S_T} = F|_{S_T}$$

such that $\operatorname{Fix} \widetilde{F} \cap S = \emptyset$.

Using the principle of map restriction (d) of Section 3.1 we obtain

$$\deg(\Phi, S) = \deg_T(\Phi, S_T) = \deg(\widetilde{\Phi}, S)$$

where $\widetilde{\Phi} = i - \widetilde{F}$, and the application of the previous result to the equivariant completely continuous multifield $\widetilde{\Phi}$ completes the proof. □

As a corollary we obtain the following result on the topological degree for equivariant condensing multifields.

Theorem 3.3.8. *Let* β *be a monotone nonsingular MNC in E and* $\Phi = i - F : S \to Kv(E)$, $\operatorname{Fix} F \cap S = \emptyset$ *be an u.s.c.* β*-condensing multifield equivariant on* S *with respect to a periodic operator* d *of period* $p \geq 2$ *such that* $\operatorname{Fix} d = \{0\}$. *Then*

$$\deg(\Phi, S) \equiv 1 \ (\mathrm{mod}\ p).$$

Proof. Take the orbit $\nu = \{d^k x\}_{k=0}^{p-1}$ of an arbitrary point $x \in S$ and consider the collection $\{T_i\}_{i \in I}$ of all fundamental sets of the multifield Φ which are d-invariant and contain ν. This collection is nonempty since it contains the whole space E. Then

$$T = \bigcap_{i \in I} T_i$$

is the essential d-invariant fundamental set of Φ.

In fact, from the equivariance condition, it follows that the set $\overline{\mathrm{co}}(F(S \cap T) \cup \nu)$ also belongs to the collection $\{T_i\}_{i \in I}$. From the minimality of T we have

$$T = \overline{\mathrm{co}}(F(S \cap T) \cup \nu)$$

and using the procedure described in proof of Theorem 2.2.1 (i) we can see that the set $S \cap T$ is compact.

We apply now the previous result to the multifield Φ and the fundamental set T. □

As a direct consequence we obtain the following fixed point principle.

Corollary 3.3.5. *Let β be a monotone nonsingular MNC in E and $\Phi = i - F : \overline{B} \to Kv(E)$,* Fix $F \cap S = \emptyset$ *be an u.s.c. β-condensing multifield equivariant on S with respect to a periodic operator d of period $p \geq 2$ such that* Fix $d = \{0\}$. *Then* Fix $F \neq \emptyset$.

Remark 3.3.1. *The case $L =$* Fix $d \neq \{0\}$. Suppose that $\Phi = i - F : S \to Kv(E)$, Fix $F \cap S = \emptyset$ is an u.s.c. 2-fundamentally restrictible multifield equivariant on S with respect to a periodic operator d of period $p \geq 2$ such that $L =$ Fix $d \neq \{0\}$ is a closed subset of E. Assume that Φ has an essential fundamental set T such that $dT \subseteq T$ and $S \cap T \cap L \neq \emptyset$.

Then we can define the multifield $\Phi_L : S \cap L \to Kv(L)$ by the formula

$$\Phi_L(x) = \Phi(x) \cap L.$$

In fact for every $x \in S \cap L$ the set $\Phi(x)$ is d-invariant and for any $y \in \Phi(x)$, $x \in S \cap L$ we have

$$\frac{1}{p} \sum_{j=0}^{p-1} d^j y \in \Phi(x) \cap L.$$

Hence the set $\Phi(x) \cap L$ is nonempty and clearly compact and convex.

It is also easy to verify that the multifield Φ_L is completely fundamentally restrictible (on the essential fundamental set $T \cap L$).

It is proved in [123], [124] that in this situation the following interesting formula holds:

$$\deg(\Phi, S) \equiv \deg(\Phi_L, S \cap L) \pmod{p}.$$

The same relation is valid for multifields condensing with respect to a monotone nonsingular MNC.

3.4 The topological degree for condensing nonconvex-valued multimaps

Our aim is to extend the topological degree theory for noncompact multimaps to the case of nonconvex-valued multimaps.

3.4.1 Vietoris pairs and multimaps

We recall some basic facts from algebraic topology (see, e.g. [208], [79] for general reference).

Given a nonempty metric space D, we denote the collection of the Čech homologies with compact carriers and with coefficients in the field of rational numbers $\mathbb{Q}$ by the symbol $\{H_k(D)\}_{k=0}^{\infty}$.

Definition 3.4.1. A nonempty space A is said to be *acyclic* if

$$H_0(A) = \mathbb{Q} \quad \text{and} \quad H_k(A) = 0, \quad k \geq 1,$$

i.e., A has the same homologies as an one-point space.

Natural examples of acyclic sets are convex or, more generally, contractible sets. Another important class of acyclic sets is formed by R_δ-sets.

Definition 3.4.2. A nonempty space is said to be an *R_δ-set* if it can be represented as the intersection of a decreasing sequence of compact, contractible sets (see [120]).

We use the following definition [102].

Definition 3.4.3. A continuous surjective map $t : Y \to X$ from a metric space Y onto a metric space X is said to be a *Vietoris map* if:

(i) t is proper, i.e., $t^{-1}(K)$ is compact for every compact $K \subset X$;

(ii) the set $t^{-1}(x)$ is acyclic for each $x \in X$.

It is easy to see that if $t : Y \to X$ is a Vietoris map and $X_0 \subset X$ is a subspace then the restriction $t_1 : t^{-1}(X_0) \to X_0$, $t_1(y) = t(y)$ is a Vietoris map too.

Let X, Y and Z be metric spaces. We will say that continuous maps $t : Y \to X$, $r : Y \to Z$ form a Vietoris pair (t, r) if t is a Vietoris map. Every Vietoris pair (t, r) defines a multimap $F_{(t,r)} : X \to K(Z)$ by the rule

$$F_{(t,r)}(x) = r(t^{-1}(x)).$$

$F_{(t,r)}$ is called *the Vietoris multimap associated with the pair* (t, r).

Using Theorem 1.1.12 it is easy to see that every Vietoris multimap is u.s.c.

The class of Vietoris multimaps is sufficiently large. Following [102] one can verify the following properties.

Proposition 3.4.1. (a) *If $F_0 : X \to K(Z_0)$, $F_1 : Z_0 \to K(Z_1)$ are Vietoris multimaps then the composition $F_1 \circ F_0 : X \to K(Z_1)$ is a Vietoris multimap.*

(b) *If $F_0 : X \to K(Z_0)$, $F_1 : X \to K(Z_1)$ are Vietoris multimaps then the Cartesian product $F_0 \times F_1 : X \to K(Z_0 \times Z_1)$ is a Vietoris multimap.*

(c) *If Z is a normed space; $F_0, F_1 : X \to K(Z)$ are Vietoris multimaps then the sum $F_0 + F_1 : X \to K(Z)$ is a Vietoris multimap.*

(d) *If Z is a normed space; $F : X \to K(Z)$ is a Vietoris multimap and $f : X \to \mathbb{R}$ is a continuous function then the product $f(x) \cdot F(x) : X \to K(Z)$ is a Vietoris multimap.*

Proof. For example, in case (a), if

$$X \xleftarrow{t_0} Y_0 \xrightarrow{r_0} Z_0 \xleftarrow{t_1} Y_1 \xrightarrow{r_1} Z_1$$

are Vietoris pairs associated with F_0 and F_1 then the Vietoris pair

$$X \xleftarrow{t} Y \xrightarrow{r} Z_1$$

for the composition $F_1 \circ F_0$ can be constructed as follows: $Y = \{(y_0, y_1) \in Y_0 \times Y_1 : r_0(y_0) = t_1(y_1)\}$; $t(y_0, y_1) = t_0(y_0)$; $r(y_0, y_1) = r_1(y_1)$ (see [102] for further details). □

We recall the following important class of Vietoris multimaps to which Vietoris pairs can be associated very naturally.

Definition 3.4.4. An upper semicontinuous multimap $F : X \to K(Z)$ is said to be *acyclic* if every value $F(x)$, $x \in X$ is an acyclic set.

An acyclic multimap F is associated with the Vietoris pair (t, r) where $t = \mathrm{pr}_X^F : \Gamma_F \to X$, $r = \mathrm{pr}_Z^F : \Gamma_F \to Z$ are the natural projections from the graph Γ_F. From Proposition 3.4.1 (a) it follows that every multimap F which can be represented as a finite composition $F = F_n \circ F_{n-1} \circ \cdots \circ F_0$ of acyclic multimaps is a Vietoris multimap.

We select an important family from this collection: a multimap $F : X \to K(Z)$ is *quasiacyclic* if it can be represented as a composition $F = g \circ F_0$ where $F_0 : X \to K(X_0)$ is an acyclic multimap and $g : X_0 \to Z$ is a continuous map. For the quasiacyclic multimap $F = g \circ F_0$ the corresponding Vietoris pair (t, s) can be taken as

$$t = \mathrm{pr}_X^{F_0} : \Gamma_{F_0} \to X, \quad r = g \circ \mathrm{pr}_{X_0}^{F_0} : \Gamma_{F_0} \to Z.$$

We will be interested in studying Vietoris pairs (t, r) only when $X \subseteq Z$. In this situation the set

$$\Pi(t, r) = \{y \in Y : t(y) = r(y)\}$$

is called *the set of coincidence points* for (t, r). It is clear that $\Pi(t, r) \neq \emptyset$ is equivalent to Fix $F_{(t,r)} \neq \emptyset$ and, moreover

$$\text{Fix } F_{(t,r)} = t(\Pi(t, r)).$$

3.4.2 The coincidence index for a Vietoris pair in a finite-dimensional space

We study here some fundamental facts concerning the coincidence index for compact Vietoris pair in a finite-dimensional space E^n (see [153] [154] for details).

Let $U \subset E^n$ be an open set, then a *the coincidence index* $I(t, r) \in \mathbb{Q}$ can be associated to every Vietoris pair (t, r):

$$\overline{U} \xleftarrow{t} Y \xrightarrow{r} E^n$$

such that the map r is compact and Fix $F_{(t,r)} \cap \partial U = \emptyset$,. This index has the following main properties.

(i) Normalization property. If $r(y) \equiv x_0$ is a constant map, then

$$I(t, r) = \begin{cases} 1, & \text{if } x_0 \in U, \\ 0, & \text{if } x_0 \in E^n \setminus \overline{U}. \end{cases}$$

(ii) *Homotopy invariance.* If (t, r) is the Vietoris pair

$$[0, 1] \times \overline{U} \xleftarrow{t} Y \xrightarrow{r} E^n,$$

with r compact and for every $\lambda \in [0, 1]$ the Vietoris pair

$$\overline{U} \xleftarrow{t_\lambda} t^{-1}(\{\lambda\} \times \overline{U}) \xrightarrow{r_\lambda} E^n$$

is such that $t_\lambda(\Pi(t_\lambda, r_\lambda)) \cap \partial U = \emptyset$ then $I(t_0, r_0) = I(t_1, r_1)$.

(iii) *Additive dependence on the domain.* If $U = \bigcup_{j=1}^{k} U_j$ where $\{U_j\}_{j=1}^{k}$ is a family of open disjoint subsets of E^n then

$$I(t, r) = \sum_{j=1}^{k} I(t_j, r_j)$$

where $(t_j, r_j) = (t, r)|_{\overline{U}_j}$, i.e., $\overline{U}_j \xleftarrow{t_j} t^{-1}(\overline{U}_j) \xrightarrow{r_j} E^n$.

(iv) *The restriction principle.* Let E^{n-1} be a subspace of E^n and $r(Y) \subset E^{n-1}$. Let $U' = U \cap E^{n-1}$; $Y' = t^{-1}(\overline{U'})$ and $t' : Y' \to \overline{U'}$, $r' : Y' \to E^{n-1}$ be the corresponding restrictions of t and r. Then

$$I(t, r) = I(t', r').$$

(v) *The coincidence point principle.* If $I(t, r) \neq 0$ then

$$\Pi(t, r) \neq \emptyset.$$

Note that if $Y = \overline{U}$, $t = \mathrm{id}_{\overline{U}}$ the coincidence index corresponds to the usual topological degree of a map r (cf. Section 3.1).

3.4.3 The topological degree for a completely fundamentally restrictible Vietoris multimap

Let E be an infinite-dimensional Banach space and U an open set. Let $F_{(t,r)} : \overline{U} \to K(E)$ be a 2-completely fundamentally restrictible Vietoris multimap associated with the Vietoris pair

$$\overline{U} \xleftarrow{t} Y \xrightarrow{r} E$$

and such that Fix $F_{(t,r)} \cap \partial U = \emptyset$. Our aim is to define the topological degree of the multifield $\Phi_{(t,r)} = i - F_{(t,r)}$. We remark that this theory will include the topological degree for acyclic and quasiacyclic multifields condensing with respect to a singular monotone measure of noncompactness.

Let T be any essential fundamental set of $F_{(t,r)}$ such that $\partial U \cap T \neq \emptyset$. Without loss of generality we can assume that T is compact (otherwise we can pass to the fundamental set $\overline{\mathrm{co}}(F_{(t,r)}(\overline{U} \cap T) \cup \{a\})$ where $a \in \partial U \cap T$ is an arbitrary point). From Lemma 3.1.1 we know that the set $\Phi_{(t,r)}(\partial U \cap T)$ is closed and therefore the distance δ of the set $\Phi_{(t,r)}(\partial U \cap T)$ from zero is positive.

For arbitrary ε, $0 < \varepsilon < \delta$ there exist a finite-dimensional subspace $E' \subset E$, $E' \cap T \neq \emptyset$ and a continuous map $\rho_\varepsilon : E \to E' \cap T$ such that

$$\| \rho_\varepsilon(y) - y) \| < \varepsilon \tag{3.4.1}$$

for every $y \in T$. Indeed, we can take ρ_ε as the composition of a retraction $\rho : E \to T$ and a Schauder projection $\mu_\varepsilon : T \to E'$. The map ρ_ε will be called *an ε-quasiretraction.*

Now denote

$$r_\varepsilon = \rho_\varepsilon \circ r, \quad U' = U \cap E', \quad Y' = t^{-1}(\overline{U'})$$

and consider the following diagram

$$\overline{U'} \xleftarrow{t'} Y' \xrightarrow{r'_\varepsilon} E' \cap T,$$

where $t'(y) = t(y)$, $r'_\varepsilon(y) = r_\varepsilon(y)$ are the corresponding restrictions. We will say that (t', r'_ε) is an *ε-approximating Vietoris pair* for the pair (t, r).

Lemma 3.4.1. Fix $F_{(t',r'_\varepsilon)} \cap \partial U' = \emptyset$.

Proof. Suppose the contrary. Then there exists $y \in Y'$ such that

$$x = t'(y) = r'_\varepsilon(y) \in \partial U' \cap T.$$

We have

$$\begin{aligned} \| t'(y) - r'_\varepsilon(y) \| &= \| x - r_\varepsilon(y) \| = \| x - r(y) - (r_\varepsilon(y) - r(y)) \| \\ &\geq \| x - r(y) \| - \| r_\varepsilon(y) - r(y) \|. \end{aligned}$$

As $x - r(y) \in \Phi_{(t,r)}(\partial U \cap T)$, it follows that $\| x - r(y) \| > \varepsilon$ and

$$\| r_\varepsilon(y) - r(y) \| = \| \rho_\varepsilon r(y) - r(y) \| < \varepsilon$$

since $r(y) \in F_{(t,r)}(\partial U \cap T) \subset T$. Therefore $\|t'(y) - r'_\varepsilon(y)\| > 0$ giving the contradiction. □

The topological degree $\deg(\Phi_{(t,r)}, \overline{U})$ can be defined in the following way.

Definition 3.4.5. Let T be an essential compact fundamental set of $F_{(t,r)}$ such that $\partial U \cap T \neq \emptyset$ and (t', r'_ε) the corresponding ε-approximating Vietoris pair of (t, r). Then

$$\deg(\Phi_{(t,r)}, \overline{U}) = I(t', r'_\varepsilon).$$

We want to prove the independence of this degree from the choice of approximating space E', quasiretraction ρ_ε and fundamental set T.

Let E', E'' be two approximating subspaces of E such that $E' \subset E''$ and $\dim E'' = \dim E' + 1$. Take an ε-quasiretraction $\rho^0_\varepsilon : E \to E' \cap T$ and set $\rho^1_\varepsilon : E \to E'' \cap T$, $\rho^1_\varepsilon(z) = \rho^0_\varepsilon(z)$, $z \in E$. From the restriction principle (iv) it follows immediately that $I(t', r'_\varepsilon) = I(t'', r''_\varepsilon)$ where t' and r'_ε are restrictions of t and $\rho^0_\varepsilon \circ r$ to $Y' = t^{-1}(\overline{U'})$ and, respectively, t'' and r''_ε are restrictions of t and $\rho^1_\varepsilon \circ r$ to $Y'' = t^{-1}(\overline{U''})$.

Lemma 3.4.2. *The degree* $\deg(\Phi_{(t,r)}, \overline{U})$ *does not depend on the choice of* T *and* ρ_ε.

Proof. Let T_1, T_2 be essential compact fundamental sets of $F_{(t,r)}$ such that $T_i \cap \partial U \neq \emptyset$ for $i = 1, 2$. Moreover, without loss of generality we can assume that $T_1 \cap T_2 \cap \partial U \neq \emptyset$. Consider the essential compact fundamental set $T_0 = T_1 \cap T_2$.

Let $\rho^0_\varepsilon : E \to E' \cap T_0$, $\rho^1_\varepsilon : E \to E' \cap T_1$ be corresponding ε-quasiretractions and $r^0_\varepsilon = \rho^0_\varepsilon \circ r$, $r^1_\varepsilon = \rho^1_\varepsilon \circ r$. Define the maps $\widetilde{r} : [0, 1] \times Y \to E$,

$$\widetilde{r}(\lambda, y) = \begin{cases} 2\lambda r(y) + (1 - 2\lambda) r^0_\varepsilon(y), & 0 \leq \lambda \leq 1/2, \\ (2 - 2\lambda) r(y) + (2\lambda - 1) r^1_\varepsilon(y), & 1/2 \leq \lambda \leq 1, \end{cases}$$

and $\widetilde{t}: [0,1] \times Y \to \overline{U}$, $\widetilde{t}(\lambda, y) = t(y)$. The restrictions of these maps to $[0,1] \times \widehat{Y}$ where $\widehat{Y} = t^{-1}(\partial U)$ have no coincidence points. In fact, if $\widetilde{t}(\lambda_0, y_0) = \widetilde{r}(\lambda_0, t_0)$ for some $0 \le \lambda_0 \le 1/2$, $y_0 \in \widehat{Y}$, then for $x_0 = t(y_0) \in \partial U$ we have $x_0 \in \overline{\mathrm{co}}(r(y_0) \cup T_0) \subset \overline{\mathrm{co}}(F_{(t,r)}(x_0) \cup T_0)$ hence, $x_0 \in T_0 \cap \partial U$, $r(y_0) \in T_0$ and the equality $t(y_0) = 2\lambda_0 r(y_0) + (1 - 2\lambda_0)\rho_\varepsilon^0 r(y_0)$ is impossible from the choice of ε. Analogously, $\widetilde{t}(\lambda, y) \neq \widetilde{r}(\lambda, y)$ when $1/2 \le \lambda \le 1$.

Consider now the continuous map $\widetilde{g} : [0,1] \times Y \to E$ defined as $\widetilde{g}(\lambda, y) = \widetilde{t}(\lambda, y) - \widetilde{r}(\lambda, y)$. The set $\widetilde{g}([0,1] \times Y_1)$ where $Y_1 = t^{-1}(\partial U \cap T_1)$ is compact. Therefore we can choose $\widetilde{\varepsilon} > 0$ less than the distance of the set $\widetilde{g}([0,1] \times Y_1)$ from zero and consider the corresponding $\widetilde{\varepsilon}$-quasiretraction $\rho_{\widetilde{\varepsilon}} : E \to E'' \cap T_1$. Without loss of generality we can assume that $E' \subset E''$.

Consider the map $h : [0,1] \times [0,1] \times Y'' \to E$,

$$h(\lambda, \tau, y) = (1 - \tau)\widetilde{r''}(\lambda, y) - \tau \rho_{\widetilde{\varepsilon}} \widetilde{r''}(\lambda, y),$$

where $U'' = U \cap E''$, $Y'' = t^{-1}(\overline{U''})$, $\widetilde{r''} = \widetilde{r}|_{[0,1]\times Y''}$.

Let $t'' = t|_{Y''}$, $\widehat{Y''} = t^{-1}(\partial U'')$, then $t''(y) \neq h(\lambda, \tau, y)$ for all $(\lambda, \tau, y) \in [0,1] \times [0,1] \times \widehat{Y''}$. In fact, if $t''(y_0) = h(\lambda_0, \tau_0, y_0)$ for some $\lambda_0, \tau_0 \in [0,1]$, $y_0 \in \widehat{Y''}$, then

$$\begin{aligned} x_0 = t''(y_0) &\in \overline{\mathrm{co}}(\widetilde{r''}(\lambda_0, y_0) \cup T_1) \\ &\subset \overline{\mathrm{co}}(r(y_0) \cup T_1) \subset \overline{\mathrm{co}}(F(x_0) \cup T_1). \end{aligned}$$

It follows that $y_0 \in Y_1$ and the equality $t''(y_0) = h(\lambda_0, \tau_0, y_0)$ is impossible from the choice of $\widetilde{\varepsilon}$.

Consider the Vietoris pair

$$[0,1] \times \overline{U''} \xleftarrow{\widetilde{t''}} [0,1] \times Y'' \xrightarrow{s} E''$$

where map s is defined as

$$s(\mu, y) = \begin{cases} h(0, 3\mu, y), & 0 \le \mu \le 1/3; \\ h(3\mu - 1, 1, y), & 1/3 \le \mu \le 2/3; \\ h(1, 3 - 3\mu, y), & 2/3 \le \mu \le 1 \end{cases}$$

and $\widetilde{t''}(\mu, y) = (\mu, t''(y))$. The map s is compact and $\widetilde{t''}(\mu, y) \neq s(\mu, y)$ for all $(\mu, y) \in [0,1] \times \widehat{Y''}$. Therefore from the homotopy invariance property (ii) it follows that $I(t'', \widetilde{r}_\varepsilon^0) = I(t'', \widetilde{r}_\varepsilon^1)$ where $\widetilde{r}_\varepsilon^0, \widetilde{r}_\varepsilon^1$ are the restrictions of $r_\varepsilon^0, r_\varepsilon^1$ on Y''.

The independence of the degree from the choice of the approximating finite-dimensional space completes the proof. □

Note also the obvious independence of the degree from the choice of ε, $0 < \varepsilon < \delta$.

The following properties of the degree can be directly deduced from the definition and corresponding properties of the coincidence index (i)–(v).

(a) *Normalization property.* If $F_{(t,r)}(x) \equiv x_0$ is a constant map, then

$$\deg(\Phi_{(t,r)}, \overline{U}) = \begin{cases} 1, & \text{if } x_0 \in U, \\ 0, & \text{if } x_0 \in E \setminus \overline{U}. \end{cases}$$

(b) *Homotopy invariance.* Let $G_{(t,r)} : [0,1] \times \overline{U} \to K(E)$ be a 2-completely fundamentally restrictible Vietoris multimap associated with the Vietoris pair

$$[0,1] \times \overline{U} \xleftarrow{t} Y \xrightarrow{r} E.$$

If $x \notin G(\lambda, x)$ for all $(\lambda, x) \in [0,1] \times \partial U$ then

$$\deg(i - G_{(t_0,r_0)}, \overline{U}) = \deg(i - G_{(t_1,r_1)}, \overline{U})$$

where

$$\overline{U} \xleftarrow{t_i} t^{-1}(\{i\} \times \overline{U}) \xrightarrow{r_i} E$$

$i = 0, 1$.

In fact, let T be a compact essential fundamental set of the family G. If $\Psi = i - G : [0,1] \times \overline{U} \to K(E)$ is the corresponding family of multifields then the closed set $\Psi([0,1] \times (\partial U \cap T))$ is at distance $\delta > 0$ from the origin. For any $0 < \varepsilon < \delta$ we can take an ε-approximating Vietoris pair (t', r'_ε) for (t, r) and note that

$$\overline{U'} \xleftarrow{t'_i} (t')^{-1}(\{i\} \times \overline{U'}) \xrightarrow{r'_i} E' \cap T$$

for $i = 0, 1$ would be the ε-approximating Vietoris pairs for (t_i, r_i), $i = 0, 1$. It remains to apply the homotopy invariance property (ii) to the pair (t', r'_ε).

(c) *Additive dependence on the domain.* If $U = \bigcup_{j=1}^k U_j$ where $\{U_j\}_{j=1}^k$ is a family of open disjoint subsets of E and the restrictions of $F_{(t,r)}$ to every $\overline{U}_j$, $j = 1, \ldots, k$ are completely fundamentally restrictible, then

$$\deg(\Phi_{(t,r)}, \overline{U}) = \sum_{j=1}^k \deg(\Phi_{(t_j,r_j)}, \overline{U}_j)$$

where $(t_j, r_j) = (t, r)|_{\overline{U}_j}$, i.e., $\overline{U}_j \xleftarrow{t_j} t^{-1}(\overline{U}_j) \xrightarrow{r_j} E$.

(d) *The fixed point principle.* If $\deg(\Phi_{(t,r)}, \overline{U}) \neq 0$ then

$$\operatorname{Fix} F_{(t,r)} \neq \emptyset.$$

In fact, to prove (d) it is sufficient to note that condition $\operatorname{Fix} F_{(t,r)} = \emptyset$ implies that the set $\Phi_{(t,r)}(\overline{U} \cap T)$ is at positive distance δ from the origin. Then for every ε-approximating Vietoris pair (t', r'_ε) of $F_{(t,r)}$ with $0 < \varepsilon < \delta$ we have $\Pi(t', r'_\varepsilon) = \emptyset$ and hence $\deg(\Phi_{(t,r)}, \overline{U}) = I(t', r'_\varepsilon) = 0$ by the coincidence property (v).

Remark 3.4.1. All the properties (a)–(d) are valid for Vietoris multimaps condensing with respect to a nonsingular monotone measure of noncompactness since they are 2-completely fundamentally restrictible (see Section 2.2).

Using the general fixed point principle (d) we can obtain the usual fixed point theorems. Consider some examples.

Theorem 3.4.1. *Let U be an open neighborhood of the origin and $F_{(t,r)} : \overline{U} \to K(E)$ a 2-completely fundamental restrictible Vietoris multimap. If for every $x \in \partial U$*

$$\mu x \notin F_{(t,r)}(x)$$

for all $\mu \geq 1$, then Fix $F_{(t,r)} \neq \emptyset$.

Proof. Let T be an essential fundamental set of $F_{(t,r)}$ containing 0 and such that $\partial U \cap T \neq \emptyset$. Then the family $G : [0,1] \times \overline{U} \to K(E)$, defined as $G(\lambda, x) = \lambda F_{(t,r)}(x)$ is fundamentally restrictible on T. At the same time

$$[0,1] \times \overline{U} \xleftarrow{t_G} [0,1] \times Y \xrightarrow{r_G} E$$

where $t_G(\lambda, y) = (\lambda, t(y))$, $r_G(\lambda, y) = \lambda r(y)$ is the Vietoris pair for G. From the hypotheses it follows that $x \notin G(\lambda, x)$ for all $(\lambda, x) \in [0,1] \times \partial U$. Therefore the homotopy invariance principle (b) gives

$$\deg(i - G_{(t_{G_0}, r_{G_0})}, \overline{U}) = \deg(i - G_{(t_{G_1}, r_{G_1})}, \overline{U})$$

where $G_0 = G(0, \cdot)$, $G_1 = G(1, \cdot)$. As $G_{(t_{G_1}, r_{G_1})} = F_{(t,r)}$ we can conclude that $G_{(t_{G_0}, r_{G_0})}(x) \equiv \{0\}$.

The normalization property (a) implies that $\deg(i - G_{(t_{G0}, r_{G0})}, \overline{U}) = 1$ and we can apply (d). □

Corollary 3.4.1. *Let U be an open convex neighborhood of the origin and $F_{(t,r)} : \overline{U} \to K(E)$ a 2-completely fundamental restrictible Vietoris multimap such that*

$$F(\partial U) \subset \overline{U}.$$

Then Fix $F_{(t,r)} \neq \emptyset$.

We can deduce from the above statement the following fixed point principle generalizing Theorems 3.1.5 and 3.3.3.

Theorem 3.4.2. *If M is a convex closed subset of E and $F_{(t,r)} : M \to K(M)$ a completely fundamental restrictible Vietoris multimap then* Fix $F \neq \emptyset$.

Proof. Let $T \subseteq M$ be a nonempty compact fundamental set of F, U a convex open bounded neighborhood of the origin containing T, $\rho : \overline{U} \to T$ a retraction. Consider the multimap $\widetilde{F} = F_T \circ \rho : \overline{U} \multimap K(\overline{U})$ where $F_T = F|_T$. From the one side $\overline{U}$ is the essential fundamental set of $\widetilde{F}$ and from the other side the multimap $\widetilde{F}$ is Vietoris: if

$$T \overset{t_T}{\longleftarrow} Y \overset{r_T}{\longrightarrow} T$$

is the Vietoris pair of F_T, then the Vietoris pair of $\widetilde{F}$ is defined as

$$\overline{U} \overset{\widetilde{t}}{\longleftarrow} \widetilde{Y} \overset{\widetilde{r}}{\longrightarrow} \overline{U}$$

where $\widetilde{Y} = \{(x, y) \in \overline{U} \times Y : \rho(x) = t_T(y)\}$, $\widetilde{t}(x, y) = x$ and $\widetilde{r}(x, y) = r_T(y)$. We can apply now the previous corollary to the multimap $\widetilde{F}$. □

Proceeding as in Section 3.3.1 we obtain the following corollaries for condensing Vietoris multimaps.

Corollary 3.4.2. *If M is a convex closed set in E and $F_{(t,r)} : M \multimap K(M)$ is a Vietoris β-condensing multimap, where β is a nonsingular MNC defined on subsets of M, then* Fix $F_{(t,r)} \neq \emptyset$.

Corollary 3.4.3. *If M is a bounded, convex closed set of E and $F_{(t,r)} : M \multimap K(M)$ is a Vietoris (k, β)-condensing multimap, where β is a real nonsingular and regular MNC defined on subsets of M, then* Fix $F_{(t,r)} \neq \emptyset$.

3.4.4 Remark on single-valued approximations of nonconvex valued multimaps

The method of single-valued approximations of multimaps described in Sections 1.2.4 and 3.1 can be also applied to the study of some nonconvex-valued multimaps. In particular, we select the following class of acyclic multimaps.

Let X and Z be normed spaces.

Definition 3.4.6. An upper semicontinuous multimap $F : X \multimap K(Z)$ is said to be an *R_δ-multimap* if every value $F(x)$, $x \in X$ is an R_δ-set (see Definition 3.4.2).

In accordance with the definition of a quasiacyclic multimap given in Section 3.4.1, we say that a multimap $F : X \multimap K(Z)$ is a *quasi R_δ-multimap* if it can be represented as a composition $F = g \circ F_1$, where $F_1 : X \multimap K(X_1)$ is an R_δ-multimap, X_1 a normed space and $g : X_1 \to Z$ is a continuous map.

The approximating theorem below follows from a result in [105].

Theorem 3.4.3. *Let $M \subset X$ be a compact convex set and $F : M \multimap K(Z)$ an R_δ-multimap. Then for every $\varepsilon > 0$ there exists a continuous single- valued approximation $f_\varepsilon : M \to Z$ (see Definition 1.2.10).*

Using ε-approximations as a tool it is possible to define the topological degree for multimaps which are a finite composition of R_δ-multimaps (see, e.g. [22], [107], [104], [151], [152]) and to prove several fixed point theorems. As an example we obtain the following generalization of the well-known asymptotic fixed point theorem of F. E. Browder, which we will use in Section 6.3.

Theorem 3.4.4. *Let B_0, B_1 and B be balls in a Banach space E; $\overline{B}_0 \subset B_1 \subset \overline{B}_1 \subset B$, and $F = g \circ F_1 : \overline{B} \to K(E)$ be a 1-completely fundamentally restrictible quasi R_δ-multimap such that g is linear and the following conditions hold true:*

(i) $\bigcup_{j=1}^{m-1} F^j(\overline{B}_1) \subset B$;

(ii) $\bigcup_{j=1}^{m-1} F^j(\overline{B}_0) \subset B_1$;

(iii) $F^m(\overline{B}_1) \subset B_0$

for some integer $m \geq 1$. Then Fix $F \cap B_0 \neq \emptyset$.

Proof. Let T be an essential fundamental set of F such that $B_{0T} = B_0 \cap T \neq \emptyset$. The above conditions are satisfied for the intersections of the balls with T:

(i_T) $\bigcup_{j=1}^{m-1} F^j(\overline{B}_{1T}) \subset B_T$;

(ii_T) $\bigcup_{j=1}^{m-1} F^j(\overline{B}_{0T}) \subset B_{1T}$;

(iii_T) $F^m(\overline{B}_{1T}) \subset B_{0T}$.

Considering the restriction $F_T = g \circ F_{1T}$ of F to $\overline{B}_T$ where $F_{1T} : \overline{B}_T \to K(X_1)$, $g : X_1 \to E$ we note that from the linearity of g it follows that every map of the form $g \circ f_\varepsilon : \overline{B}_T \to E$ where $f_\varepsilon : \overline{B}_T \to X_1$ is an ε-approximation of F_1 acts into $\overline{\text{co}}\, F_T(\overline{B}_T) \subset T$. Now using condition ($\mathrm{iii}_T$) we see that the set $F^m(\overline{B}_{1T})$ is compact and hence there exists $\delta > 0$ such that

$$W_\delta(F^m(\overline{B}_{1T})) \subset B_{0T}.$$

Then it is clear that by choosing an ε-approximation $f_\varepsilon : \overline{B}_T \to X_1$ of F_1 for $\varepsilon > 0$ small enough, condition (iii_T) is fulfilled for the composition $g \circ f_\varepsilon : \overline{B}_T \to E$. Analogously, conditions (i_T) and (ii_T) for $g \circ f_\varepsilon$ can be guaranteed for $\varepsilon > 0$ small enough.

Then $\mathrm{Fix}(g \circ f_\varepsilon) \cap \overline{B}_{0T} \neq \emptyset$ by Yu. G. Borisovich's generalization of the F. E. Browder's asymptotic fixed point theorem (see [53], [34]). Taking sequences $\varepsilon_n \to 0$ and $x_n \in \mathrm{Fix}(g \circ f_{\varepsilon_n})$ one can easily show that the limit point x_* of the sequence $\{x_n\}$ is the fixed point of F_T. The inclusion $x_* \in B_0$ follows from condition (iii). □

3.5 Some properties of the fixed points set

In this section we describe some properties of the fixed points set of condensing multimaps. First of all we mention the following simple but very significant facts.

Theorem 3.5.1. *Let X be a closed bounded subset of a Banach space E and $F : X \to K(E)$ a closed fundamentally restrictible multimap. Then the set* Fix F *is compact.*

Proof. If F is not completely fundamentally restrictible, then Fix $F = \emptyset$ and the case is trivial. If F is completely fundamentally restrictible, then let T be any essential fundamental set of T. As the multimap F is closed it follows that Fix F is a closed subset of a relatively compact set $F(X \cap T)$. □

For condensing multimaps we have the following refinement

Proposition 3.5.1. *Let X be a closed subset of a Banach space E and $F : X \to K(E)$ a closed multimap, which is β-condensing on every bounded subset of X, where β is a monotone MNC in E. If the fixed points set* Fix F *is bounded then it is compact.*

Proof. It is sufficient to note that from the obvious inclusion

$$F(\text{Fix } F) \supseteq \text{Fix } F$$

follows

$$\beta(F(\text{Fix } F)) \geq \beta(\text{Fix } F)$$

implying that Fix F is compact. □

Consider now the question of the continuous dependence of the fixed point set on a parameter. We need the following notion.

Definition 3.5.1. Let X be a closed subset of a Banach space E and Λ a metric space. A multimap $G : \Lambda \times X \to K(E)$ is said to be *completely fundamentally restrictible in the second variable* if for any compact subset $K \subseteq \Lambda$ the restriction $G|_{K\times X}$ is completely fundamentally restrictible. If β is a (real) MNC in E then a multimap $G : \Lambda \times X \to K(E)$ is said to be *β-condensing* (resp., *(k, β)-condensing*) *in the second variable* if for any compact subset $K \subseteq \Lambda$ the restriction $G|_{K\times X}$ is β-condensing (resp. (k, β)-condensing).

From Theorem 2.2.1 (b) if follows that if either (i) G is u.s.c. and β-condensing in the second variable with respect to a monotone nonsingular MNC β, or (ii) X is bounded and G is (k, β)-condensing in the second variable with respect to a real, monotone, regular and nonsingular MNC β then G is completely fundamentally restrictible in the second variable.

Theorem 3.5.2. *Let X be a closed subset of a Banach space E, Λ a metric space and $G : \Lambda \times X \to K(E)$ a closed multimap which is completely fundamentally restrictible in the second variable and such that*

$$\mathcal{F}(\lambda) := \operatorname{Fix} G(\lambda, \cdot) \neq \emptyset \tag{3.5.1}$$

for every $\lambda \in \Lambda$. Then the multimap $\mathcal{F} : \Lambda \to P(E)$ is u.s.c.

Proof. Supposing the contrary we can find a point $\lambda_0 \in \Lambda$ and sequences $\{\lambda_n\}_{n=1}^{\infty} \subset \Lambda$, $\lambda_n \to \lambda_0$, $\{x_n\}_{n=1}^{\infty} \subset X$ such that $x_n \in \mathcal{F}(\lambda_n) \setminus W$ for all n where W is a neighborhood of a set $\mathcal{F}(\lambda_0)$. Let T be an essential fundamental set of $G|_{K \times X}$ for the compact set $K = \overline{\{\lambda_n\}_{n=1}^{\infty}}$. Then the sequence $\{x_n\}_{n=1}^{\infty}$ lies in the relatively compact set $G(K \times (X \cap T))$, so we can assume, without loss of generality, that $x_n \to x_0 \in X$. Since G is closed we obtain $x_0 \in G(\lambda_0, x_0)$, i.e., $x_0 \in \mathcal{F}(\lambda_0)$ giving the contradiction. □

For condensing multimaps we can obtain the following result which is slightly more general than a corollary of the above theorem.

Proposition 3.5.2. *Let X be a closed subset of a Banach space E, β a monotone MNC in E, Λ a metric space and $G : \Lambda \times X \to K(E)$ a closed multimap which is β-condensing in the second variable and such that* (3.5.1) *holds for every $\lambda \in \Lambda$. Then the multimap $\mathcal{F} : \Lambda \to P(E)$ is u.s.c.*

Proof. Again supposing the contrary, we can find a point $\lambda_0 \in \Lambda$ and sequences $\{\lambda_n\}_{n=1}^{\infty} \subset \Lambda$, $\lambda_n \to \lambda_0$, $\{x_n\}_{n=1}^{\infty} \subset X$ such that $x_n \in \mathcal{F}(\lambda_n) \setminus W$ for all n where W is a neighborhood of a set $\mathcal{F}(\lambda_0)$. For a compact set $K = \overline{\{\lambda_n\}_{n=1}^{\infty}}$ and $\Omega = \{x_n\}_{n=1}^{\infty}$ we have $\Omega \subseteq G(K \times \Omega)$. Therefore

$$\beta(\Omega) \leq \beta(G(K \times \Omega)),$$

hence the set Ω is relatively compact and we can assume without loss of generality that $x_n \to x_0 \in X$ obtaining the same contradiction as in the previous case. □

To verify relation (3.5.1) we can use any of the fixed point principles from Section 3.3. One more possibility follows from the following theorem on the stability of the topological degree.

Theorem 3.5.3. *Let U be an open subset of a Banach space E, β a monotone and invariant with respect to the union with compact sets MNC in E, Λ a metric space. Let $G : \Lambda \times \partial U \to K(E)$ be an u.s.c. β-condensing in the second variable multimap such that for some $\lambda_0 \in \Lambda$ we have* $\operatorname{Fix} G(\lambda_0, \cdot) \cap \partial U = \emptyset$. *Then for all $\lambda \in \Lambda$ close enough to λ_0, the degree*

$$\deg(i - G(\lambda, \cdot), \partial U)$$

is well-defined and coincides with $\deg(i - G(\lambda_0, \cdot), \partial U)$.

Proof. Let $\widehat{\Psi} = i - \widehat{G} : \partial U \to Kv(E)$ be a compact homotopy approximation of the multifield $\Psi_0 = i - G(\lambda_0, \cdot)$ on ∂U and the joining β-homotopy $\Psi_0 \underset{\beta}{\sim} \widehat{\Psi}$ has the form

$$\Phi(\nu, x) = (1 - \nu)\Psi_0(x) + \nu\widehat{\Psi}(x)$$

(see Definition 3.2.4 and Lemma 3.2.1).

Let us prove that for all $\lambda \in \Lambda$ close enough to λ_0, the multimap $\mathcal{F}_\lambda : [0, 1] \times \partial U \to Kv(E)$,

$$\mathcal{F}_\lambda(\mu, x) = \mu G(\lambda, x) + (1 - \mu)\widehat{G}(x),$$

defines a β-homotopy

$$i - G(\lambda, \cdot) \underset{\beta}{\sim} \widehat{\Psi}.$$

In fact, it is easy to see that $\mathcal{F}_\lambda$ is u.s.c. and β-condensing (see Proposition 2.2.1 (ii)). So it remains to demonstrate only that

$$\operatorname{Fix} \mathcal{F}_\lambda(\mu, \cdot) \cap \partial U = \emptyset, \quad 0 \le \mu \le 1$$

for all $\lambda \in \Lambda$ close enough to λ_0. Supposing the contrary we would have sequences $\{\lambda_n\}_{n=1}^\infty \subset \Lambda$, $\lambda_n \to \lambda_0$; $\{\mu_n\}_{n=1}^\infty \subset [0, 1]$, $\mu_n \to \mu_0$ and $\{x_n\}_{n=1}^\infty \subset \partial U$ such that

$$x_n \in \mathcal{F}_{\lambda_n}(\mu_n, x_n).$$

Then

$$\{x_n\} \subset \operatorname{co}\left(G\left(\overline{\{\lambda_n\}} \times \{x_n\}\right) \cup \widehat{G}\left(\{x_n\}\right)\right)$$

and from the properties of the MNC β it follows that

$$\beta\left(\{x_n\}\right) \le \beta\left(G\left(\overline{\{\lambda_n\}} \times \{x_n\}\right)\right).$$

The condition that G is β-condensing in the second variable implies that the sequence $\{x_n\}$ is relatively compact and we can assume, without loss of generality, that $x_n \to x_0 \in \partial U$. It is easy to verify (see Section 1.2.3) that the multimap $\mathcal{G} : \Lambda \times [0, 1] \times \partial U \to Kv(E)$, $\mathcal{G}(\lambda, \mu, x) = \mathcal{F}_\lambda(\mu, x)$ is u.s.c. and hence closed. So $x_0 \in \mathcal{F}_{\lambda_0}(\mu_0, x_0)$, i.e.,

$$x_0 = \mu_0 G(\lambda_0, x_0) + (1 - \mu_0)\widehat{G}(x_0)$$

contradicting the relation $0 \notin \Phi([0, 1] \times \partial U)$. □

As a corollary we obtain the following implicit multifunction theorem.

Corollary 3.5.1. *Let U be an open subset of a Banach space E. β a monotone and invariant with respect to the union with compact sets MNC in E, Λ a metric space. Let $G : \Lambda \times \overline{U} \to Kv(E)$ be an u.s.c. multimap β-condensing in the second variable such that*

$$\emptyset \neq \operatorname{Fix} G(\lambda_0, \cdot) \subset U$$

for some $\lambda_0 \in \Lambda$. *Moreover, let* $\operatorname{Ind}(G(\lambda_0, \cdot), \mathcal{M}) \neq 0$ *for a certain isolated component* $\mathcal{M}$ *of the set* $\operatorname{Fix} G(\lambda_0, \cdot)$ *(see Section* 3.3.1*). Then there exists a neighborhood* $W(\lambda_0)$ *of* λ_0 *such that for all* $\lambda \in W(\lambda_0)$

$$\operatorname{Fix} G(\lambda, \cdot) \neq \emptyset$$

and the multimap $\mathcal{F} : W(\lambda_0) \to P(E)$,

$$\mathcal{F}(\lambda) = \operatorname{Fix} G(\lambda, \cdot)$$

is u.s.c.

3.6 Solvability of systems of inclusions with condensing multioperators

In this section we study the existence of solutions for a system of inclusions of the form

$$\begin{cases} y \in \mathcal{F}(x, y) \\ x \in \mathcal{G}(x, y) \end{cases}$$

in Banach spaces X, Y. Here $\mathcal{F} : \overline{U} \times \overline{V} \to P(Y)$, $\mathcal{G} : \overline{U} \times \overline{V} \to P(X)$ are multimaps and $U \subseteq X$, $V \subseteq Y$ are open sets. It is clear that every solution of the above system can be considered as a fixed point of a multimap $\mathcal{G} \times \mathcal{F} : \overline{U} \times \overline{V} \to P(X \times Y)$ but this circumstance is of a little use since in the situations considered below this multimap does not belong to any of the classes for which the fixed point principles are directly adaptable.

As an application we consider the problem of optimal controllability for a system governed by a neutral functional differential equation. Other applications will be given in Section 6.3.1.

3.6.1 Preliminaries

We introduce the following definitions taken from [66].

Definition 3.6.1. Let X and Y be Hausdorff topological spaces. An u.s.c. multimap $M : X \to K(Y)$ is said to be a *weighted carrier* (*w-carrier*) if for any $x \in X$, to any piece C (i.e., open and closed subset) of $M(x)$ is assigned a weight or multiplicity $m(C, M(x)) \in \mathbb{Z}$ in such a way that the following properties hold:

(a) $m(\cdot\,, M(x))$ is an additive function, i.e.,

$$m(C_1 \cup C_2, M(x)) = m(C_1, M(x)) + m(C_2, M(x)),$$

whenever $C_1 \cap C_2 = \emptyset$;

(b) if U is open in Y with $\partial U \cap M(x) = \emptyset$, then

$$m(M(x) \cap U, M(x)) = m(M(x') \cap U, M(x'))$$

whenever x' is close enough to x.

Definition 3.6.2. If $U \subseteq Y$ is an open set, $M(x) \cap \partial U = \emptyset$, then the number

$$i(M(x), U) = m(M(x)) \cap U, M(x))$$

is called the *index* or *multiplicity* of $M(x)$ in U.

If X is connected the number $i(M(x), Y)$ does not depend on $x \in X$. In this case this number will be called the *index of the w-carrier M* and we denote it by $i(M)$.

We mention the following properties of w-carriers (see [66]).

(1) Let $M : X \to K(Y)$ be an u.s.c. multimap with connected values. Then M becomes a w-carrier by assigning a multiplicity m to $M(x)$. In particular every continuous single-valued map f can be considered as w-carrier with a prescribed index m. In this case we will denote it by $mf(x)$.

(2) Let $M : X \to K(Y)$ be a w-carrier and $f : Z \to X$ (resp. $f' : Y \to W$) be a continuous single-valued map. Then

$$M \circ f : Z \to K(Y) \quad (\text{resp. } f' \circ M : X \to K(W))$$

is a w-carrier.

(3) Let $M : X \to K(Y)$ be a w-carrier and $f : Z \to W$ be a continuous single-valued map. Then the multimap

$$M \times f : X \times Z \to K(Y \times W)$$

is w-carrier.

(4) Let $M : X \to K(Y)$ be a w-carrier, then the graph multimap $\Gamma_F : X \to K(X \times Y)$,

$$\Gamma_F(x) = \{x\} \times M(x)$$

is a w-carrier.

By a *weighted multimap* (*w-multimap*) we mean any w- carrier $M : X \to K(Y)$ such that $M(x)$ is a finite subset of Y for every $x \in X$.

As before (see Section 3.4), let

$$\{H_k(D)\}_{k=0}^{\infty}$$

be the graded Čech homology module of a space D with rational coefficients.

Definition 3.6.3. A w-carrier $M : X \to K(Y)$ is said to be

a) *pseudo-acyclic* if for every $x \in X$

$$H_k(M(x)) = 0$$

for all $k > 0$;

b) a *Lefschetz w-carrier* if it can be factorized in the form $M = r \circ Q$ where $Q : X \to K(Z)$ is a pseudo-acyclic w- carrier, Z is a metric ANR-space (absolute neighborhood retract, see e.g. [48]) and $r : Z \to Y$ is a continuous map.

It is known (see [66]) that, if X is a finite polyhedron and Z is a metric ANR-space, every pseudo-acyclic w-carrier $Q : X \to K(Z)$ induces homomorphisms

$$Q_{*k} : H_k(X) \to H_k(Z), \quad k \geq 0$$

which are invariant with respect to the corresponding homotopies. So if X is a finite polyhedron and $M : X \to K(Y)$ is a Lefschetz w-carrier of the form $M = r \circ Q$ we can define homomorphisms $M_{*k} : H_k(X) \to H_k(Y)$, $k \geq 0$ as the compositions $M_{*k} = r_{*k} \circ Q_{*k}$,

$$H_k(X) \xrightarrow{Q_{*k}} H_k(Z) \xrightarrow{r_{*k}} H_k(Y).$$

From the method of construction of homomorphisms Q_{*k} (see [66]) the following properties of homomorphisms M_{*k} can be deduced.

Proposition 3.6.1.

(a) *If two Lefschetz w-carriers $M_0, M_1 : X \to K(Y)$ are homotopic, $M_0 \sim M_1$, in the sense that there exists a Lefschetz w-carrier $N : [0, 1] \times X \to K(Y)$ with $N(0, \cdot) = M_0$ and $N_1(\cdot) = M_1$ then*

$$(M_0)_{*k} = (M_1)_{*k}, \quad k \geq 0.$$

(b) *If X is connected, Y is a metric ANR-space and $M(x) = m \cdot f(x)$ where $f : X \to Y$ is a continuous single valued map, $m = i(M)$, then*

$$M_{*k} = m \cdot f_{*k}, \quad k \geq 0$$

where $f_{*k} : H_k(X) \to H_k(Y)$, $k \geq 0$ are the homomorphisms induced by f.

Definition 3.6.4. Let X be a topological vector space and $A \subset X$ a symmetric set with respect to the origin. A multimap $L : A \to P(X)$ is said to satisfy the *Borsuk–Ulam property on A (B.U.-property)* if for each $x \in A$ the sets $L(x)$ and $L(-x)$ are strictly separated by a hyperplane; that is, there exists a continuous functional $x^* \in X^*$, the dual of X, such that

$$x^*(y) > 0 \quad \text{for all } y \in L(x)$$

and

$$x^*(y) < 0 \quad \text{for all } y \in L(-x).$$

We can prove the following result on the existence of a singular point for a Lefschetz w-carrier generalizing Theorem 2.8 of [167].

Theorem 3.6.1. *Let $B \subset \mathbb{R}^n$ be an open ball with center at the origin, $n \geq 2$ and $M : B \to K(\mathbb{R}^n)$ a Lefschetz w-carrier with $i(M) \neq 0$. If M satisfies the B.U.-property on $S = \partial B$ then there exists $x \in B$ such that $0 \in M(x)$.*

Proof. Denote by $\widetilde{M} : S \to K(\mathbb{R}^n \setminus 0)$ the restriction of M to S. From the B.U.-property it follows that for each $x \in S$ there exists $y \in S$ such that

$$\begin{aligned} \langle y, z\rangle > 0 \quad &\text{for all} \quad z \in \widetilde{M}(x) \\ \langle y, z\rangle < 0 \quad &\text{for all} \quad z \in \widetilde{M}(-x) \end{aligned} \tag{3.6.1}$$

For $y \in S$, let $V_y = \{x \in S : (3.6.1) \text{ holds}\}$. From the upper semicontinuity of $\widetilde{M}$ it follows that every V_y, $y \in S$ is an open subset of S and the B.U.-property implies that $\{V_y\}_{y \in S}$ is a covering of S. Let $\{V_{y_i}\}_{i=1}^m$ be a finite subcovering of $\{V_y\}_{y\in S}$ and let $p_i : S \to [0, 1]$, $i = 1, \ldots, m$ be the partition of the unity subordinated to $\{V_{y_i}\}_{i=1}^m$.

Define $f : S \to \mathbb{R}^n$ as an odd continuous map:

$$f(x) = \sum_{i=1}^m (p_i(x) - p_i(-x))y_i.$$

Consider the multimap $H : [0, 1] \times S \to K(\mathbb{R}^n)$ defined as the composition

$$[0,1] \times S \xrightarrow{\ \text{id} \times \Delta\ } [0,1] \times S \times S \xrightarrow{\ \text{id} \times \widetilde{M} \times f\ } [0,1] \times \mathbb{R}^n \times \mathbb{R}^n \xrightarrow{\ g\ } \mathbb{R}^n$$

where $\Delta : S \to S \times S$, $\Delta(x) = (x, x)$ and

$$g(\lambda, x, y) = \lambda x + (1 - \lambda)y.$$

H is a Lefschetz w-carrier. Let us prove that H acts into $\mathbb{R}^n \setminus 0$. In fact, if $0 \in H(\lambda_*, x_*)$ for some $(\lambda_*, x_*) \in [0, 1] \times S$ then there exists $z_* \in \widetilde{M}(x_*)$ such that

$$\lambda_* z_* = -(1 - \lambda_*) f(x_*) = -(1 - \lambda_*)\Big(\sum_{i=1}^m (p_i(x_*) - p_i(-x_*))y_i\Big).$$

Clearly $\lambda_* \neq 1$. Then for $\mu_* = -\lambda_*/(1 - \lambda_*) < 0$ we have that

$$\mu_* z_* = \sum_{i=1}^m p_i(x_*)y_i - \sum_{i=1}^m p_i(-x_*)y_i$$

hence

$$\mu_* \| z_* \|^2 = \sum_{i=1}^m p_i(x_*)\langle y_i, z_*\rangle - \sum_{i=1}^m p_i(-x_*)\langle y_i, z_*\rangle. \tag{3.6.2}$$

But if $\langle y_i, z_* \rangle < 0$ then $x_* \notin V_{y_i}$ and so $p_i(x_*) = 0$ and if $\langle y_i, z_* \rangle > 0$ then $-x_* \notin V_{y_i}$ and so $p_i(-x_*) = 0$. Therefore the right-hand side of equality (3.6.2) is always positive, a contradiction.

So $H : [0,1] \times S \to K(\mathbb{R}^n \setminus 0)$ defines a homotopy between $H(0,x) = i(M) \cdot f(x)$ and $H(1,x) = \widetilde{M}(x)$. Since f is odd, the homomorphism $\widetilde{f}_{(n-1)*} : H_{n-1}(S) \to H_{n-1}(\mathbb{R}^n \setminus 0)$ is non-zero and hence the homomorphism $\widetilde{M}_{(n-1)*} : H_{n-1}(S) \to H_{n-1}(\mathbb{R}^n \setminus 0)$, $\widetilde{M}_{(n-1)*} = i(M) \cdot f_{(n-1)*}$ is non-trivial too.

If we suppose that $0 \notin M(x)$ for all $x \in B$ then the multimap $G : [0,1] \times S \to K(\mathbb{R}^n \setminus 0)$, $G(\lambda, x) = M(\lambda x)$ induces a homotopy between $G(1,x) = \widetilde{M}(x)$ and the constant multimap $G(0,x) = \{M(0)\}$. If $M = r \circ Q$ is the factorization of M then $H_{n-1}(Q(0)) = 0$ and so $\widetilde{M}_{(n-1)*} = G(0, \cdot)_{(n-1)*} = 0$. □

We also use the following "relative" version of the above theorem.

Theorem 3.6.2. *Let $\mathcal{R} \subset \mathbb{R}^n$ be a convex, compact set, symmetric with respect to the origin,* $\dim \mathcal{R} \geq 2$ *and let $\overline{B}_{\mathcal{R}} = \overline{B}_r(0) \cap \mathcal{R}$. Assume that $M : \overline{B}_{\mathcal{R}} \to K(\mathbb{R}^n)$ is a Lefschetz w-carrier such that $i(M) \neq 0$ and $N(\overline{B}_{\mathcal{R}}) \subset \mathcal{R}$, where $N(x) = x - M(x)$. If M verifies the B.U.-property on $\partial B_{\mathcal{R}} = \partial B_r(0) \cap \mathcal{R}$, then there exists $x \in B_{\mathcal{R}}$ such that $0 \in M(x)$.*

Proof. Passing if necessary to the linear hull of the set $\mathcal{R}$, we can assume, without loss of generality, that $\dim \mathcal{R} = n$. Consider the retraction $\rho : \mathbb{R}^n \to \mathcal{R}$ defined as

$$\rho(x) = \begin{cases} x, & \text{if } x \in \mathcal{R}, \\ \nu x, \ \nu = \sup\{\nu' : \nu' x \in \mathcal{R}\}, & \text{if } x \notin \mathcal{R}. \end{cases}$$

It is clear that the set $\rho^{-1}(\overline{B}_{\mathcal{R}})$ is symmetric with respect to the origin and connected. Now, let $B_1 = B_{r_1}(0) \subset \mathbb{R}^n$ be an open ball containing the set $\mathcal{R}$. Consider on the set $D = \overline{B}_1 \cap \rho^{-1}(\overline{B}_{\mathcal{R}})$ the multimap $\widetilde{M} : D \to K(\mathbb{R}^n)$, given by $\widetilde{M}(x) = x - N \circ \rho(x)$. The multimap $\widetilde{M}$ is a Lefschetz w-carrier since it can be decomposed into the form

$$D \stackrel{\Gamma}{\multimap} \mathbb{R}^n \times \mathbb{R}^n \stackrel{d}{\longrightarrow} \mathbb{R}^n,$$

where $\Gamma = \Gamma_{N \circ \rho}$ is the graph multimap and $d(x,y) = x - y$. Moreover, $i(\widetilde{M}) = i(M) \neq 0$.

We show that $\widetilde{M}$ verifies the B.U.-property on ∂D. In fact, if $x \in \partial D$ and $x \in \rho^{-1}(\partial B_{\mathcal{R}})$ the assertion follows from the fact that $\widetilde{M}|_{\partial B_{\mathcal{R}}} = M|_{\partial B_{\mathcal{R}}}$ and the assumption of the theorem.

If $x \in \partial B_1 \cap \rho^{-1}(\overline{B}_{\mathcal{R}})$, we show that

$$0 \notin \Lambda(\overline{\text{co}}\, \widetilde{M}(x)) - \overline{\text{co}}\, \widetilde{M}(-x),$$

where $\Lambda(A) = \{\lambda x : \lambda \in [0,1],\ x \in A\}$. Assume, to the contrary, that there exist $\lambda_0 \in [0,1]$, $y_1 \in \overline{\text{co}}(N \circ \rho(x)) \subset \mathcal{R}$ and $y_2 \in \overline{\text{co}}(N \circ \rho(-x)) \subset \mathcal{R}$ such that $\lambda_0(x - y_1) = -x - y_2$. Then

$$(\lambda_0 + 1)\| x \| = \| \lambda y_1 - y_2 \| < (\lambda + 1) r_1.$$

This contradicts the fact that $x \in \partial B_1$. Analogously

$$0 \notin \Lambda(\overline{\mathrm{co}}\,\widetilde{M}(-x)) - \overline{\mathrm{co}}\,\widetilde{M}(x).$$

From the Hahn–Banach separation property it follows that there exist $u, v \in \mathbb{R}^n$ and $\alpha, \beta \in \mathbb{R}$ such that

$$\langle u, \Lambda(\overline{\mathrm{co}}\,\widetilde{M}(x))\rangle \subset (\alpha, +\infty), \quad \langle u, \overline{\mathrm{co}}\,\widetilde{M}(-x)\rangle \subset (-\infty, \alpha)$$

and

$$\langle v, \Lambda(\overline{\mathrm{co}}\,\widetilde{M}(-x))\rangle \subset (\beta, +\infty), \quad \langle v, \overline{\mathrm{co}}\,\widetilde{M}(x)\rangle \subset (-\infty, \beta).$$

Since $0 \in \Lambda(\overline{\mathrm{co}}\,\widetilde{M}(x)) \cap \Lambda(\overline{\mathrm{co}}\,\widetilde{M}(-x))$ it follows that $\alpha, \beta < 0$. Then it is easy to see that the sets $\widetilde{M}(x)$ and $\widetilde{M}(-x)$ are strictly separated by the hyperplane defined by the vector $w = \alpha v - \beta u$.

Since the set D is homeomorphic to the ball if follows from the previous theorem that there exists $x \in D$ such that $0 \in \widetilde{M}(x)$, that is $x \in N \circ \rho(x) \subset \mathcal{R}$, hence $\rho(x) = x$ and $0 \in M(x)$. □

3.6.2 Systems of inclusions

We can study now our main problem on the solvability of systems of inclusions.

Let X and Y be Banach spaces and $U \subseteq X$, $V \subseteq Y$ open sets. Assume that $\mathcal{F} : \overline{U} \times \overline{V} \to K(Y)$ is a multimap and let $\Phi(x, y) = y - \mathcal{F}(x, y)$. We introduce the notation

$$\begin{aligned} S^{\mathcal{F}} &= \{(x, y) \in \overline{U} \times \overline{V} : y \in \mathcal{F}(x, y)\}, \\ D^{\mathcal{F}} &= \{x \in X : S^{\mathcal{F}}_x \subset U \times V\}, \end{aligned}$$

where $S^{\mathcal{F}}_A = S^{\mathcal{F}} \cap (A \times Y)$ for any $A \subset X$.

We study the existence of solutions for the system

$$\begin{cases} y \in \mathcal{F}(x, y), \\ x \in \mathcal{G}(x, y), \end{cases} \tag{3.6.3}$$

or equivalently

$$\begin{cases} 0 \in y - \mathcal{F}(x, y) = \Phi(x, y), \\ 0 \in x - \mathcal{G}(x, y) = \Psi(x, y). \end{cases}$$

Theorem 3.6.3. *Let $U \subseteq X$ be an open neighborhood of the origin and $V \subseteq Y$ an open set. Let $\mathcal{F} : \overline{U} \times \overline{V} \to Kv(Y)$ be a closed completely fundamentally restrictible in the second variable multimap and $\mathcal{G} : \overline{U} \times \overline{V} \to K(X)$ a quasiacyclic multimap (see Sections* 3.5 *and* 3.4.1*).*

(i) *Suppose that there exists* $r > 0$ *such that* $\overline{B}_r(0) \subset \overline{U} \cap D^{\mathcal{F}}$ *and*

$$\deg(\Phi(0,\cdot), \overline{V}) \neq 0.$$

(ii) *Suppose that the multimap* $\widehat{\mathcal{G}} : \overline{B}_r(0) \to K(X)$, *given by* $\widehat{\mathcal{G}}(x) = \mathcal{G}(x, \overline{V})$ *is closed and completely fundamentally restrictible on a compact essential fundamental set* $\mathcal{K}$ *which is symmetric with respect to the origin and such that* $\dim \mathcal{K} \geq 2$.

Let $T = i - Q : \overline{B}_r(0) \to P(X)$ *be the multifield corresponding to the multimap* $Q(x) = \mathcal{G}(x, S(x))$ *where* $S(x) = \{y \in \overline{V} : y \in \mathcal{F}(x, y)\}$.

(iii) *Suppose that the multifield* T *satisfies the B.U.-condition on* $\partial B_r(0)$.

Then the system (3.6.3) *has a solution.*

Remark 3.6.1. The condition (ii) is fulfilled if the closed multimap $\widehat{\mathcal{G}}$ is (k, ψ)-condensing with respect to a real MNC ψ which is regular, monotone, nonsingular, semiadditive and invariant with respect to reflection at the origin. In fact, from the proof of Theorem 3.3.5 we know that $\widehat{\mathcal{G}}$ possesses an essential fundamental set $\mathcal{K}$ which is symmetric with respect to the origin. On the other hand, from Theorem 2.2.1 it follows that we can assume, without loss of generality, that $\mathcal{K}$ is compact and contains any prescribed finite subset from the ball $\overline{B}_r(0)$ so the condition $\dim \mathcal{K} \geq 2$ can also be ensured.

The proof of Theorem 3.6.3 needs several preliminary remarks and results.

Remark 3.6.2. The multimap Q (and hence the multifield T) is correctly defined, i.e., it has nonempty compact values and it is closed. In fact, from assumption (i) it follows that for any $x \in \overline{B}_r(0)$ we have $S(x) = \operatorname{Fix} \mathcal{F}(x, \cdot) \cap \partial V = \emptyset$. Further, the fundamental restrictibility of the multimap $\mathcal{F}$ in the second variable implies that for each $x \in \overline{B}_r(0)$ the multimap $\mathcal{F}' : [0, 1] \times \overline{V} \to Kv(Y)$,

$$\mathcal{F}'(\lambda, y) = \mathcal{F}(\lambda x, y)$$

defines a homotopy connecting the multifields $\Phi(0, \cdot)$ and $\Phi(x, \cdot)$ and therefore $\deg(\Phi(x, \cdot), \overline{V}) = \deg(\Phi(0, \cdot), \overline{V}) \neq 0$. So for every $x \in \overline{B}_r(0)$ the set $S(x)$ is nonempty and compact (Theorem 3.5.1). Finally we can apply Theorem 3.5.2 to conclude that the multimap S is u.s.c. and we can use the continuity properties of multimaps described in Theorems 1.2.12 and 1.2.9.

Let $\overline{B}_{\mathcal{K}} = \overline{B}_r(0) \cap \mathcal{K}$ and let $\mathcal{M} \subset Y$ be an essential fundamental set of the family $\mathcal{F}|_{\overline{B}_{\mathcal{K}} \times \overline{V}}$.

From condition (i) it follows that $\emptyset \neq S(0) = \operatorname{Fix} \mathcal{F}(0, \cdot) \subset V \cap \mathcal{M}$ and so $V_{\mathcal{M}} = V \cap \mathcal{M} \neq \emptyset$. Let $\overline{V}_{\mathcal{M}}$, $\partial V_{\mathcal{M}}$ denote the closure and the boundary of $V_{\mathcal{M}}$ in the relative topology of the space $\mathcal{M}$. Since Fix $\mathcal{F}(0,\cdot) \subset V_{\mathcal{M}} \cap \mathcal{F}(0, V_{\mathcal{M}})$, we can assume, without loss of generality, that the set $\mathcal{M}$ is compact. (Otherwise we can take the compact essential fundamental set $\overline{\text{co}}\,\mathcal{F}(\overline{B}_{\mathcal{K}} \times \overline{V}_{\mathcal{M}})$).

From the condition (iii) it follows that multifield T has no zeros on $\partial B_{\mathcal{K}}$ and so $S^{\mathcal{F}}_{\overline{B}_{\mathcal{K}}} \subset B_{\mathcal{K}} \times V_{\mathcal{M}}$. First of all we need the following extension of Lemma 2.2 of [63].

Lemma 3.6.1. *For each neighborhood $W \subset B_{\mathcal{K}} \times V_{\mathcal{M}}$ (in the relative topology) of the set $S^{\mathcal{F}}_{\overline{B}_{\mathcal{K}}}$, there exists $\varepsilon > 0$ such that if $f : \overline{B}_{\mathcal{K}} \times \overline{V}_{\mathcal{M}} \to Y$ is an ε-approximation of $\mathcal{F}|_{\overline{B}_{\mathcal{K}} \times \overline{V}_{\mathcal{M}}}$ then $S^{f}_{\overline{B}_{\mathcal{K}}} \subset W$.*

Proof. Since the multimap $\mathcal{F}$ is closed the set $S^{\mathcal{F}}_{\overline{B}_{\mathcal{K}}}$ is compact. Let $W_1 \subset B_{\mathcal{K}} \times V_{\mathcal{M}}$ be an δ_1-neighborhood of $S^{\mathcal{F}}_{\overline{B}_{\mathcal{K}}}$ such that $W_1 \subset W$ and $\partial W_1 \cap \partial W = \emptyset$. Let

$$\delta_2 = \text{dist}(\partial W, \partial W_1) = \min\{d(w, w_1) : W \in \partial W,\ w_1 \in \partial W_1\}$$

and

$$A = (\overline{B}_{\mathcal{K}} \times \overline{V}_{\mathcal{M}}) \setminus W_1.$$

Then $\mathcal{F}(A) \subset Y$ is a closed set (see Theorem 1.1.6), $0 \notin \mathcal{F}(A)$ and let $\delta_3 > 0$ be such that $\overline{B}_{\delta_3}(0) \cap \mathcal{F}(A) = \emptyset$. Hence, if $(x, y) \in A$, then $\text{dist}(y, \mathcal{F}(x, y)) > \delta_3$.

Let $\delta = \min\{\delta_1, \delta_2, \delta_3\}$ and let $f : \overline{B}_{\mathcal{K}} \times \overline{V}_{\mathcal{M}} \to Y$ be an ε-approximation of $\mathcal{F}_{\overline{B}_{\mathcal{K}} \times \overline{V}_{\mathcal{M}}}$ where $\varepsilon = \delta/2$. Let $(x, y) \in (\overline{B}_{\mathcal{K}} \times \overline{V}_{\mathcal{M}}) \setminus W$ be such that $y = f(x, y)$, that is $(x, y) \in S^{f}_{\overline{B}_{\mathcal{K}}}$. Then there exists $(\overline{x}, \overline{y}) \in \overline{B}_{\mathcal{K}} \times \overline{V}_{\mathcal{M}}$ such that

$$\text{dist}[(x, y), (\overline{x}, \overline{y})] + \| f(x, y) - z \| < 2\varepsilon = \delta$$

for some $z \in \mathcal{F}(\overline{x}, \overline{y})$. From $\text{dist}[(x, y), (\overline{x}, \overline{y})] < \delta$ it follows that $(\overline{x}, \overline{y}) \notin S^{\mathcal{F}}_{\overline{B}_{\mathcal{K}}}$. Since $y = f(x, y)$ we get

$$\| \overline{y} - z \| \le \| y - \overline{y} \| + \| y - z \| \le \text{dist}[(x, y), (\overline{x}, \overline{y})] + \| f(x, y) - z \| < \delta.$$

Hence $(\overline{x}, \overline{y}) \notin A$. Thus $(\overline{x}, \overline{y}) \in W_1 \setminus S^{\mathcal{F}}_{\overline{B}_{\mathcal{K}}}$. This is a contradiction, since $(x, y) \in (\overline{B}_{\mathcal{K}} \times \overline{V}_{\mathcal{M}}) \setminus W$ and $\text{dist}[(x, y), (\overline{x}, \overline{y})] < \delta$ imply that $(\overline{x}, \overline{y}) \notin W_1$. □

Lemma 3.6.2. *Let $X = \mathbb{R}^n$, $Y = \mathbb{R}^m$. There exists an $\varepsilon_0 > 0$ such that for all ε, $0 < \varepsilon < \varepsilon_0$, there exists a continuous map $f : \overline{U} \times \overline{V} \to \mathbb{R}^m$ such that $f|_{\overline{B}_{\mathcal{K}} \times \overline{V}_{\mathcal{M}}} : \overline{B}_{\mathcal{K}} \times \overline{V}_{\mathcal{M}} \to \mathcal{M}$ is an ε-approximation of $\mathcal{F}|_{\overline{B}_{\mathcal{K}} \times \overline{V}_{\mathcal{M}}}$ and*

(a) *for every $x \in \overline{B}_{\mathcal{K}}$ the set $S^f(x) = \{y \in Y : y = f(x, y)\}$ is a finite subset of V;*

(b) $\deg(\Phi(0,\cdot), \overline{V}) = \deg_{\mathcal{M}}(i - f(0,\cdot), \overline{V}_{\mathcal{M}})$.

Proof. Let $\varepsilon_0 > 0$ be such that every ε-approximation f of $\mathcal{F}|_{\overline{B}_{\mathcal{K}}\times\overline{V}_{\mathcal{M}}}$ with $0 < \varepsilon < \varepsilon_0$ satisfies property (b). Fix any particular ε, $0 < \varepsilon < \varepsilon_0$ and let $f_0 : \overline{B}_{\mathcal{K}} \times \overline{V}_{\mathcal{M}} \to \mathcal{M}$ be any $(\varepsilon/2)$-approximation of $\mathcal{F}|_{\overline{B}_{\mathcal{K}}\times\overline{V}_{\mathcal{M}}}$. From Lemma 2.4 of [167] it follows that there exists a continuous map $f_1 : \overline{B}_{\mathcal{K}} \times \overline{V}_{\mathcal{M}} \to \mathcal{M}$ such that

1) $\| f_1(x, y) - f_0(x, y) \| < \varepsilon/2$ for all $(x, y) \in \overline{B}_{\mathcal{K}} \times \overline{V}_{\mathcal{M}}$;

2) for every $x \in \overline{B}_{\mathcal{K}}$ the set $S^{f_1}(x) = \{y \in \overline{V}_{\mathcal{M}} : y = f_1(x, y)\}$ is a finite subset of V;

3) $\deg_{\mathcal{M}}(i - f_1(0, \cdot), \overline{V}_{\mathcal{M}}) = \deg_{\mathcal{M}}(i - f_0(0, \cdot), \overline{V}_{\mathcal{M}})$.

Then any continuous extension f to f_0 on $\overline{U} \times \overline{V}$ is the required map. □

Lemma 3.6.3. *Let X, Y and $\mathcal{F}$ satisfy conditions in Theorem* 3.6.3. *Let us suppose that for all $x \in \overline{B}_{\mathcal{K}}$ the inclusion $y \in \mathcal{F}(x, y)$ has only isolated solutions. Then the multimap $S : \overline{B}_{\mathcal{K}} \to K(V_{\mathcal{M}})$ is a w-multimap and*

$$i(S) = \deg(\Phi(0, \cdot), \overline{V}).$$

Proof. If $y \in S(x)$ then y is an isolated fixed point of $\mathcal{F}(x, \cdot)$ and we can put $m(y, S(x)) = \operatorname{ind}(i - \mathcal{F}(x, \cdot), y)$, the topological index of y (see Section 3.3.1). Now, let Ω be an open set in V such that $S(x) \cap \partial\Omega_{\mathcal{M}} = \emptyset$, where $\Omega_{\mathcal{M}} = \Omega \cap \mathcal{M}$. From the upper semicontinuity of S it follows that there exists a ball $\overline{B}_\delta(x)$ such that $S(\overline{B}_\delta(x) \cap \mathcal{K}) \cap \partial\Omega_{\mathcal{M}} = \emptyset$. Then, for every $x' \in \overline{B}_\delta(x) \cap \mathcal{K}$, the homotopy $H : [0, 1] \times \overline{\Omega}_{\mathcal{M}} \to Kv(\mathcal{M})$, given by

$$H(\lambda, y) = \mathcal{F}(\lambda x + (1 - \lambda)x', y)$$

ensures that

$$\deg_{\mathcal{M}}(\Phi(x, \cdot), \overline{\Omega}) = \deg_{\mathcal{M}}(\Phi(x', \cdot), \overline{\Omega}).$$

Therefore

$$\begin{aligned} m(S(x) \cap \Omega, S(x)) &= \sum_{y \in S(x)\cap\Omega} \operatorname{ind}(\mathcal{F}(x, \cdot), y) \\ &= \deg_{\mathcal{M}}(\Phi(x, \cdot), \overline{\Omega}) = \deg_{\mathcal{M}}(\Phi(x', \cdot), \overline{\Omega}) \\ &= \sum_{y' \in S(x')\cap\Omega} \operatorname{ind}(\mathcal{F}(x, \cdot), y') = m(S(x') \cap \Omega, S(x')). \end{aligned}$$

Thus S is a w-multimap and

$$i(S) = i(S(0), V_{\mathcal{M}}) = \sum_{y \in S(0)} \operatorname{ind}(\mathcal{F}(0, \cdot), y) = \deg(\Phi(0, \cdot), \overline{V}).$$ □

We are now in a position to prove Theorem 3.6.3.

Proof of Theorem 3.6.3. Let $T_{\mathcal{K}} = T|_{\overline{B}_{\mathcal{K}}}$. Assume to the contrary that system (3.6.3) does not have solutions. Then $0 \notin T_{\mathcal{K}}(x)$ for all $x \in \overline{B}_{\mathcal{K}}$. Since $T_{\mathcal{K}}$ is an u.s.c. multimap, this assumption implies the existence of an $\varepsilon_1 > 0$ such that $B_{\varepsilon_1}(0) \cap T_{\mathcal{K}}(\overline{B}_{\mathcal{K}}) = \emptyset$.

On the other hand it is easy to verify (see Proposition 2.13 of [167]) that there exists $\varepsilon_2 > 0$ such that every multimap $T' : \overline{B}_{\mathcal{K}} \to P(X)$ with the property that $\Gamma_{T'} \subset W_{\varepsilon_2}(\Gamma_T)$ satisfies the B.U.-property on $\partial B_{\mathcal{K}}$.

Let $\delta = \min\{\varepsilon_1, \varepsilon_2\}$ and let $W \subset \overline{B}_{\mathcal{K}} \times V_{\mathcal{M}}$ be given by

$$W = \{(x, y) \in B_{\mathcal{K}} \times V_{\mathcal{M}} : (x, \Psi(x, y)) \in W_\delta(\Gamma_{T_{\mathcal{K}}})\}.$$

It is clear that W is an open neighborhood (in the relative topology) of the set $S^{\mathcal{F}}_{\overline{B}_{\mathcal{K}}}$.

We will divide the rest of the proof into three parts.

First step. $X = \mathbb{R}^n$, $n \geq 2$ and $Y = \mathbb{R}^m$.

Let ε^* be the one in Lemma 3.6.1, i.e., every ε^*-approximation f of $\mathcal{F}|_{\overline{B}_{\mathcal{K}} \times \overline{V}_{\mathcal{M}}}$ has the property that $S^f_{\overline{B}_{\mathcal{K}}} \subset W$.

Let $\varepsilon_0 > 0$ be the one in Lemma 3.6.2 and let $\varepsilon' = \min\{\varepsilon^*, \varepsilon_0, \delta\}$. By Lemmas 3.6.2 and 3.6.3 there exists an ε'-approximation f' of $\mathcal{F}|_{\overline{B}_{\mathcal{K}} \times \overline{V}_{\mathcal{M}}}$ such that the multimap $S' : \overline{B}_{\mathcal{K}} \to P(V_{\mathcal{M}})$ defined by $S'(x) = \{y : y = f'(x, y)\}$ is a w-multimap. Its index can be evaluated in the following way:

$$i(S') = \deg\left(f'(0, \cdot), \overline{W(0)}\right) = \deg\left(\Phi(0, \cdot), \overline{W(0)}\right) = \deg\left(\Phi(0, \cdot), \overline{V}\right) \neq 0.$$

We show now that the multimap $Q'(x) = G(x, S'(x))$ is a Lefschetz w-carrier. Indeed, since the multimap G is quasiacyclic it can be represented as $G = q \circ \mathcal{G}$ where $\mathcal{G} : \overline{B}_{\mathcal{K}} \to K(Z)$ is an acyclic multimap and $q : Z \to \mathbb{R}^n$ is a continuous map. Then Q' can be decomposed as

$$\overline{B}_{\mathcal{K}} \overset{\Lambda}{\multimap} Z \times \mathbb{R}^m \xrightarrow{\mathrm{pr}_1} Z \xrightarrow{q} \mathbb{R}^n,$$

where $\Lambda(x) = \bigcup_{y \in S'(x)} (Q(x, y) \times \{y\})$ is an acyclic w-carrier and pr_1 is the natural projection. Moreover, for $T' = i - Q'$ we have

$$i(T') = i(Q') = i(S') \neq 0.$$

Since $S^{f'}_{\overline{B}_{\mathcal{K}}} \subset W$ we have $\Gamma_{T'} \subset W_\delta(\Gamma_{T_{\mathcal{K}}})$, i.e., T' is a δ-approximation of $T_{\mathcal{K}}$. From our choice of δ it follows that T' satisfies the B.U.-condition on $\partial B_{\mathcal{K}}$. Hence, from Theorem 3.6.2 it follows that there exists $x \in B_{\mathcal{K}}$ such that $0 \in T'(x)$. Then $0 \in W_\delta(T_{\mathcal{K}}(B_\delta(x))) \subseteq W_{\varepsilon_1}(T_{\mathcal{K}}(B_{\varepsilon_1}(x)))$. This contradiction establishes the result.

Second step. $X = \mathbb{R}^n$, Y is a Banach space.

Let $\varepsilon' > 0$ be as in the first step and f_1 be an ε'-approximation of $\mathcal{F}|_{\overline{B}_{\mathcal{K}} \times \overline{V}_{\mathcal{M}}}$ such that its range $f_1(\overline{B}_{\mathcal{K}} \times \overline{V}_{\mathcal{M}})$ is contained in a finite-dimensional subspace $Y_1 \subset Y$.

Let $V^1_{\mathcal{M}} = V_{\mathcal{M}} \cap Y_1$, then $S^{f_1}_{\overline{B}_{\mathcal{K}}} \subset W \cap (B_{\mathcal{K}} \times V^1_{\mathcal{M}}) = W_1$; let $f_2 = f_1|_{\overline{B}_{\mathcal{K}} \times \overline{V}^1_{\mathcal{M}}}$, then from the properties of the topological degree (see Section 3.1) we have

$$\begin{aligned}\deg(f_2(0,\cdot), \overline{W_1(0)}) &= \deg(f_1(0,\cdot), \overline{W(0)}) \\ &= \deg(\Phi(0,\cdot), \overline{W(0)}) = \deg(\Phi(0,\cdot), \overline{V}) \neq 0.\end{aligned}$$

Applying the first step to the map f_2 and the multimap $\mathcal{G}_2 = \mathcal{G}|_{\overline{B}_{\mathcal{K}} \times \overline{V}^1_{\mathcal{M}}}$, we conclude that the multimap $T_2 : \overline{B}_{\mathcal{K}} \to P(Y_1)$, given by $T_2(x) = \Psi_2(x, S^{f_2}(x))$, $\Psi_2(x, y) = x - \mathcal{G}_2(x, y)$, has a zero singular point $x \in B_{\mathcal{K}}$. As $S^{f_2}_{\overline{B}_{\mathcal{K}}}(x) \subset W$ we have $\Gamma_{T_2} \subset W_\delta(\Gamma_{T_{\mathcal{K}}})$, contradicting the fact that $0 \notin W_{\varepsilon_1}(T_{\mathcal{K}}(B_{\varepsilon_1}(x)))$.

Third step. X, Y are Banach spaces.

Let $\mu_\varepsilon : \mathcal{K} \to \mathcal{K}_1 = \mathcal{K} \cap X_1$ be a Schauder ε-projection, X_1 a finite-dimensional subspace of X. Let $\mathcal{G}_3 : \overline{B}_{\mathcal{K}} \times \overline{V}_{\mathcal{M}} \to K(\mathcal{K}_1)$, $\mathcal{G}_3 = \mu_\varepsilon \circ \mathcal{G}$, $\Psi_3(x, y) = x - \mathcal{G}_3(x, y)$ and $\mathcal{F}_3 = \mathcal{F}|_{\overline{B}_{\mathcal{K}} \times \overline{V}_{\mathcal{M}}}$. It is clear that $\mathcal{G}_3$ is quasiacyclic and if $T' : \overline{B}_{\mathcal{K}_1} \to K(X_1)$ is defined as $T'(x) = \mathcal{G}_3(x, S(x))$ then $\Gamma_{T'} \subset W_\varepsilon(\Gamma_{\widehat{T}})$ where $\widehat{T} = T|_{\overline{B}_{\mathcal{K}_1}}$ and hence, if $\varepsilon > 0$ is sufficiently small, T' satisfies the B.U.-condition on $\partial B_{\mathcal{K}_1}$. By the second step we conclude that T' has a zero on $B_{\mathcal{K}_1}$, contradicting the fact that $0 \notin W_\varepsilon(T(B_\varepsilon(x)))$ for $\varepsilon > 0$ sufficiently small. □

Theorem 3.6.4. *Let X, Y be Banach spaces, $U \subset X$, $V \subset Y$ be open sets. Let $\mathcal{F} : \overline{U} \times \overline{V} \to Kv(Y)$ be a closed completely fundamentally restrictible in the second variable multimap. Let $N \subset \overline{U}$ be a nonempty compact convex set such that $N \subset D^{\mathcal{F}}$ and* $\deg(\mathcal{F}(x,\cdot), \overline{V}) \neq 0$ *for some (and hence for all) $x \in N$. If $\mathcal{G} : \overline{U} \times \overline{V} \to K(X)$ is a quasiacyclic multimap such that $\mathcal{G}(N, \overline{V}) \subseteq N$ then there exists a solution $(x, y) \in N \times V$ of the system* (3.6.3).

Proof. Let $B_r(0) \subset X$ be any ball containing the set N and $\rho : B_r(0) \to N$ any retraction. We can now apply Theorem 3.6.3 to the multimaps $\mathcal{F}_1 : \overline{B}_r(0) \times \overline{V} \to Kv(Y)$ and $\mathcal{G}_1 : \overline{B}_r(0) \times \overline{V} \to K(X)$ defined as $\mathcal{F}_1(x, y) = \mathcal{F}(\rho(x), y)$, $\mathcal{G}_1(x, y) = \mathcal{G}(\rho(x), y)$. In fact, it is not difficult to see that all the conditions in this theorem are satisfied except the closedness of the multimap $\widehat{\mathcal{G}}$, but here this condition is unnecessary since $\widehat{\mathcal{G}}$ is compact.

So there exists $(x, y) \in B_r(0) \times V$ such that $y \in \mathcal{F}_1(x, y)$, $x \in \mathcal{G}_1(x, y)$. Since $x \in N$ we conclude that (x, y) is a solution of the system (3.6.3). □

In the following statement we do not assume any additional condition concerning the structure of the values of the multimap $\mathcal{F}$. Instead, we assume the acyclicity of the solution multimap S. Later on it will be shown that this hypothesis is natural in applications (see Section 6.3).

The next result is a direct consequence of the u.s.c. dependence of the fixed points set (Section 3.5) and the topological degree theory constructed in Section 3.4.

Theorem 3.6.5. *Let X, Y be Banach spaces, $U \subset X$, $V \subset Y$ be open sets. Let $\mathcal{F} : \overline{U} \times \overline{V} \to K(Y)$ be a closed completely fundamentally restrictible in the second variable multimap and let $\mathcal{G} : \overline{U} \times \overline{V} \to K(X)$ be a Vietoris multimap (see Section 3.4.1). Suppose that there exists a ball $B_r(0) \subset X$ such that for every $x \in \overline{B}_r(0)$ the set $S(x) = \{y \in \overline{V} : y \in \mathcal{F}(x, y)\}$ is nonempty and acyclic. Then*

(a) *$Q : \overline{B}_r(0) \to K(X)$, $Q(x) = G(x, S(x))$ is a Vietoris multimap;*

(b) *if Q is a 2-completely fundamentally restrictible and $\deg(i - Q, \overline{B}_r(0)) \neq 0$ then the system (3.6.3) has a solution in $B_r(0) \times \overline{V}$.*

Proof. From Theorems 3.5.1 and 3.5.2 it follows that the multimap S is compact-valued and u.s.c. Then Proposition 3.4.1 (a) implies that Q is a Vietoris multimap. From the fixed point principle for Vietoris multimaps (see Section 3.4.3) it follows that there exists $x \in B_r(0)$, $x \in Q(x)$ and taking any $y \in S(x)$ we obtain the required solution (x, y). □

Corollary 3.6.1. *Let $\mathcal{F}$, $\mathcal{G}$ and S satisfy the conditions of Theorem 3.6.5. Assume that the multimap Q is 2-completely fundamentally restrictible and for every $x \in \partial B_r(0)$*

$$\mu x \notin Q(x)$$

for all $\mu \geq 1$. Then the system (3.6.3) has a solution in $B_r(0) \times \overline{V}$.

Proof. From Theorem 3.4.1 we know that $\deg(i - Q, \overline{B}_r(0)) = 1$ and we can apply above result. □

3.6.3 Application: optimal control for a neutral functional differential equation

We apply Theorem 3.6.3 to an optimal control problem for a system governed by a neutral functional differential equation. We want to find a control, $u \in \mathbb{R}^n$, in such a way that, starting from a given convex, compact set $C \subset \mathbb{R}^n$, the corresponding state vector y, reaches a variable target set $H(y(0), u) \subset \mathbb{R}^n$ at a specified final time, and the pair (y, u) minimizes a given functional. To this goal consider the following problem.

$$\begin{cases} y'(t) = f(t, y(t), y(\theta(t)), y'(\theta(t)), u), \quad t \in [0, 1]; \\ y(0) \in C; \\ y(1) \in H(y(0), u); \\ \int_0^1 f_0(s, y(s), y(\theta(s)), y'(\theta(s)), u)\, ds \to \min . \end{cases} \tag{3.6.4}$$

Here C is a given convex, compact subset of $\mathbb{R}^n$; the multimap $H : C \times \mathbb{R}^n \to K(\mathbb{R}^n)$ is acyclic and compact; $\theta \in C([0, 1]; \mathbb{R})$ satisfies $0 \leq \theta(t) \leq t$ for any $t \in [0, 1]$ and $\theta(0) = 0$; $f_0 : [0, 1] \times \mathbb{R}^{4n} \to \mathbb{R}$ is a continuous functional.

Denoting by $|\cdot|$ the Euclidean norm in $\mathbb{R}^n$, we assume the following conditions on the dynamics $f : [0,1] \times \mathbb{R}^{4n} \to \mathbb{R}^n$:

(I) the map f is continuous;

(II) there exist positive constants a, b, c, d and k, with $0 \le k < 1$, such that

$$|f(t, v_1, v_2, v_3, u)| \le a + b|v_1| + c|v_2| + k|v_3| + d|u|$$

for all $(t, v_1, v_2, v_3, u) \in [0,1] \times \mathbb{R}^{4n}$;

(III) $f(t, v_1, v_2, v_3, 0) = 0$ for any $(t, v_1, v_2, v_3) \in [0,1] \times \mathbb{R}^{3n}$;

(IV) there exists $N > 0$ such that

$$\liminf_{|u|\to\infty} \frac{\langle u, f(t, v_1, v_2, v_3, u)\rangle}{|u|^{1+\delta}} \ge N$$

for some $0 < \delta \le 1$, for any $t \in [0,1]$ and for every (v_1, v_2, v_3) belonging to any bounded subset of $\mathbb{R}^{3n}$;

(V) the inequality $|f(t, v_1, v_2, v_3, u) - f(t, v_1, v_2, v_3', u)| \le k|v_3 - v_3'|$ holds for any $(t, v_1, v_2, v_3, v_3', u) \in [0,1] \times \mathbb{R}^{5n}$, where the constant k is the same as in assumption (f2).

Let $X = \mathbb{R}^n$ and $Y = C^1([0,1]; \mathbb{R}^n)$ be the space of continuously differentiable functions $x : [0,1] \to \mathbb{R}^n$ with norm $\| x \|_{C^1} = \| x \|_C + \| x' \|_C$. Under assumptions (I)–(II) the problem

$$\begin{cases} y'(t) = f(t, y(t), y(\theta(t)), y'(\theta(t)), u), \\ y(0) \in C, \end{cases} \tag{3.6.5}$$

can be rewritten in the form

$$y \in \mathcal{F}(x, y)$$

where $x = u \in X$, $y \in Y$ and the multimap $\mathcal{F} : X \times Y \to Kv(Y)$ is defined by

$$\mathcal{F}(x, y) = C + \int_0^t f(s, y(s), y(\theta(s)), y'(\theta(s)), x)\, ds$$

$$= \left\{ \tilde{y} \in Y : \tilde{y}(t) = c + \int_0^t f(s, y(s), y(\theta(s)), y'(\theta(s)), x)\, ds, \quad c \in C \right\}.$$

Further, the multimap $\mathcal{G} : X \times Y \to K(X)$ defined by

$$\mathcal{G}(x, y) = x + H(y(0), x) - y(1)$$

is clearly acyclic.

Now the problem of finding $u \in \mathbb{R}^n$ for which the corresponding solution of (3.6.5) satisfies $y(1) \in H(y(0), u)$ is equivalent to solving the system

$$\begin{cases} y \in \mathcal{F}(x, y), \\ x \in \mathcal{G}(x, y). \end{cases}$$

We show that this system has a solution by using Theorem 3.6.3. To this aim we need some preliminary results.

Lemma 3.6.4. *Assume* (I)–(II). *Then for any $r > 0$ there exists a constant $M(r) > 0$ such that any solution y of* (3.6.5) *corresponding to some control $x = u \in B_r(0)$ satisfies $\| y \|_Y < M(r)$.*

Proof. Let $y \in Y$ be a solution of the problem (3.6.5) corresponding to $x = u \in B_r(0)$. By (II) we have

$$|y'(t)| \le a_r + b|y(t)| + c|y(\theta(t))| + k|y'(\theta(t))| \tag{3.6.6}$$

for any $t \in [0, 1]$, where $a_r = a + dr$.

Define now a C^1-function $z : [0, 1] \to \mathbb{R}$ as the solution of the Cauchy problem

$$\begin{cases} z'(t) = a_r + (b + c)z(t) + kz'(t) + 1, \\ z(0) = 1 + |C|, \end{cases}$$

where $|C| = \max\{|\widetilde{c}| : \widetilde{c} \in C\}$. Equivalently,

$$\begin{cases} z'(t) = \dfrac{1 + a_r}{1 - k} + \dfrac{b + c}{1 - k} z(t), \\ z(0) = 1 + |C|. \end{cases}$$

It is easy to see that $z(t)$ is a positive and increasing function on $[0, 1]$, hence $z'(t)$ is also increasing on $[0, 1]$ and so

$$z'(t) > a_r + bz(t) + cz(\theta(t)) + kz'(\theta(t)) \tag{3.6.7}$$

for any $t \in [0, 1]$. We want to show that (3.6.6) and (3.6.7) imply

$$\| y \|_Y < \| z \|_{C^1([0,1];\mathbb{R})}.$$

For this, let

$$\tau = \sup\{t \in [0, 1] : |y(s)| < z(s) \text{ and } |y'(s)| < z'(s), \ 0 \le s < t\}.$$

Since $|y(0)| < z(0)$ and $\theta(0) = 0$, from (3.6.6) and (3.6.7) we obtain $|y'(0)| < z'(0)$. Therefore $\tau > 0$. We claim that $\tau = 1$. In fact, assume to the contrary that

$|y(\tau)| = z(\tau)$, $\tau < 1$. Integrating (3.6.7) between 0 and τ and using (3.6.6), the definition of τ and the fact that $0 \leq \theta(t) \leq t$ we get

$$z(\tau) - z(0) > \int_0^\tau |y'(s)|\, ds.$$

But $\frac{d}{dt}|y(t)| \leq |y'(t)|$, thus

$$z(\tau) > z(0) + |y(\tau)| - |y(0)| = 1 + |C| - |y(0)| + |y(\tau)| \geq 1 + |y(\tau)|.$$

Furthermore, $z(\tau) > |y(\tau)|$ implies, by virtue of (3.6.6) and (3.6.7) that $z'(\tau) > |y'(\tau)|$.

So, $\| y \|_Y < \| z \|_{C^1([0,1];\mathbb{R})}$ and we can take $M(r) = \| z \|_{C^1([0,1];\mathbb{R})}$. □

As a direct consequence of this definition we have $D^F = \mathbb{R}^n$. Indeed, the multimap

$$x \multimap S(x) = \{y \in Y : y \text{ is a solution of (3.6.5)}\}$$

sends bounded sets into bounded sets, thus, for any $r > 0$, we can take $U = B_r(0) \subset X$ and $V = B_{M(r)}(0) \subset Y$ as open sets in Theorem 3.6.3.

Define the MNC ψ_Y in the space Y by the formula

$$\psi_Y(\Omega) = \alpha_C(\Omega')$$

where $\Omega \subset Y$ is a bounded set, $\Omega' = \{y' : y \in \Omega\}$ and α_C is the Kuratowski MNC in the space $C([0, 1]; \mathbb{R}^n)$ (see Section 2.1.1). The real MNC ψ_Y is monotone, nonsingular and regular.

We can now prove the following.

Lemma 3.6.5. *Assume that conditions* (I), (II) *and* (V) *are satisfied. Then the multimap* $\mathcal{F} : X \times Y \to Kv(Y)$ *is* (k, ψ_Y)*-condensing in the second variable.*

Proof. Let $K \subset X$ be a compact set, $\Omega \subset Y$ a bounded set. We want to show that

$$\psi_Y(\mathcal{F}(K, \Omega)) \leq k\psi_Y(\Omega).$$

For this, let the operators $h : Y \times Y \to C([0, 1]; \mathbb{R}^n)$ and

$$q : X \times C([0, 1]; \mathbb{R}^n) \times C([0, 1]; \mathbb{R}^n) \to C([0, 1]; \mathbb{R}^n)$$

be defined as follows

$$h(x, y)(t) = f(t, y(t), y(\theta(t)), y'(\theta(t)), x)$$

and

$$q(x, u, v)(t) = f(t, u(t), u(\theta(t)), v(\theta(t)), x).$$

Since $h(K,\Omega) \subseteq q(K,\Omega,\Omega')$ it suffices to estimate $\alpha(q(K,\Omega,\Omega'))$. The set $\mathcal{T} \subset \mathbb{R}^n$,

$$\mathcal{T} = \{z : z = y(t) \text{ or } z = y'(t),\ \ y \in \Omega,\ t \in [0,1]\}$$

is bounded. Therefore, from the uniform continuity of f on $[0,1] \times \mathcal{T} \times K^3$ it follows that for a given $\varepsilon > 0$ there exists $\eta > 0$ such that

$$|f(t,u_1,u_2,v,x) - f(t,\widetilde{u}_1,\widetilde{u}_2,v,\widetilde{x})| < \varepsilon$$

whenever $|u_1 - \widetilde{u}_1| + |u_2 - \widetilde{u}_2| + |x - \widetilde{x}| < 2\eta$, $t \in [0,1]$; $u_1, \widetilde{u}_1, u_2, \widetilde{u}_2, v \in \mathcal{T}$; $x, \widetilde{x} \in K$.

As Ω is a bounded equicontinuous subset of $C([0,1];\mathbb{R}^n)$ there is an η-net $\{u_1,\dots,u_m\}$ for Ω. Let $\{k_1,\dots,k_p\}$ be an η-net of K. Then

$$q(K,\Omega,\Omega') \subset \bigcup_{\substack{1\le i\le m\\ 1\le j\le p}} W_\varepsilon(q(k_j,u_i,\Omega')).$$

From the properties of the Kuratowski MNC (see Section 2.1.1) it then follows that

$$\alpha(q(K,\Omega,\Omega')) \le \max\{\alpha(q(k_j,u_i,\Omega') + 2\varepsilon : \ 1 \le i \le m,\ 1 \le j \le p\}.$$

The operators $q(k_j,u_i,\cdot)$, $1 \le i \le m$, $1 \le j \le p$ are k-nonexpansive. In fact, if $v_1, v_2 \in C([0,1];\mathbb{R}^n)$ then

$$\begin{aligned}
&\|q(k_j,u_i,v_1) - q(k_j,u_i,v_2)\| \\
&\quad = \sup_{t\in[0,1]} |f(t,u_i(t),u_i(\theta(t)),v_1(t),k_j) - f(t,u_i(t),u_i(\theta(t)),v_2(t),k_j)| \\
&\quad \le k \sup_{t\in[0,1]} |v_1(t) - v_2(t)| = k\|\,v_1 - v_2\,\|.
\end{aligned}$$

Therefore

$$\alpha(q(k_j,u_i,\Omega')) \le k\alpha(\Omega')$$

and, since $\varepsilon > 0$ is arbitrary, it follows that

$$\alpha(q(K,\Omega,\Omega')) \le k\alpha(\Omega').$$

Hence

$$\psi_Y(F(K,\Omega)) = \alpha(h(K,\Omega)) \le \alpha(q(K,\Omega,\Omega')) \le k\alpha(\Omega') = k\psi_Y(\Omega),$$

completing the proof. □

Consider the multimap $Q : X \to P(X)$,

$$\begin{aligned}
Q(x) &= G(x,S(x)) \\
&= x - \bigcup_{y\in S(x)} \left\{ \int_0^1 f(s,y(s),y(\theta(s)),y'(\theta(s)),x)\,ds + C - H(y(0),x) \right\}
\end{aligned}$$

and the multifield $T = i - Q : X \to P(X)$,

$$T(x) = \bigcup_{y \in S(x)} \left\{ \int_0^1 f(s, y(s), y(\theta(s)), y'(\theta(s)), x)\, ds + C - H(y(0), x) \right\}.$$

Lemma 3.6.6. *Assume* (I)–(V). *Then there exists* $r > 0$ *such that:*

(a) *the multifield* T *satisfies the B.U.-property on* $\partial B_r(0)$*;*

(b) $\deg(i - \mathcal{F}(0, \cdot), \overline{B}_{M(r)}(0)) \neq 0$.

Proof. a) By (IV) and Lemma 3.6.4 for $r > 0$ sufficiently large we have

$$\frac{\langle u, f(t, v_1, v_2, v_3, u)\rangle}{|u|^{1+\delta}} + \inf_{y \in C - L} \frac{\langle u, y\rangle}{|u|^{1+\delta}} \geq \frac{N}{2},$$

where $L = \overline{H(C, X)}$, for any $t \in [0, 1]$, for any u, $|u| \geq r$, and any $(v_1, v_2, v_3) \in \mathbb{R}^{3n}$ such that $|v_1| + |v_2| + |v_3| \leq 2M(r)$. In fact, note that

$$\lim_{|u| \to +\infty} \inf_{y \in C - L} \frac{\langle u, y\rangle}{|u|^{1+\delta}} = 0.$$

It is now easy to verify that for any $x \in \partial B_r(0)$ the sets $T(x)$ and $T(-x)$ are strictly separated by the hyperplane defined by $x^* \in X^*$, $x^*(\cdot) = \langle x, \cdot\rangle$.

b) From the condition (III) it follows that

$$\deg(i - \mathcal{F}(0, \cdot), \overline{B}_{M(r)}(0)) = \deg(i - \widetilde{C}, \overline{B}_{M(r)}(0)) = 1$$

where $\widetilde{C} = \{\tilde{c} \in Y : \tilde{c}(t) \equiv c \in C\}$ and $\widetilde{C} \subset B_{M(r)}(0)$. □

Now we are in a position to prove the existence of solutions for the problem (3.6.4).

Theorem 3.6.6. *If* (I)–(V) *hold, then problem* (3.6.4) *has a solution* $(u, y) \in X \times Y$.

Proof. From the proof of Lemma 3.6.6 it follows that all the solutions of the system (3.6.3) are contained in $B_r(0) \times \overline{B}_{M(r)}(0) \subset X \times Y$ and the above lemmas show that for $\mathcal{F}$ and $\mathcal{G}$ on $\overline{B}_r(0) \times \overline{B}_{M(r)}(0)$ all the assumptions of Theorem 3.6.3 are satisfied.

The conditions of Theorem 3.6.3 imply that the set $\{(x, y)\}$ of solutions of system (3.6.3) is a nonempty compact subset of $X \times Y$. Hence there exists a pair $(u, y) \in X \times Y$ giving a solution of the problem (3.6.4). □

Bibliographic notes

The idea of relative topological degree (or rotation) of completely continuous single-valued vector fields was introduced and investigated by Yu. G. Borisovich [32]–[34]. It was extended in the paper of Ya. A. Izrailevich [122] to completely continuous multifields in locally convex spaces and studied in detail in [40] and [41].

The first construction of the topological degree for condensing multifields in Fréchet spaces was by V. V. Obukhovskii [175]. It was based on the concepts of fundamental set and relative topological degree. Subsequently different approaches to topological degree theory for condensing multimaps were developed in the works of W. V. Petryshyn and P. M. Fitzpatrick [196], J. R. L. Webb [218], V. V. Obukhovskii and E. V. Gorokhov [181], A. Vanderbauwhede [217] and others (we again refer the reader to the surveys [40], [41], [43], [44] and [95] for the details). Concerning recent trends in this direction see the publications of Yu. G. Borisovich [35]–[38].

We also mention that the problem of defining the topological degree for condensing multimaps on the boundary of a domain was considered in the dissertation of V. V. Obukhovskii (1974) and in the work of E. De Pascale and R. Guzzardi [74]. A generalization of Yu. I. Sapronov's theorem [200] on the existence of a compact approximation for a condensing multifield was proved in the paper of V. V. Obukhovskii and E. V. Gorokhov [181], and on that basis the topological degree was defined.

The method of construction of relative topological degree theory for fundamentally restrictible multifields presented in Chapter 3, is apparently new and extend the definitions mentioned above. We also mention that the topological degree for equivariant condensing multifields, presented in Section 3.3.2, was developed in the paper of Ya. A. Izrailevich and V. V. Obukhovskii [124], but the approach used in this section, based on the construction of an equivariant single-valued approximation, is new. (Note that independently a similar construction was suggested recently, using a different method, by Z. Dzedzej [83]).

The topological degree theory for condensing acyclic multimaps was suggested by V. V. Obukhovskii [179] (see also the survey [43]). This paper was preceded by works where some fixed point theorems for multimaps of the same type were considered (see [55], [62], [88], [91]). Among other work in this direction let us mention the papers of L. Górniewicz and Z. Kucharski [106] and D. Gabor and W. Kryszewski [92]. The topological degree theory given in the Section 3.4.3 for a completely fundamentally restrictible Vietoris multimap is new and its presentation is based on a recent paper of G. Conti, V. V. Obukhovskii and P. Zecca [65].

The first work concerning single-valued approximations of nonconvex-valued multimaps (Section 3.4.4) is the paper of A. D. Myshkis [170] where some fixed point theorems were proved. Further development of this approach to fixed point theory, and the construction of various topological invariants for multimaps of such type, was carried out by G. Anichini [8], G. Anichini, G. Conti and P. Zecca [9]–[11], Yu. G. Borisovich and Yu. E. Gliklih [45], R. Bader and W. Kryszewski [22], Yu. E. Gliklih [102],

L. Górniewicz, A. Granas and W. Kryszewski [105], L. Górniewicz and M. Lassonde [107], W. Kryszewski [151], [152] and others.

The asymptotic fixed point result (Theorem 3.4.4) is a generalization of a result in the work of V. V. Obukhovskii and P. Zecca [183]. It is based on Yu. G. Borisovich's generalization of the F. E. Browder's fixed point theorem (see [53], [33]) and extends a number of asymptotic fixed point theorems for multivalued maps (see, for example, [175], [178], [40], [41]).

The results of Section 3.5 on the properties of the fixed point set seem to be new. The statements about continuous dependence of the fixed point set on a parameter generalize results of J. Hale [112], Do Hong Tan [78] and M. I. Kamenskii, V. V. Obukhovskii and P. Zecca [136].

Section 3.6.1 contains some concepts on system of inclusions and a description of their properties taken from the work of G. Conti and J. Pejsachowicz [66] and also extensions of results from a paper of I. Massabo, P. Nistri and J. Pejsachowicz [167].

The results of Section 3.6.2 on the solvability of systems of inclusions generalize theorems in the work of P. Nistri, V. V. Obukhovskii and P. Zecca [172]. We mention that solvability of systems of inclusions was also studied in the work of G. Conti, W. Kryszewski and P. Zecca [61], G. Conti, P. Nistri and P. Zecca [63], D. Gabor and W. Kryszewski [92] and others. The application to the optimal control problem (Section 3.6.3) was described in the paper [172]. In the proofs of Lemmas 3.6.4 and 3.6.5 some methods from the paper [209] are used.

Chapter 4

Semigroups and measures of noncompactness

In this chapter we give a summary of facts from the theory of strongly continuous or C_0-semigroups, necessary to our study of semilinear inclusions. We continue also the study of measures of noncompactness in some function spaces. Conditions under which a semigroup is condensing are given.

4.1 Semigroups: general facts

In this section we present some results from the theory of semigroups of linear operators in Banach spaces. For a complete treatment see, for example [195].

We recall first the following basic concepts.

Definition 4.1.1. A family of bounded linear operators $T(t),\ 0 \leq t < \infty$ in a Banach space E forms a C_0-*semigroup* if

(i) $T(0) = I$ (I the identity operator on E);

(ii) $T(t+s) = T(t)T(s) = T(s)T(t)$ for every $t, s \geq 0$;

(iii) the function $T(\cdot)x : [0, \infty) \to E,\ t \to T(t)x$ is continuous for every $x \in E$.

Definition 4.1.2. Let $T(t)$ be a C_0-semigroup of linear operators in a Banach space E. Denote by A the linear operator defined by the relations

$$D(A) = \left\{x \in E : \lim_{\Delta t \to +0} \frac{1}{\Delta t}(T(\Delta t) - I)x \text{ exists}\right\}$$

$$Ax = \lim_{\Delta t \to +0} \frac{1}{\Delta t}(T(\Delta t) - I)x\,,\ x \in D(A).$$

The operator A is called the *infinitesimal generator* of the semigroup $T(t)$ with *domain* $D(A)$.

We say that A *generates* the semigroup $T(t)$ and write

$$T(t) = \exp\{At\}\,.$$

It can be shown that the infinitesimal generator of a C_0-semigroup is a closed operator, and $D(A)$ is dense in E.

We recall also that *the resolvent set* $\Lambda(A)$ of a linear operator A consists of all complex numbers λ for which the linear operator $\lambda I - A$ is invertible, i.e., $(\lambda I - A)^{-1}$ is a bounded linear operator in E. The operator

$$R(\lambda, A) = (\lambda I - A)^{-1}\ ,\ \lambda \in \Lambda(A)$$

is called the *resolvent* of A.

The criterion for a linear operator to be the infinitesimal generator of a C_0-semigroup is given by the following theorem of Hille–Phillips–Miyadera (see, for example, [149], [195]).

Theorem 4.1.1. *The closed densely defined linear operator A is the infinitesimal generator of a C_0-semigroup $T(t)$ if and only if there exist a real number $\sigma_0 = \sigma_0(A)$ and a constant $C \in R_+$ such that the resolvent set $\Lambda(A)$ contains the half-plane $\Omega_A = \{\lambda \in \mathbb{C} : \operatorname{Re}\lambda \geq \sigma_0(A)\}$ and for all $\lambda \in \Omega_A$ the resolvent $R(\lambda, A)$ satisfies the inequalities*

$$\left\| (R(\lambda, A))^i \right\| \leq \frac{C}{(\operatorname{Re}\lambda - \sigma_0)^i}, \quad i = 1, 2, \ldots .$$

It is well known (see, for example [195]) that in the case when A is the infinitesimal generator of the C_0-semigroup $\exp\{At\}$ and $x_0 \in D(A)$ the function $x(t) = \exp\{At\}x_0$ is the smooth solution of the Cauchy problem

$$\begin{cases} x'(t) = Ax(t), \\ x(0) = x_0. \end{cases}$$

Moreover,

$$x'(t) = A\exp\{At\}x_0 = \exp\{At\}Ax_0.$$

Consider the problem

$$x'(t) = Ax(t) + f(t), \tag{4.1.1}$$

$$x(0) = x_0, \tag{4.1.2}$$

where $f(\cdot) \in L^1([0, d]; E)$.

Definition 4.1.3. An absolutely continuous function $x(\cdot) \in C([0, d]; E)$ satisfying equation (4.1.1) a.e. on $[0, d]$ with initial condition (4.1.2) is said to be a *Carathéodory solution* of the problem (4.1.1), (4.1.2).

The following statement is well known.

Theorem 4.1.2. *Every Carathéodory solution $x(\cdot)$ of the problem* (4.1.1), (4.1.2) *satisfies the integral equation*

$$x(t) = \exp\{At\}x_0 + \int_0^t \exp\{A(t-s)\} \cdot f(s)\, ds. \tag{4.1.3}$$

Sufficient conditions under which the solution of the equation (4.1.3) is a Carathéodory solution of the problem (4.1.1), (4.1.2) are given in the following statement.

Theorem 4.1.3. *Let A be the infinitesimal generator of a C_0-semigroup in a reflexive Banach space E, $x_0 \in D(A)$, and assume that*

(i) $f(t) \in D(A)$ *for a.e.* $t \in [0, d]$;

(ii) *the function* $Af(\cdot)$ *is summable on* $[0, d]$.

Then every solution of the integral equation (4.1.3) *is a Carathéodory solution of the problem* (4.1.1), (4.1.2).

Proof. Let $x(\cdot) \in C([0, d]; E)$ be a solution of the integral equation (4.1.3). Then for any $t \in [0, d]$ and $h > 0$ small enough we have that

$$\begin{aligned}\frac{x(t+h)-x(t)}{h} &= \frac{1}{h}[\exp\{A(t+h)\}x_0 - \exp\{At\}x_0] \\ &\quad + \frac{1}{h}\left[\int_0^{t+h} \exp\{A(t+h-s)\}f(s)\,ds - \int_0^t \exp\{A(t-s)\}f(s)\,ds\right] \\ &= \frac{\exp\{At\}(\exp\{Ah\}x_0 - x_0)}{h} \\ &\quad + \frac{1}{h}\int_0^t (\exp\{A(t+h-s)\} - \exp\{A(t-s)\})f(s)\,ds \\ &\quad + \frac{1}{h}\int_t^{t+h} \exp\{A(t+h-s)\}f(s)\,ds.\end{aligned}$$

Consider the limit of each term as $h \to 0$. Since the semigroup $\exp\{At\}$ is C_0, $x_0 \in D(A)$, $\exp\{At\}$ is differentiable, the limit of the first term is equal to $A\exp\{At\}x_0$.

The second term can be rewritten as

$$\int_0^t \exp\{A(t-s)\}\frac{\exp\{Ah\} - I}{h} f(s)\,ds. \tag{4.1.4}$$

Taking into consideration conditions (i) and (ii) of the theorem, the integrand in (4.1.4) can be estimated by

$$\left\| \exp\{A(t-s)\}\frac{\exp\{Ah\} - I}{h} f(s)\right\| \le M\left\|\frac{1}{h}\int_0^h \exp(A\xi)Af(s)\,d\xi\right\| \le M^2\|Af(s)\|$$

where $M = \sup_{t\in[0,d]} \|\exp\{At\}\|$.

Applying the Lebesgue Theorem we conclude that the limit of (4.1.4) is equal to

$$\int_0^t \exp\{A(t-s)\}Af(s)\,ds = A\int_0^t \exp\{A(t-s)\}f(s)\,ds.$$

(The equality follows from the fact that A commutes with $\exp\{At\}$ on $D(A)$.)
Finally, for the third term we have

$$\lim_{h\to 0}\frac{1}{h}\int_t^{t+h} \exp\{A(t+h-s)\}f(s)\,ds = f(t)$$

for a.e. $t \in [0, d]$. In fact, by Theorem 4.1.1 the operator $(A - nI)^{-1}$ is bounded for some integer n. We can write

$$\frac{1}{h}\int_t^{t+h} \exp\{A(t+h-s)\}f(s)\,ds = \frac{1}{h}\int_t^{t+h} \exp\{(A-nI)(t+h-s)\}e^{n(t+h-s)}f(s)\,ds.$$

Now changing the variables ($t + h - s = \xi$) we have

$$\begin{aligned}
&\frac{1}{h}\int_t^{t+h} \exp\{A(t-nI)(t+h-s)\}e^{n(t+h-s)}f(s)\,ds\\
&= \frac{1}{h}\int_0^h (A-nI)^{-1}\exp\{(A-nI)\xi\}e^{n\xi}(A-nI)f(t+h-\xi)\,d\xi\\
&= \frac{1}{h}\int_0^h \Big[\int_0^\xi \exp\{(A-nI)\tau\}\,d\tau + (A-nI)^{-1}\Big]e^{n\xi}(A-nI)f(t+h-\xi)\,d\xi\\
&= \int_0^h \frac{1}{h}\Big(\int_0^\xi \exp\{(A-nI)\tau\}\,d\tau\Big)f(t+h-\xi)\,d\xi\\
&\quad + \frac{1}{h}\int_0^h f(t+h-\xi)\,d\xi + \frac{1}{h}\int_0^h (e^{n\xi}-1)f(t+h-\xi)\,d\xi\\
&= I_1(t,h) + I_2(t,h) + I_3(t,h).
\end{aligned}$$

Two of these terms can be estimated as follows:

$$\| I_1(t,h) \| \le M\int_0^h \| f(t+h-\xi) \|\,d\xi = M\int_t^{t+h} \| f(s) \|\,ds \underset{h\to 0}{\longrightarrow} 0$$

$$\| I_3(t,h) \| \leq \frac{e^{nh}-1}{h} \int_0^h \| f(t+h-\xi) \| \, d\xi = e^n \int_t^{t+h} \| f(s) \| \, ds \underset{h \to 0}{\longrightarrow} 0.$$

Finally, since the space E is reflexive the term

$$I_2(t,h) = \frac{1}{h} \int_t^{t+h} f(s) \, ds$$

tends to $f(t)$ as $h \to 0$ for a.e. $t \in [0,d]$.

So, finally, we have

$$\begin{aligned} x'(t) &= A \exp\{At\} x_0 + A \int_0^t \exp\{A(t-s)\} f(s) \, ds + f(t) \\ &= Ax(t) + f(t), \end{aligned}$$

concluding the proof. □

Condition (i) of Theorem 4.1.3 is rather restrictive. To avoid this difficulty, and to establish an equivalence between the solutions of the initial value problem (4.1.1), (4.1.2) and the integral equation (4.1.3), equality (4.1.1) can be considered in a rather weakened form.

Let A^* denote the adjoint of A and $\langle \, , \, \rangle$ the pairing between E and its dual space E^*.

Definition 4.1.4. A function $x(\cdot) \in C([0,d]; E)$ is a *weak solution* of the Cauchy problem (4.1.1), (4.1.2) if it satisfies the initial condition (4.1.2) and for every $v \in D(A^*)$ the function $\langle x(t), v \rangle$ is absolutely continuous on $[0,d]$ and the equation

$$\frac{d}{dt} \langle x(t), v \rangle = \langle x(t), A^* v \rangle + \langle f(t), v \rangle \tag{4.1.5}$$

holds for a.e. $t \in [0,d]$.

Theorem 4.1.4. *A function $x(\cdot) \in C([0,d]; E)$ is a weak solution of the Cauchy problem* (4.1.1), (4.1.2) *if and only if it is a solution of the integral equation* (4.1.3).

Proof. We will only prove the "if" part since we will not use the converse. The full proof can be found in [24] or [23, Theorem 4.8.3].

Let $x(\cdot)$ be a solution of the integral equation (4.1.3). For a C_0-semigroup $\exp\{At\}$ generated by A, let $\max_{t \in [0,d]} \| \exp\{At\} \| = M$. Note that for every $u \in E$ and $v \in D(A^*)$ the function $\langle \exp\{At\} u, v \rangle$ is differentiable with respect to t and

$$\frac{d}{dt} \langle \exp\{At\} u, v \rangle = \langle \exp\{At\} u, A^* v \rangle.$$

This equality is clear when $u \in D(A)$ and holds for arbitrary $u \in E$ since $\overline{D(A)} = E$ and $\exp\{At\}$ is strongly continuous.

For each $v \in D(A^*)$ and $t \in [0, d]$ we have

$$\langle x(t), v\rangle = \langle \exp\{At\}x_0, v\rangle + \int_0^t \langle \exp\{A(t-s)\} \cdot f(s), v\rangle\, ds.$$

Suppose first that $f(\cdot)$ belongs to $C([0, d]; E)$. Since $(t, u) \to \exp\{At\}u$ is jointly continuous on $[0, d] \times E$ we have

$$\frac{d}{dt}\int_0^t \langle \exp\{A(t-s)\} \cdot f(s), v\rangle\, ds = \langle f(t), v\rangle + \int_0^t \langle \exp\{A(t-s)\} \cdot f(s), A^*v\rangle\, ds$$

and therefore the function $\langle x(t), v\rangle$ satisfies (4.1.5).

If $f(\cdot) \in L^1([0, d]; E)$, approximate f by a sequence of continuous functions $\{f_n\}_{n=1}^{\infty} \subset C([0, d]; E)$ with $f_n \to f$ in $L^1([0, d]; E)$. We consider the corresponding sequence

$$x_n(t) = \exp\{At\}x_0 + \int_0^t \exp\{A(t-s)\} \cdot f_n(s)\, ds, \quad t \in [0, d].$$

Then we have that

$$\| x_n(t) - x(t) \| \le M \int_0^d \| f_n(s) - f(s) \|\, ds,$$

so that $x_n \to x$ in $C([0, d]; E)$.

As was proved above, each function x_n for every $v \in D(A^*)$ satisfies (4.1.5) and therefore we have the following representation for $\langle x_n(t), v\rangle$:

$$\langle x_n(t), v\rangle = \langle x_0, v\rangle + \int_0^t [\langle x_n(s), A^*v\rangle + \langle f_n(s), v\rangle]\, ds, \quad t \in [0, d].$$

Passing to the limit as $n \to \infty$ we obtain that $x(\cdot)$ satisfies (4.1.5). □

Definition 4.1.5. A semigroup $\exp\{At\}$ is called *decreasing* if

$$\| \exp\{At\} \| \le Ce^{-at}, \quad t \ge 0 \tag{4.1.6}$$

where C and a are positive constants. If $C = 1$ then the semigroup $\exp\{At\}$ is called *strongly contractive*.

In the study of the averaging principle we need the following result.

Theorem 4.1.5. *Let A be the infinitesimal generator of a decreasing C_0-semigroup. Then, the family of operators $\Pi(\varepsilon) : C([0,T], E) \to E$, defined by*

$$\Pi(\varepsilon)y = \varepsilon\left[I - \exp\{\varepsilon AT\}\right]^{-1} \int_0^T \exp\{\varepsilon A(T-s)\}\, y(s)\, ds,$$

is bounded in norm, uniformly with respect to ε, and

$$\Pi(\varepsilon)y \to -A^{-1}\frac{1}{T}\int_0^T y(s)\, ds$$

as $\varepsilon \to 0$ for every $y \in C([0,T]; E)$.

Proof. We will use C to denote generic constants. Since the semigroup $\exp\{At\}$ satisfies the estimate (4.1.6) we have the following bound for the spectral radius:

$$r\big(\exp\{\varepsilon AT\}\big) \le e^{-\varepsilon a T},$$

hence the operators $\Pi(\varepsilon)$ are well defined. Moreover, we have

$$\left[I - \exp\{\varepsilon AT\}\right]^{-1} = \sum_{k=0}^{\infty} \exp\{\varepsilon k AT\}. \tag{4.1.7}$$

Therefore

$$\left\| [I - \exp \varepsilon AT]^{-1} \right\| \le \sum_{k=0}^{\infty} Ce^{-\varepsilon kaT} \le C\left(1 - e^{-\varepsilon aT}\right)^{-1}. \tag{4.1.8}$$

Then we have the following estimate:

$$\|\Pi(\varepsilon)\| \le C\varepsilon\left(1 - e^{-\varepsilon aT}\right)^{-1} \| y \|_C \le C\| y \|_C\,.$$

We can now prove that

$$\varepsilon\left(I - \exp\{\varepsilon AT\}\right)^{-1} x \underset{\varepsilon\to 0}{\longrightarrow} -A^{-1}\frac{1}{T}x \tag{4.1.9}$$

for every $x \in E$. In fact, by (4.1.8) we have

$$\left\| \varepsilon\left[I - \exp\{\varepsilon AT\}\right]^{-1} \right\| \le C$$

therefore it is sufficient to verify (4.1.9) only for x belonging to $D(A)$ which is dense in E. From Definition 4.1.2 we have

$$\frac{\exp\{\varepsilon AT\} - I}{\varepsilon T} x - Ax \underset{\varepsilon \to 0}{\longrightarrow} 0 .$$

It follows that

$$-A^{-1}\frac{1}{T}x - \varepsilon\left[I - \exp\{\varepsilon AT\}\right]^{-1}x$$
$$= \varepsilon\left[I - \exp\{\varepsilon AT\}\right]^{-1} A^{-1}\left(\frac{\exp\{\varepsilon AT\} - I}{\varepsilon T}x - Ax\right) \underset{\varepsilon \to 0}{\longrightarrow} 0.$$

Note now that the set $\{y(t) : t \in [0, T]\}$ is obviously compact and, since $\exp\{AT\}$ is a C_0-semigroup we obtain

$$\exp\{\varepsilon A(T - s)\}\, y(s) \underset{\varepsilon \to 0}{\longrightarrow} y(s) ,$$

uniformly with respect to $s \in [0, T]$, hence we can pass to the limit $\varepsilon \to 0$ under the integral in the expression for $\Pi(\varepsilon)$. □

4.2 Measures of noncompactness in function spaces: measurability and integrability

In this section we consider some properties of measures of noncompactness generated by the Hausdorff MNC. These properties are related to estimates of their values on sequences of functions and multivalued integrals.

Let E be a (not necessarily separable) Banach space.

In this part by measurable we mean strongly measurable. We begin with the following statement.

Theorem 4.2.1. *Let the sequence of functions $\{f_n\}_{n=1}^{\infty} \subset L^1([0, d]; E)$ be integrably bounded:*

$$\| f_n(t) \|_E \le \nu(t) \quad \textit{for all } n = 1, 2, \ldots \textit{ and a.e. } t \in [0, d], \tag{4.2.1}$$

where $\nu \in L^1([0, d])$. Assume that

$$\chi(\{f_n(t)\}_{n=1}^{\infty}) \le q(t) \tag{4.2.2}$$

for a.e. $t \in [0, d]$ where $q \in L^1([0, d])$. Then for every $\delta > 0$ there exist a compact set $K_\delta \subset E$, a set $m_\delta \subseteq [0, d]$, meas $m_\delta < \delta$ *and a set of functions $G_\delta \subset L^1([0, d]; E)$ with values in K_δ such that for every $n \ge 1$ there exists $g_n \in G_\delta$ for which*

$$\| f_n(t) - g_n(t) \| \le 2q(t) + \delta, \quad t \in [0, d] \setminus m_\delta. \tag{4.2.3}$$

Proof. From the measurability of functions f_n, $n \geq 1$, it follows that there exist a separable subspace $Y \subset E$ and a set $m_0 \subset [0, d]$, meas $m_0 = 0$ such that

$$\{f_n(t)\}_{n=1}^{\infty} \subset Y$$

for $t \in [0, d] \setminus m_0$. Without loss of generality we can assume that the inequality (4.2.2) holds for all $t \in [0, d] \setminus m_0$. Choose $\delta_1 \in (0, \delta/3)$. Now let $M > 0$ be a constant such that for the set

$$m_1 = \{t : t \in [0, d],\ \nu(t) > M\}$$

we have

$$\operatorname{meas} m_1 < \delta_1.$$

The functions f_n satisfy the inequality

$$\| f_n(t) \|_E \leq M, \quad n = 1, 2, \dots,$$

for $t \in [0, d] \setminus m_1$. Using inequality (4.2.2) we have

$$\chi_Y\big(\{f_n(t)\}_{n=1}^{\infty}\big) \leq 2\chi\big(\{f_n(t)\}_{n=1}^{\infty}\big) \leq 2q(t) \tag{4.2.4}$$

for $t \in [0, d] \setminus (m_0 \cup m_1)$.

In accordance with the formula for the Hausdorff MNC in separable spaces (see Section 2.1.2) we have

$$\chi_Y\big(\{f_n(t)\}_{n=1}^{\infty}\big) = \lim_{k \to \infty} d_*\big(\{f_n(t)\}_{n=1}^{\infty}, Y_k\big) \tag{4.2.5}$$

where d_* is the deviation of the set $\{f_n(t)\}_{n=1}^{\infty}$ from the space Y_k and $\{Y_k\}_{k=1}^{\infty}$ is a sequence of k-dimensional subspaces of Y such that

$$Y_{k+1} \supseteq Y_k \quad \text{and} \quad \overline{\bigcup_{k \geq 1} Y_k} = Y.$$

As already noted (see Section 2.1.2) in formula (4.2.5) we can take the balls $\overline{B}^k = \overline{B}^k_{2M}(0) \subset Y_k$ instead of the whole of Y_k.

Since the functions f_n are measurable, the functions $\alpha_{nk} : [0, d] \to \mathbb{R}_+$,

$$\alpha_{nk}(t) = d\big(f_n(t), \overline{B}^k\big), \quad t \in [0, d] \setminus (m_0 \cup m_1)$$

are also measurable, and therefore the functions $\beta_k : [0, d] \to \mathbb{R}_+$

$$\beta_k(t) = \sup_n \alpha_{nk}(t)$$

are measurable as well.

Therefore, by the Egorov Theorem there exist a set $m_2 \subset [0, d] \setminus (m_0 \cup m_1)$, $\operatorname{meas} m_2 < \delta_1$ and an integer k_0 such that

$$\beta_k(t) < 2q(t) + \delta/3$$

for $t \in [0, d] \setminus (m_0 \cup m_1 \cup m_2)$ and $k \geq k_0$. It follows that

$$d(f_n(t), \overline{B}^k) < 2q(t) + \delta/3$$

for $n = 1, 2, \ldots;$ $t \in [0, d] \setminus (m_0 \cup m_1 \cup m_2)$ and $k \geq k_0$. Since the functions f_n are measurable on $[0, d] \setminus (m_0 \cup m_1 \cup m_2)$ we conclude that each of them is the pointwise limit of step functions. Hence there exist a set $m_3 \subset [0, d] \setminus (m_0 \cup m_1 \cup m_2)$, $\operatorname{meas} m_3 < \delta_1$ and step functions $h_n : [0, d] \to E$ such that

$$\| f_n(t) - h_n(t) \|_E < \delta/3$$

for $n = 1, 2, \ldots; t \in [0, d] \setminus m_3$. Then

$$d\big(h_n(t), \overline{B}^k\big) < 2q(t) + 2\delta/3$$

for $n = 1, 2, \ldots;$ $k \geq k_0$ and $t \in [0, d] \setminus m$ where $m = m_0 \cup m_1 \cup m_2 \cup m_3$, i.e., there exist step functions $g_n : [0, d] \to \overline{B}^k$ such that

$$\| f_n(t) - g_n(t) \|_E < 2q(t) + \delta \tag{4.2.6}$$

for the same n, t and k. Note that $\operatorname{meas} m < 3\delta_1 < \delta$. □

Remark 4.2.1. From the proof it is clear that we can assume without loss of generality that $g|_{m_\delta} = 0$ for all $g \in G_\delta$.

Remark 4.2.2. From the proof of Theorem 4.2.1 it follows also that the compact set K_δ can be considered as lying in a certain finite-dimensional subspace of the space E. By (4.2.1) and (4.2.6) the functions g_n satisfy the estimate

$$\| g_n(t) \|_E < \nu(t) + 2q(t) + \delta \quad \text{for a.e. } t \in [0, d]$$

and hence the sequence $\{g_n\}_{n=1}^{\infty}$ appearing in the conclusion of Theorem 4.2.1 is weakly compact (see [81], p. 294).

Corollary 4.2.1. *If the space E is separable the estimate* (4.2.3) *has the form*

$$\| f_n(t) - g_n(t) \| \leq q(t) + \delta, \quad t \in [0, d] \setminus m_\delta. \tag{4.2.7}$$

Proof. In this case we can take $Y = E$ and hence the coefficient 2 in (4.2.3) can be omitted. □

We select the following important class of sequences satisfying the conditions of Theorem 4.2.1.

Definition 4.2.1. The sequence $\{f_n\}_{n=1}^{\infty} \subset L^1([0,d];E)$ is *semicompact* if it is integrably bounded and the set $\{f_n(t)\}_{n=1}^{\infty}$ is relatively compact for almost every $t \in [0,d]$.

Using Theorem 4.2.1 we can write the following corollaries describing the properties of semicompact sequences.

Corollary 4.2.2. *Let $\{f_n\}_{n=1}^{\infty} \subset L_1([0,d];E)$ be a semicompact sequence. Then for every $\delta > 0$ there exist a set $m_\delta \subset [0,d]$ and a compact set $K_\delta \subset E$ such that* $\operatorname{meas} m_\delta < \delta$ *and* $\operatorname{dist}(f_n(t), K_\delta) < \delta$ *for all n and $t \in [0,d] \setminus m_\delta$.*

Proof. It is sufficient to note that for a semicompact sequence the function $q(t)$ in (4.2.2) is equal to zero and to apply Theorem 4.2.1. □

Corollary 4.2.3. *If the sequence $\{f_n\}_{n=1}^{\infty} \subset L^1([0,d];E)$ is semicompact then for every $\delta > 0$ there exist a compact set $K_\delta \subset E$ and a δ-net of the sequence $\{f_n\}_{n=1}^{\infty}$ formed by the functions with values in K_δ.*

Proof. The functions f_n, $n = 1, 2, \dots$ satisfy the inequality (4.2.1). Choose $\delta_1 \in (0, \delta/2d)$ such that for every set $m \subset [0,d]$, $\operatorname{meas} m < \delta_1$ we have

$$\int_m \nu(s)\, ds < \delta/2.$$

Take now a set $m_{\delta_1} \subset [0,d]$, a compact set $K_{\delta_1} \subset E$ and $G_{\delta_1} \subset L^1([0,d];E)$ as in Theorem 4.2.1. Then taking into account Remark 4.2.1 for the function f_n and the corresponding function g_n we have that

$$\| f_n - g_n \|_{L^1} \le \int_{[0,d]\setminus m} \| f_n(s) - g_n(s) \| \, ds + \int_m \| f_n(s) \| \, ds < \delta_1 d + \delta/2 < \delta. \quad \square$$

This statement implies the following important sufficient condition for the weak compactness of a set in $L^1([0,d];E)$.

Proposition 4.2.1. *Assume that $\Omega \subset L^1([0,d];E)$ is integrably bounded and the sets $\Omega(t)$ are relatively compact for a.e. $t \in [0,d]$. Then Ω is weakly compact in $L^1([0,d];E)$.*

Proof. By the Eberlein–Šmulian Theorem (see [81, p. 430]) it is sufficient to show the sequential weak compactness of the set Ω.

Let $\{f_n\}_{n=1}^{\infty} \subset \Omega$ be a sequence. By Definition 4.2.1 it is semicompact. From Corollary 4.2.3 and Remark 4.2.2 it follows that for any $\delta > 0$ the sequence $\{f_n\}_{n=1}^{\infty}$ has a weakly compact δ-net. Then by the Grothendieck Theorem (see [56], Lemma V-6) the weak closure of the sequence $\{f_n\}_{n=1}^{\infty}$ is weakly compact. Applying the Eberlein–Šmulian Theorem once more we obtain the weak compactness of the sequence $\{f_n\}_{n=1}^{\infty}$. □

We consider now the problem of exchanging MNC and integral. For convenience we will treat the general case of an abstract operator $S : L^1([0,d]; E) \to C([0,d]; E)$ satisfying the following conditions:

(S1) there exists $D \geq 0$ such that

$$\| Sf(t) - Sg(t) \|_E \leq D \int_0^t \| f(s) - g(s) \|_E \, ds$$

for every $f, g \in L^1([0,d]; E),\ 0 \leq t \leq d$;

(S2) for any compact $K \subset E$ and sequence $\{f_n\}_{n=1}^{\infty} \subset L^1([0,d]; E)$ such that $\{f_n(t)\}_{n=1}^{\infty} \subset K$ for a.e. $t \in [0,d]$ the weak convergence $f_n \underset{w}{\to} f_0$ implies $Sf_n \to Sf_0$.

Note that condition (S1) implies that the operator S satisfies the Lipschitz condition

(S1′) $\| Sf - Sg \|_C \leq D \| f - g \|_{L^1}$.

To present an example of an operator satisfying these requirements we introduce the following notion.

Definition 4.2.2. Let A be the infinitesimal generator of a C_0-semigroup $\exp\{At\}$. Then the operator $G : L^1([0,d]; E) \to C([0,d]; E)$ defined as

$$Gf(t) = \int_0^t \exp\{A(t-s)\} f(s) \, ds$$

is called the *Cauchy operator*.

Remark 4.2.3. Note that by taking $A = 0$ we obtain the usual integral operator $G_I : L^1([0,d]; E) \to C([0,d]; E)$,

$$G_I f(t) = \int_0^t f(s) \, ds$$

as a particular case of the Cauchy operator.

Lemma 4.2.1. *The Cauchy operator G satisfies the properties* (S1) *and* (S2).

Proof. Since $\exp\{At\}$ is a C_0-semigroup there exists a constant $D > 0$ such that

$$\| \exp\{At\} \|_E \le D, \quad t \in [0, d] \tag{4.2.8}$$

then property (S1) easily follows. To prove property (S2) we note that for every compact $K \subset E$ the functions $s \to \exp\{A(t-s)\}x, s \in [0, t], \ x \in K$ are equicontinuous in s with respect to x on $[0, t]$. Therefore the set $Q \subset E$,

$$Q = \bigcup_{s \in [0,t]} \exp\{A(t-s)\}K$$

is relatively compact. For every sequence $\{f_n\}_{n=1}^{\infty} \subset L^1([0, d]; E)$ such that $\{f_n(s)\}_{n=1}^{\infty} \subset K$ for a.e. $s \in [0, d]$ we have

$$\{Gf_n(t)\}_{n=1}^{\infty} \subset tQ$$

and hence the sequence $\{Gf_n(t)\}_{n=1}^{\infty} \subset E$ is relatively compact for every $t \in [0, d]$.

We show now that the sequence of functions $\{Gf_n\}_{n=1}^{\infty} \subset C([0, d]; E)$ is equicontinuous. Since $f_n(t) \in K$ for $n \ge 1$ and a.e. $t \in [0, d]$, there exists a constant $N > 0$ such that

$$\| f_n(t) \| \le N \quad \text{for } n \ge 1 \text{ and a.e. } t \in [0, d].$$

Take $\varepsilon > 0$ and choose $\delta_1 > 0$ such that the inequality $DN\delta_1 < \varepsilon/2$ holds, where D is the constant in (4.2.8).

Since $\exp\{At\}$ is a C_0-semigroup it is possible to choose $0 < \delta < \delta_1$ such that

$$\| (\exp\{A\tau\} - I)x \|_E < \frac{\varepsilon}{2Dd}$$

for all $0 < \tau < \delta, \ x \in K$.

Then for $t_1, t_2 \in [0, d], \ t_2 > t_1$ and $t_2 - t_1 < \delta$ we have that

$$\begin{aligned}
\|Gf_n(t_2) - Gf_n(t_1)\|_E &\le \left\| \int_{t_1}^{t_2} \exp\{A(t_2 - s)\} \, f_n(s) \, ds \right\|_E \\
&\quad + \left\| \int_0^{t_1} (\exp\{A(t_2 - s)\} - \exp\{A(t_1 - s)\}) \, f_n(s) \, ds \right\|_E \\
&\le DN(t_2 - t_1) + \left\| \int_0^{t_1} \exp\{A(t_1 - s)\} (\exp\{A(t_2 - t_1)\} - I) \cdot f_n(s) \, ds \right\|_E \\
&\le \frac{\varepsilon}{2} + Dd \frac{\varepsilon}{2Dd} = \varepsilon
\end{aligned}$$

The relative compactness of the sequence $\{Gf_n\}_{n=1}^{\infty}$ follows from the well-known Arzela–Ascoli criterion.

Condition (S1) implies that G is a bounded linear operator from the space $L^1([0,d];E)$ into $C([0,d];E)$. Therefore it is continuous if these spaces are endowed with the topology of weak sequential convergence and $f_n \underset{w}{\to} f_0$ implies $Gf_n \underset{w}{\to} Gf_0$. But the relative compactness of the sequence $\{Gf_n\}_{n=1}^{\infty}$ implies that this last convergence is in the norm of the space $C([0,d];E)$. □

Theorem 4.2.2. *Let the sequence of functions* $\{f_n\}_{n=1}^{\infty} \subset L^1([0,d];E)$ *satisfy the conditions of Theorem* 4.2.1. *If* $S : L^1([0,d];E) \to C([0,d];E)$ *is an operator satisfying properties* (S1) *and* (S2) *then*

$$\chi(\{Sf_n(t)\}_{n=1}^{\infty}) \le 2D \int_0^t q(s)\,ds \tag{4.2.9}$$

for all $t \in [0,d]$, *where* $D \ge 0$ *is the constant in condition* (S1).

Proof. For any $\varepsilon > 0$ choose $\delta \in (0,\varepsilon)$ such that for every set $m \subset [0,d]$, meas $m < \delta$ it is

$$\int_m \nu(s)\,ds < \varepsilon.$$

Take m_δ and G_δ from Theorem 4.2.1. Then any sequence $\{g_n\}_{n=1}^{\infty}$, $g_n \in G_\delta$ is weakly compact by Remark 4.2.2. Therefore from condition (S2) it follows that the sequence $\{Sg_n\}_{n=1}^{\infty}$ is relatively compact in $C([0,d];E)$. This means that SG_δ is relatively compact in $C([0,d];E)$.

By condition (S1) and using Remark 4.2.1 we have that f_n and corresponding function g_n satisfy

$$\begin{aligned}
\| Sf_n(t) - Sg_n(t) \|_E &\le D \int_0^t \| f_n(s) - g_n(s) \|_E\,ds \\
&\le D \int_{[0,t]\setminus m_\delta} \| f_n(s) - g_n(s) \|_E\,ds + D \int_{[0,t]\cap m_\delta} \| f_n(s) \|_E\,ds \\
&\le D \int_{[0,t]\setminus m} (2q(s) + \delta)\,ds + D\varepsilon \le 2D \int_0^t q(s)\,ds + D\varepsilon(d+1).
\end{aligned}$$

Therefore the relatively compact set $SG_\delta(t)$ forms a $\big(2D\int_0^t q(s)\,ds + D\varepsilon(d+1)\big)$-net for the set $\{Sf_n(t)\}_{n=1}^{\infty}$, proving the theorem due to the arbitrary choice of ε. □

Corollary 4.2.4. *If the space E is separable the estimate* (4.2.9) *has the form*

$$\chi\big(\{Sf_n(t)\}_{n=1}^{\infty}\big) \le D\int_0^t q(s)\,ds. \tag{4.2.10}$$

Proof. It is sufficient in the proof of Theorem 4.2.2 to change the reference from Theorem 4.2.1 to Corollary 4.2.1. □

Corollary 4.2.5. *For any sequence of functions* $\{f_n\}_{n=1}^{\infty} \subset L^1([0,d];E)$ *satisfying the conditions of Theorem* 4.2.1 *we have the estimate*

$$\chi\left(\left\{\int_0^t f_n(s)\,ds\right\}_{n=1}^{\infty}\right) \le 2\int_0^t q(s)\,ds$$

for $t \in [0,d]$ *in the general case and*

$$\chi\left(\left\{\int_0^t f_n(s)\,ds\right\}_{n=1}^{\infty}\right) \le \int_0^t q(s)\,ds$$

for $t \in [0,d]$ *in the case of a separable space* E.

Proof. Take $A = 0$ and apply Lemma 4.2.1, Theorem 4.2.2 and the corollary above. □

We recall the following important property.

Lemma 4.2.2. *Let E be a separable Banach space.* $\{f_n\}_{n=1}^{\infty} \subset L^1([0,d];E)$ *an integrably bounded sequence of functions. Then the function* $\varphi : [0,d] \to \mathbb{R}_+$,

$$\varphi(t) = \chi\big(\{f_n(t)\}_{n=1}^{\infty}\big) \quad \textit{for a.e. } t \in [0,d]$$

is measurable and hence integrable.

Proof. Since the space E is separable and the sequence $\{f_n\}_{n=1}^{\infty}$ is integrably bounded we know from Section 2.1.2 that

$$\varphi(t) = \lim_{k\to\infty} d_*\big(\{f_n(t)\}_{n=1}^{\infty}, E_k\big) \quad \text{for a.e. } t \in [0,d]$$

where $\{E_k\}_{k=1}^{\infty}$ is a sequence of k-dimensional subspaces of E such that

$$E_k \subset E_{k+1} \quad \text{and} \quad \overline{\bigcup_{k\ge 1} E_k} = E.$$

We note that functions $t \to d(f_n(t), E_k)$, $n = 1, 2, \dots$ are measurable and hence the functions

$$\tau_k(t) = d_*\big(\{f_n(t)\}_{n=1}^{\infty}, E_k\big) = \sup_n d(f_n(t), E_k), \quad k = 1, 2, \dots$$

are measurable too. For a.e. $t \in [0, d]$ the sequence $\{\tau_k(t)\}_{n=1}^{\infty}$ is monotonically increasing and converges to $\chi\big(\{f_n(t)\}_{n=1}^{\infty}\big)$, hence the function φ is measurable. □

We can prove now the following result about χ-estimates for a multivalued integral.

Theorem 4.2.3. *Let E be a separable Banach space and $G : [0, d] \to P(E)$ an integrable, integrably bounded multifunction such that*

$$\chi(G(t)) \le q(t)$$

for a.e. $t \in [0, d]$ where $q(\cdot) \in L^1_+([0, d])$. Then

$$\chi\Big(\int_0^t G(s)\,ds\Big) \le \int_0^t q(s)\,ds$$

for all $t \in [0, d]$. In particular, if the multifunction $G : [0, d] \to C(E)$ is measurable and integrably bounded then the function $\chi(G(\cdot))$ is integrable and

$$\chi\Big(\int_0^t G(s)\,ds\Big) \le \int_0^t \chi(G(s))\,ds$$

for all $t \in [0, d]$.

Proof. Since E is separable, the space $L^1([0, d]; E)$ is separable. Therefore the set of integrable selections S^1_G contains a dense denumerable subset $\{f_n\}_{n=1}^{\infty}$. Hence the set

$$\Big\{\int_0^t f_n(s)\,ds\Big\}_{n=1}^{\infty}$$

is dense in the set $\int_0^t G(s)\,ds$.

From the properties of the Hausdorff MNC χ (see Section 2.1.1) it follows that

$$\chi\Big(\int_0^t G(s)\,ds\Big) = \chi\Big(\Big\{\int_0^t f_n(s)\,ds\Big\}_{n=1}^{\infty}\Big)$$

and

$$\chi(\{f_n(t)\}_{n=1}^{\infty}) \le \chi(G(t)) \le q(t) \quad \text{a.e. } t \in [0, d].$$

Then from Corollary 4.2.5 it follows that

$$\chi\left(\left\{\int_0^t f_n(s)\,ds\right\}_{n=1}^{\infty}\right) \le \int_0^t q(s)\,ds$$

concluding the proof of the first part of the theorem.

When the multifunction $G : [0, d] \to C(E)$ is measurable and integrably bounded, without loss of generality, it is possible to take as a Castaing representation of G the above family $\{f_n\}_{n=1}^{\infty}$ (see Remark 1.3.1). Then Lemma 4.2.2 gives that

$$\chi(G(t)) = \chi\left(\{f_n(t)\}_{n=1}^{\infty}\right) \quad \text{a.e. } t \in [0, d]$$

is integrable and we can apply the first part of the theorem. □

We consider now the problem of a χ-estimate for a set of integrals taken over a bounded family of continuous functions.

Theorem 4.2.4. *Let E be a separable Banach space and $\Omega \subset C([0, d]; E)$ a bounded subset. Then the function*

$$\varphi_\Omega : [0, d] \to \mathbb{R}_+, \quad \varphi_\Omega(t) = \chi(\Omega(t))$$

where $\Omega(t) = \{y(t) : y \in \Omega\}$, is measurable and hence integrable and

$$\chi\left(\left\{\int_0^t y(s)\,ds : y \in \Omega\right\}\right) \le \int_0^t \varphi_\Omega(s)\,ds$$

for all $t \in [0, d]$.

Proof. Since the space E is separable we can choose a dense countable subset W in Ω. It is easy to see that the sets $W(t)$ are dense in $\Omega(t)$, $t \in [0, d]$ and hence from the property of a measure of noncompactness (see Section 2.1.1) we know that

$$\varphi_\Omega(t) = \chi(W(t)).$$

The integrability of the function φ_Ω follows from Lemma 4.2.2.

From the same property for the MNC it follows that

$$\chi\left(\left\{\int_0^t y(s)\,ds : y \in \Omega\right\}\right) = \chi\left(\left\{\int_0^t w(s)\,ds : w \in W\right\}\right)$$

and we apply Corollary 4.2.5 to obtain the desired estimate. □

Theorem 4.2.5. *Let E be a separable Banach space, $\Omega \subset C([0,d];E)$ a bounded set and assume that the function φ_Ω is zero a.e. on $[0,d]$. Then for every $\delta > 0$ there exist a set $J \subseteq [0,d]$, a compact set $K \subset E$ such that* meas $J < \delta$ *and*

$$\sup_{x\in\Omega} \operatorname{dist}(x(t), K) < \delta \quad \textit{for } t \in [0,d] \setminus J.$$

Proof. From the formula for the Hausdorff MNC in a separable Banach space (see Section 2.1.2) it follows that the function φ_Ω is a pointwise limit of the functions

$$y_m(t) = \sup_{x\in\Omega} \operatorname{dist}\big(x(t), B^m_{2M}(0)\big),$$

where $m = 1, 2, \dots$. Recall that $M > 0$ is chosen so that Ω is contained in $B^C_M(0)$, a ball in the space $C([0,d];E)$, and $B^m_{2M}(0)$ denotes a ball in the space E_m.

Note that the functions $y_m(t)$ are measurable. In fact, if Ω is an open subset in $C([0,d];E)$ then

$$y_m(t) = \sup_{x\in\Omega\cap H} \operatorname{dist}\big(x(t), B^m_{2M}(0)\big),$$

where H is a dense denumerable subset of the space $C([0,d];E)$, i.e., the supremum is taken over a denumerable set of continuous functions and hence y_m are measurable functions.

Now, if Ω is an arbitrary set let us send $\varepsilon_n \to 0$ and then

$$y_m(t) = \lim_{n\to\infty} \sup_{x\in W_{\varepsilon_n}(\Omega)} \operatorname{dist}\big(x(t), B^m_{2M}(0)\big),$$

where $W_{\varepsilon_n}(\Omega)$ is an ε_n-neighborhood of Ω in the space $C([0,d];E)$. Therefore the function y_m is measurable as a pointwise limit of a sequence of measurable functions.

We apply the Egorov Theorem to the sequence of functions $\{y_m\}$ and deduce that for every $\delta > 0$ there exists a set $J' \subset [0,d]$ such that meas $J' < \delta$ and

$$\sup_{t\in[0,d]\setminus J'} |y_m(t) - \varphi_\Omega(t)| < \delta.$$

Finally, join to the set J', if necessary, the points at which the function φ_Ω does not vanish, and denote the resulting set by J. We have

$$\sup_{t\in[0,d]\setminus J} |y_m(t)| < \delta,$$

or

$$\Big|\sup_{x\in\Omega} \operatorname{dist}\big(x(t), B^m_{2M}(0)\big)\Big| < \delta,$$

i.e., we can take the ball $B^m_{2M}(0)$ as the compact set K. □

4.3 Condensing semigroups

In this section we focus our attention on χ*-decreasing semigroups*, that is on semigroups satisfying the estimate

$$\| \exp\{At\} \|^{(\chi)} \leq Ce^{-\gamma t} \tag{4.3.1}$$

where γ and C are positive constants (see Section 2.1.1 for the definition of the χ-norm $\| \cdot \|^{(\chi)}$). An important particular case is when $C = 1$ and hence the estimate 4.3.1 has the form

$$\| \exp\{At\} \|^{(\chi)} \leq e^{-\gamma t} \tag{4.3.2}$$

In this case the semigroup $\exp\{At\}$ is called χ*-strongly contractive.*

Theorem 4.3.1. *Let E be a separable Banach space, A an infinitesimal generator of a decreasing C_0-semigroup* $\exp\{At\}$ *and B a bounded linear operator such that*

$$\| B \|^{(\chi)} \leq b\,,$$

with $b < a/C$, where a and C are the constants from (4.1.6). *Then the operator $A + B$ is the infinitesimal generator of a χ-decreasing C_0-semigroup satisfying the estimate* (4.3.1) *with constant $\gamma = a - Cb$.*

Proof. It is known that the operator $A + B$ is the infinitesimal generator of a C_0-semigroup (see e.g. [139]). To estimate $\| \exp\{(A + B)t\} \|^{(\chi)}$ in accordance with Definition 2.1.2, consider the value $\chi\big(\exp\{(A + B)t\}\mathcal{B}\big)$, where $\mathcal{B}$ is the unit ball in E. For every $u_0 \in \mathcal{B}$ the function $u(t) = \exp\{(A + B)t\}u_0$ is the solution of the integral equation

$$u(t) = \exp\{At\}u_0 + \int_0^t \exp\{A(t - s)\}Bu(s)\,ds. \tag{4.3.3}$$

Note that solutions of this equation belong to the space $C([0, d]; E)$ for arbitrary $d > 0$. Denote by Ω the set of solutions of the equation (4.3.3) obtained when u_0 varies over $\mathcal{B}$. It is clear that the set Ω is bounded in $C([0, d], E)$ and from Theorem 4.2.4 it follows that the function $y(t) = \chi(\Omega(t))$ is measurable and so we have

$$y(t) \leq Ce^{-at} + \int_0^t Ce^{-a(t-s)} \| B \|^{(\chi)} y(s)\,ds = Ce^{-at} + b\int_0^t Ce^{-a(t-s)} y(s)\,ds.$$

Applying Gronwall's lemma we obtain

$$y(t) \leq Ce^{-(a-Cb)t}\,,$$

hence

$$\chi\big(\exp\{(A + B)t\}\mathcal{B}\big) = \chi(\Omega(t)) \leq Ce^{-(a-Cb)t}. \qquad \square$$

Definition 4.3.1. A χ-strongly contractive C_0-semigroup $\exp\{At\}$ is said to be *sequentially condensing* with exponent γ if for every sequence of positive numbers $\{t_m\}$ converging to $t_0 > 0$ and every bounded sequence $\{x_m\} \subset E$ we have

$$\chi\left(\{\exp\{At_m\}x_m\}\right) \le e^{-\gamma t_0}\chi\left(\{x_m\}\right). \tag{4.3.4}$$

Remark 4.3.1. If a space E is separable then the estimate (4.3.2) follows from the inequality (4.3.4).

Remark 4.3.2. If a semigroup $\exp\{At\}$ is χ-strongly contractive and continuous in the operator norm topology for $t > 0$ then it is sequentially condensing with the exponent γ from (4.3.2). To see this it is sufficient to note that $\exp\{At_m\}x_m - \exp\{At_0\}x_m \to 0$ as $m \to \infty$, hence

$$\chi\left(\{\exp\{At_m\}x_m\}\right) = \chi\left(\{\exp\{At_0\}x_m\}\right) \le e^{-at_0}\chi\left(\{x_m\}\right).$$

Theorem 4.3.2. *Let E be a separable Banach space, A the infinitesimal generator of a strongly contractive C_0-semigroup* $\exp\{At\}$ *(see Definition* 4.1.5*) and B a linear compact operator. Then the operator $A + B$ is the infinitesimal generator of a C_0-semigroup which is sequentially condensing with exponent $\gamma = a$ where a comes from the estimate* (4.1.6).

Proof. Using Theorem 4.3.1 and the estimate (2.1.1) we have

$$\chi\left(\{\exp\{(A+B)t_0\}x_m\}\right) \le e^{-\gamma t_0}\chi\left(\{x_m\}\right).$$

Given $\varepsilon > 0$, we construct a compact $(e^{-\gamma t_0}\chi(\{x_m\}) + \varepsilon)$-net for the set $\{\exp\{(A+B)t_m\}x_m\}$. Note that

$$\exp\{(A+B)t_m\}x_m = \exp\{At_m\}x_m + \int_0^{t_m} \exp\{A(t_m - s)\}Bu^m(s)\,ds,$$

where u^m are the solutions of the equation (4.3.3) with $u_0 = x_m$.

Choosing $d > 0$ so that $\{t_m\} \subset [0, d]$ we conclude that the sequence $\{u^m\}$ is bounded in the space $C([0,d]; E)$. Therefore, the set $\{Bu^m(s) : s \in [0,d],\ m = 1, 2, \dots\}$ is relatively compact and, since $\exp\{At\}$ is a C_0-semigroup we have

$$\left(\exp\{A(t_m - s)\} - \exp\{A(t_0 - s)\}\right) Bu^m(s) \to 0$$

uniformly with respect to $s \in [0,d]$ as $m \to \infty$. It follows that

$$\int_0^{t_m} \exp\{A(t_m - s)\}Bu^m(s)\,ds - \int_0^{t_0} \exp\{A(t_0 - s)\}Bu^m(s)\,ds \underset{m\to\infty}{\longrightarrow} 0. \tag{4.3.5}$$

Since the set $\left\{ \int_0^{t_0} \exp\{A(t_0 - s)\} B u^m(s)\, ds \right\}$ is relatively compact, from relation (4.3.5) we deduce that the set $\left\{ \int_0^{t_m} \exp\{A(t_m - s)\} B u^m(s)\, ds \right\}$ is also relatively compact.

Suppose the set $\{z_i\}_{i=1}^n$ forms a ρ-net of the set $\{x_m\}$. Then the set

$$\{\exp\{At_m\} z_i\}_{i=1}^{n} {}_{m=1}^{\infty}$$

is relatively compact.

Since for given $\varepsilon > 0$, for every m large enough it is possible to find z_i for which the inequality

$$\| \exp\{At_m\} z_i - \exp\{At_m\} x_m \| < e^{-\gamma t_0} \rho + \varepsilon$$

holds, we obtain

$$\chi\left(\{\exp\{(A+B)t_m\} x_m\}\right) \le e^{-\gamma t_0} \rho + \varepsilon,$$

concluding the proof. □

Example 4.3.1. Conditions on the infinitesimal generator of a C_0-semigroup of Theorem 4.3.2, hold for the following operator appearing in the study of a nonlinear transmission line problem (see Example 5.2.1 below).

Consider the Hilbert space $E = L^2[0,l] \times L^2[0,l] \times \mathbb{R}$ with norm:

$$|(\varphi, \psi, b)|^2 = \int_0^l \varphi^2(x)\, dx + \int_0^l \psi^2(x)\, dx + b^2, \tag{4.3.6}$$

and the linear operator $A : D(A) \subset E \to E$ given by

$$A \begin{pmatrix} \varphi \\ \psi \\ b \end{pmatrix} = \begin{pmatrix} -a_{11} & -a_{12} D_x & 0 \\ -a_{21} D_x & -a_{22} & 0 \\ a_{31} \Gamma & 0 & 0 \end{pmatrix} \begin{pmatrix} \varphi \\ \psi \\ b \end{pmatrix} \tag{4.3.7}$$

where $a_{ij} > 0$, $D_x = \frac{d}{dx}$, and $\Gamma[\varphi] = \varphi(l)$. The domain of A is defined as

$$D(A) = \left\{ (\varphi, \psi, b) \in E \;:\; \varphi, \psi \in H^1[0,l],\ \psi(0) + R_0 \varphi(0) = 0,\ b = \psi(l) \right\} \tag{4.3.8}$$

where $R_0 > 0$ and H^1 denotes the Sobolev space. We represent A as the sum $A = A_0 + B$ where

$$A_0 = \begin{pmatrix} -a_{11} & -a_{12} D_x & 0 \\ -a_{21} D_x & -a_{22} & 0 \\ a_{31} \Gamma & 0 & -N \end{pmatrix} \tag{4.3.9}$$

with $N > \min\{a_{11}, a_{12}\}$ and

$$B = \begin{pmatrix} 0 & 0 & 0 \\ 0 & 0 & 0 \\ 0 & 0 & N \end{pmatrix}. \tag{4.3.10}$$

The operator B is one dimensional and hence compact.

Define an equivalent norm $\|\cdot\|_*$, in the space E, via the inner product

$$\langle(\varphi_1, \chi_1, b_1), (\varphi_2, \chi_2, b_2)\rangle \tag{4.3.11}$$
$$= \alpha \int_0^l \varphi_1(x)\, \varphi_2(x)\, dx + \beta \int_0^l \psi_1(x)\, \psi_2(x)\, dx + \delta\, b_1 b_2.$$

Assume that the positive coefficients $\alpha,\ \beta,\ \delta$ satisfy the relation

$$\alpha\, a_{11} = \beta\, a_{21} = \delta\, a_{13}. \tag{4.3.12}$$

Then for any $(\varphi,\ \psi,\ b) \in D(A)$ we have

$$\begin{aligned}\langle A_0(\varphi,\ \psi,\ b), (\varphi,\ \psi,\ b)\rangle &= \alpha \int_0^l \left(-a_{11}\varphi^2 - a_{12}\psi'\varphi\right) dx \\ &\quad + \beta \int_0^l \left(-a_{21}\psi\varphi' - a_{22}\psi^2\right) dx \\ &\quad + \delta\, a_{31} b\, \varphi(l) - \delta\, N\, b^2 \qquad (4.3.13)\\ &= -a_{11}\alpha \int_0^l \varphi^2\, dx - a_{22}\,\beta \int_0^l \psi^2\, dx \\ &\quad - N\,\delta\, b^2 - \alpha\, a_{12} R_0 \varphi^2(0) \\ &\le -\gamma\, \langle(\varphi,\ \psi,\ b), (\varphi,\ \psi,\ b)\rangle .\end{aligned}$$

where $\gamma = \min\{a_{11},\ a_{22}\}$.

Therefore (see [195]) the semigroup generated by the operator A_0 satisfies the estimate

$$\|\exp\{A_0 t\}\|_* \le e^{-\gamma t}.$$

So conditions of Theorem 4.3.2 are fulfilled for the operator A and hence the semigroup $\exp\{At\}$ is sequentially condensing with exponent γ.

Theorem 4.3.3. *Let E be a separable Banach space, A the infinitesimal generator of a strongly contractive C_0-semigroup* $\exp\{At\}$ *which is continuous with respect to the operator norm topology for $t > 0$ and B a continuous linear operator satisfying $\|B\|^{(\chi)} \le b$, and moreover let $a - b > 0$. Then the operator $A + B$ is the infinitesimal generator of a sequentially condensing C_0-semigroup with exponent $\gamma = a - b$.*

To prove this statement it is sufficient to note that the operator $A + B$ is the infinitesimal generator of a C_0-semigroup which is continuous with respect to the operator norm for $t > 0$, and then apply Theorem 4.3.1 and Remark 4.3.2.

Proposition 4.3.1. *Let E be a separable space and A the infinitesimal generator of a χ-strongly contractive C_0-semigroup. Then for a sequence of functions $\{f_n\}_{n=1}^{\infty} \subset L^1([0,d];E)$ satisfying the conditions of Theorem* 4.2.1 *we have the estimate*

$$\chi\left(\left\{\int_0^t \exp\{A(t-s)\} f_n(s)\, ds\right\}_{n=1}^{\infty}\right) \leq \int_0^t e^{-\gamma(t-s)} q(s)\, ds$$

for $t \in [0,d]$.

Proof. This inequality follows from Corollary 4.2.4 and the estimate

$$\chi\left(\{\exp\{A(t-s)\} f_n(s)\}_{n=1}^{\infty}\right) \leq e^{-\gamma(t-s)} q(s), \quad s \in [0,t]$$

(see (2.1.1)). □

Bibliographic notes

For a detailed treatment of the theory of semigroups for linear operators in Banach spaces we refer the reader to the monographs of A. Pazy [195], E. Hille and R. S. Phillips [116], S. Goldberg [100], S. G. Krein [150], R. Martin [166], A. V. Balakrishman [23], V. Barbu [26], N. H. Pavel [194] and others.

Theorem 4.1.3 is taken from a paper of M. I. Kamenskii, V. V. Obukhovskii and P. Zecca [137] and it follows the ideas of V. Barbu [26] and R. Martin [166] about the regularity of mild solutions. Theorem 4.1.4 was proved in the paper of J. M. Ball [24]. A result similar to Theorem 4.1.5 was obtained for the case of analytic semigroups by I. N. Gourova [110].

In Section 4.2 the most part of the facts connected with the estimate of the values of the Hausdorff MNC on sequences of summable functions are new. Some of them were given in the paper of J. F. Couchouron and M. I. Kamenskii [67]. We mention the condition of weak compactness of a subset of integrable functions (Proposition 4.2.1), closely related to the condition of weak compactness of J. Diestel as given in [75] and [76] (see also [186] for the case of a separable space). Lemma 4.2.2 is due to M. Kisielewicz [141] and the result on the χ-estimation of the multivalued integral (Theorem 4.2.3) was announced in the paper of V. V. Obukhovskii [180].

Theorem 4.3.1 giving a sufficient condition for a semigroup to be χ-condensing was proved in the papers of M. I. Kamenskii, V. V. Obukhovskii and P. Zecca [137], M. I. Kamenskii [127], [130]. Linear operators like the one in Example 4.1.1 arise in the study of a nonlinear transmission line problem (see [59] and [114]).

Chapter 5

Semilinear differential inclusions: initial problem

In this chapter we use the method of integral multioperators to study classical questions concerning semilinear inclusions with unbounded linear part generating a noncompact semigroup in a Banach space. We give local and nonlocal existence theorems, investigate the topological structure of the solutions set and prove a result on continuous dependence of solutions on parameters and initial data. Special attention is paid to the foundation of the averaging principle. In the last part of this chapter we consider semilinear differential inclusions with lower semicontinuous nonlinearities.

5.1 Integral multioperator

In this section we construct the integral multioperator whose fixed points are mild solutions of the Cauchy problem for a semilinear differential inclusion in a Banach space. We give sufficient conditions under which this multioperator is condensing with respect to the MNC ν given in Example 2.1.4.

We recall from Definition 4.2.1 that the sequence $\{f_n\}_{n=1}^{\infty} \subset L^1([0, d]; E)$ is said to be semicompact if it is integrably bounded and the set $\{f_n(t)\}_{n=1}^{\infty}$ is relatively compact for almost every $t \in [0, d]$.

Theorem 5.1.1. *Let $S : L^1([0, d]; E) \to C([0, d]; E)$ be an operator satisfying the conditions* (S1′) *and* (S2) *(see Section* 4.2*). Then for every semicompact sequence $\{f_n\}_{n=1}^{\infty} \subset L^1([0, d]; E)$ the sequence $\{Sf_n\}_{n=1}^{\infty}$ is relatively compact in $C([0, d]; E)$ and, moreover, if $f_n \underset{w}{\to} f_0$ then*

$$Sf_n \to Sf_0.$$

Proof. From Corollary 4.2.3 we know that for every $\varepsilon > 0$ there exist a compact $K_\varepsilon \subset E$ and ε-net $\mathcal{R}$ of the sequence $\{f_n\}_{n=1}^{\infty}$ formed by the functions with values in K_ε. By property (S2) $S\mathcal{R}$ is a relatively compact set in $C([0, d]; E)$ and by (S1′) $S\mathcal{R}$ is an εD-net of the sequence $\{Sf_n\}_{n=1}^{\infty}$ in $C([0, d]; E)$.

To prove the second part of the theorem assume the contrary. Then, there exist $\varepsilon_0 > 0$, two sequences of indices $n_k \to +\infty$ and $m_k \to +\infty$ such that

$$\| Sf_{n_k} - Sf_{m_k} \| > \varepsilon_0. \tag{5.1.1}$$

We note that $f_n \underset{w}{\to} f_0$ implies $\kappa(\cdot)f_n(\cdot) \underset{w}{\to} \kappa(\cdot)f_0(\cdot)$ where κ is the indicatrix function of any measurable subset of $[0, d]$. Choose $\delta > 0$ such that for every set

$m \subset [0, d]$, meas $m < \delta$ we have that

$$\int_m \mu(s)\, ds < \frac{\varepsilon_0}{4D}$$

where μ is the function integrally bounding the sequence $\{f_n\}_{n=1}^{\infty}$. Let us construct, for the sequence $\{f_n\}_{n=1}^{\infty}$ and the number δ, the set m_δ and the compact set K_δ appearing in Theorem 4.2.1. Denote by κ_δ the indicatrix function of the set $[0, d] \setminus m_\delta$. By the property (S1′) of the operator S we have

$$\| Sf_n - S\kappa_\delta f_n \| \le D \int_{m_\delta} \| f_n(s) \| \, ds < \frac{\varepsilon_0}{4}.$$

Therefore from inequality (5.1.1) it follows that

$$\left\| S\kappa_\delta f_{n_k} - S\kappa_\delta f_{m_k} \right\| > \frac{\varepsilon_0}{2}. \tag{5.1.2}$$

For functions $\kappa_\delta f_{n_k}$ and $\kappa_\delta f_{m_k}$, $k = 1, 2, \ldots$, there exist (see Theorem 4.2.1 and Remark 4.2.1) functions g_{n_k} and g_{m_k} with values in the set $K_\delta \cup \{0\}$ such that

$$\begin{aligned} \left\| \kappa_\delta f_{n_k}(t) - g_{n_k}(t) \right\| &< \delta, \\ \left\| \kappa_\delta f_{m_k}(t) - g_{m_k}(t) \right\| &< \delta. \end{aligned} \tag{5.1.3}$$

By Remark 4.2.2 we can select subsequences

$$g_{n_{k_p}} \underset{w}{\rightarrow} x, \quad g_{m_{k_p}} \underset{w}{\rightarrow} y.$$

Using property (S2) of the operator S we obtain that

$$Sg_{n_{k_p}} \rightarrow Sx, \quad Sg_{m_{k_p}} \rightarrow Sy.$$

Now choose p large enough to ensure

$$\left\| Sg_{n_{k_p}} - Sx \right\| < \delta D, \quad \left\| Sg_{m_{k_p}} - Sy \right\| < \delta D. \tag{5.1.4}$$

Then the inequalities (5.1.3) give

$$\begin{aligned} \| \kappa_\delta f_0 - x \|_{L^1} &< \delta d, \\ \| \kappa_\delta f_0 - y \|_{L^1} &< \delta d. \end{aligned} \tag{5.1.5}$$

Hence by property (S1′) of the operator S it follows from the inequalities (5.1.3)-(5.1.5) that

$$\left\| S\kappa_\delta f_{n_{k_p}} - S\kappa_\delta f_{m_{k_p}} \right\| < 6\,\delta d D.$$

Corollary 5.1.1. *If $G : L^1([0,d]; E) \to C([0,d]; E)$ is the Cauchy operator (see Definition 4.2.2) then for every semicompact sequence $\{f_n\}_{n=1}^{\infty} \subset L^1([0,d]; E)$ the sequence $\{Gf_n\}_{n=1}^{\infty}$ is relatively compact in $C([0,d]; E)$ and, moreover, if $f_n \underset{w}{\to} f_0$ then*

$$Gf_n \to Gf_0.$$

Proof. See Lemma 4.2.1. □

Our aim now is to describe the principal properties of the integral multioperator appearing in the investigation of differential inclusions.

Recall that $Kv(E)$ denotes the collection of all nonempty convex compact subsets of E.

Let $F : [0,d] \times E \to Kv(E)$ be a multimap satisfying the following properties:

(F1) the multifunction $F(\cdot, x)$ has a strongly measurable selection for every $x \in E$;

(F2) the multimap $F(t, \cdot) : E \to Kv(E)$ is u.s.c. for a.e. $t \in [0,d]$;

(F3) for any nonempty bounded subset $\Omega \subset E$ there exists a function $\mu_\Omega \in L^1_+([0,d])$ such that

$$\| F(t,x) \| \leq \mu_\Omega(t)$$

for all $x \in \Omega$ and a.e. $t \in [0,d]$.

The next assumption will be called the *χ-regularity condition*:

(F4) there exists a function $k(\cdot) \in L^1_+([0,d])$ such that

$$\chi(F(t,D)) \leq k(t) \cdot \chi(D) \quad \text{a.e. } t \in [0,d]$$

for every bounded $D \subset E$ where χ is the Hausdorff MNC in E.

Note that from assumptions (F1)–(F3) it follows that the superposition multioperator $\mathcal{P}_F : C([0,d]; E) \to P(L^1([0,d]; E))$,

$$\mathcal{P}_F(x) = S^1_{F(\cdot, x(\cdot))}$$

generated by F is correctly defined (see Section 1.3.3). The superposition multioperator $\mathcal{P}_F$ is weakly closed in the following sense.

Lemma 5.1.1. *Assume the sequences*

$$\{x_n\}_{n=1}^{\infty} \subset C([0,d]; E), \quad \{f_n\}_{n=1}^{\infty} \subset L^1([0,d]; E),$$

$f_n \in \mathcal{P}_F(x_n)$, $n \geq 1$ are such that $x_n \to x_0$, $f_n \underset{w}{\to} f_0$. Then $f_0 \in \mathcal{P}_F(x_0)$.

Proof. According to Mazur's lemma (see, for example [84]) the weak convergence of $\{f_n\}_{n=1}^{\infty}$ to f_0 implies the existence of a double sequence of nonnegative numbers $\{\alpha_{ik}\}_{i=1}^{\infty}{}_{k=1}^{\infty}$ such that: (i) $\sum_{k=i}^{\infty} \alpha_{ik} = 1$ for all $i = 1, 2, \ldots$; (ii) for every $i = 1, 2, \ldots$ there exists a number $k_0(i)$ such that $\alpha_{ik} = 0$ for all $k \geq k_0(i)$; (iii) the sequence $\{\widetilde{f}_i\}_{i=1}^{\infty}$,

$$\widetilde{f}_i(t) = \sum_{k=i}^{\infty} \alpha_{ik} f_k(t)$$

converges to f_0 with respect to the norm of the space $L^1([0, d]; E)$. Passing if necessary to a subsequence we can assume that $\{\widetilde{f}_i\}_{i=1}^{\infty}$ converges to f_0 almost everywhere on $[0, d]$.

From (F2) it follows that for almost all $t \in [0, d]$, for a given $\varepsilon > 0$, there exists an integer $i_0 = i_0(\varepsilon, t)$ such that

$$F(t, x_i(t)) \subset W_\varepsilon(F(t, x_0(t)) \quad \text{for all } i \geq i_0,$$

where W_ε denotes the ε-neighborhood of a set. Then

$$f_i(t) \in W_\varepsilon(F(t, x_0(t)))$$

for $i \geq i_0$, and hence also

$$\widetilde{f}_i(t) \in W_\varepsilon(F(t, x_0(t))) \quad \text{for } i \geq i_0.$$

Therefore

$$f_0(t) \in F(t, x_0(t)) \quad \text{for a.e. } t \in [0, d],$$

i.e., $f_0 \in \mathcal{P}_F(x_0)$, concluding the proof. □

Theorem 5.1.2. *Let $\mathcal{P}_F : C([0, d]; E) \to P(L^1([0, d]; E))$ be the superposition multioperator generated by a multimap $F : [0, d] \times E \to Kv(E)$ satisfying the properties* (F1)–(F4) *and $S : L^1([0, d]; E) \to C([0, d]; E)$ be an operator satisfying* (S1′) *and* (S2). *Then the composition*

$$S \circ \mathcal{P}_F : C([0, d]; E) \to P(C([0, d]; E))$$

is a closed multioperator with compact values.

Proof. Let $\{x_n\}_{n=1}^{\infty}$, $\{z_n\}_{n=1}^{\infty} \subset C([0, d]; E)$, $x_n \to x_0$, $z_n \in S \circ \mathcal{P}_F(x_n)$, $n \geq 1$; and $z_n \to z_0$. Take an arbitrary sequence $\{f_n\}_{n=1}^{\infty} \subset L^1([0, d]; E)$ such that $f_n \in \mathcal{P}_F(x_n)$, $z_n = S(f_n)$, $n \geq 1$. From assumption (F3) it follows that the sequence $\{f_n\}_{n=1}^{\infty}$ is integrably bounded. Hypothesis (F4) implies that

$$\chi(\{f_n(t)\}_{n=1}^{\infty}) \leq k(t)\chi\left(\{x_n(t)\}_{n=1}^{\infty}\right) = 0$$

for a.e. $t \in [0, d]$, i.e., the set $\{f_n(t)\}_{n=1}^{\infty}$ is relatively compact for a.e. $t \in [0, d]$, hence $\{f_n\}_{n=1}^{\infty}$ is a semicompact sequence.

From Remark 4.2.2 it follows that the sequence $\{f_n\}_{n=1}^{\infty}$ is weakly compact in $L^1([0,d];E)$ so we can assume, without loss of generality, that $f_n \underset{w}{\rightarrow} f_0$. Theorem 5.1.1 implies that $z_n = Sf_n \rightarrow Sf_0 = z_0$. On the other hand, applying Lemma 5.1.1 we have $f_0 \in \mathcal{P}_F(x_0)$ and therefore $z_0 \in S \circ \mathcal{P}_F(x_0)$, i.e., the multioperator $S \circ \mathcal{P}_F$ is closed.

If $x \in C([0,d];E)$, the χ-regularity property (F4) implies that every sequence $\{f_n\}_{n=1}^{\infty}$ chosen in $\mathcal{P}_F(x)$ is semicompact and the same argument as above implies that the sequence $\{Sf_n\}_{n=1}^{\infty} \subset C([0,d];E)$ is relatively compact. The compactness of the set $S \circ \mathcal{P}_F(x)$ follows from its closedness. □

Corollary 5.1.2. *Under the conditions of Theorem* 5.1.2 *the multioperator $S \circ \mathcal{P}_F$ is u.s.c.*

Proof. Let us consider a convergent sequence $\{x_n\}_{n=1}^{\infty} \subset C([0,d];E)$ and an arbitrary sequence $\{f_n\}_{n=1}^{\infty} \subset L^1([0,d];E)$ such that $f_n \in \mathcal{P}_F(x_n)$, $n \geq 1$. From (F3) and (F4) it follows that the sequence $\{f_n\}_{n=1}^{\infty}$ is semicompact. From Theorem 5.1.1 it follows that the sequence $\{Sf_n\}_{n=1}^{\infty} \subset C([0,d];E)$ is relatively compact and therefore the multioperator $S \circ \mathcal{P}_F$ is quasicompact (see Definition 1.1.5 (c)). The statement follows from Theorem 1.1.12. □

Consider the measure of noncompactness

$$\nu(\Omega) = \max_{\mathcal{D} \in \Delta(\Omega)} \left(\gamma(\mathcal{D}), \operatorname{mod}_C(\mathcal{D})\right),$$

in the space $C([0,d];E)$, where

$$\gamma(\mathcal{D}) = \sup_{t \in [0,d]} e^{-Lt} \chi(\mathcal{D}(t))$$

(see Example 2.1.4). The constant $L > 0$ is chosen so that

$$q = \sup_{t \in [0,d]} \left[2D \int_0^t e^{-L(t-s)} k(s)\, ds \right] < 1, \tag{5.1.6}$$

where the constant D is the one in condition (S1) and $k(\cdot)$ is the function appearing in the χ-regularity condition (F4).

Theorem 5.1.3. *Let $\mathcal{P}_F : C([0,d];E) \rightarrow P(L^1([0,d];E))$ be a superposition multioperator generated by a multimap $F : [0,d] \times E \rightarrow Kv(E)$ satisfying properties* (F1)–(F4) *and $S : L^1([0,d];E) \rightarrow C([0,d];E)$ be an operator satisfying* (S1) *and* (S2). *Then the multioperator $S \circ \mathcal{P}_F$ is ν-condensing on bounded sets.*

Proof. Let $\Omega \subset C([0,d];E)$ be a bounded set such that

$$\nu(S \circ \mathcal{P}_F(\Omega) \geq \nu(\Omega) \tag{5.1.7}$$

where the inequality is taken in the sense of the order in $\mathbb{R}^2$ induced by the positive cone $\mathbb{R}^2_+$. We will show that (5.1.7) implies that Ω is relatively compact.

Let the maximum on the left-hand side of the inequality be achieved for the countable set $\mathcal{D}' = \{g_n\}_{n=1}^\infty$. Then

$$g_n(t) = Sf_n(t), \quad f_n \in \mathcal{P}_F(x_n), \ n \geq 1$$

where $\{x_n\}_{n=1}^\infty \subset \Omega$.

From inequality (5.1.7) it follows that

$$\gamma\big(\{g_n\}_{n=1}^\infty\big) \geq \gamma\big(\{x_n\}_{n=1}^\infty\big). \tag{5.1.8}$$

We can now give an upper estimate for $\gamma(\{g_n\}_{n=1}^\infty)$.

The χ-regularity property (F4) implies

$$\begin{aligned}\chi\big(\{f_n(s)\}_{n=1}^\infty\big) &\leq k(s)\cdot\chi\big(\{x_n(s)\}_{n=1}^\infty\big)\\ &= e^{Ls}k(s)e^{-Ls}\cdot\chi\big(\{x_n(s)\}_{n=1}^\infty\big)\\ &\leq e^{Ls}k(s)\cdot \sup_{\xi\in[0,d]} e^{-L\xi}\chi\big(\{x_n(\xi)\}_{n=1}^\infty\big) = e^{Ls}k(s)\cdot\gamma\big(\{x_n\}_{n=1}^\infty\big).\end{aligned}$$

Applying Theorem 4.2.2 we have

$$\chi\big(\{Sf_n(t)\}_{n=1}^\infty\big) \leq 2D\gamma\big(\{x_n\}_{n=1}^\infty\big)\int_0^t e^{Ls}k(s)\,ds. \tag{5.1.9}$$

Now from the inequalities (5.1.8), (5.1.9) it follows that

$$\gamma\big(\{x_n\}_{n=1}^\infty\big) \leq 2D\sup_{t\in[0,d]}\int_0^t e^{-L(t-s)}k(s)\,ds\cdot\gamma\big(\{x_n\}_{n=1}^\infty\big) \leq q\cdot\gamma\big(\{x_n\}_{n=1}^\infty\big),$$

hence

$$\gamma\big(\{x_n\}_{n=1}^\infty\big) = 0,$$

and therefore,

$$\chi\big(\{x_n(t)\}_{n=1}^\infty\big) = 0$$

for all $t \in [0, d]$. From properties (F3), (F4) we see that the sequence $\{f_n\}_{n=1}^\infty$ is semicompact. It follows now, from Theorem 5.1.1, that the sequence $\{g_n\}_{n=1}^\infty$ is relatively compact, therefore

$$\gamma\big(\{g_n\}_{n=1}^\infty\big) = 0$$

and

$$\mathrm{mod}_C(\{g_n\}) = 0,$$

i.e., $\nu(S\circ\mathcal{P}_F(\Omega)) = (0,0)$, and from (5.1.7) we also have

$$\nu(\Omega) = (0,0),$$

concluding the proof. □

We can now apply the above abstract results to the case of the multioperator appearing in the study of the Cauchy problem for a differential inclusion. In fact, we consider the multioperator $\Gamma : C([0,d];E) \to Kv(C([0,d];E))$ defined as

$$\Gamma x = \{y : y(t) = \exp\{At\}x_0 + z(t),\ z \in G \circ \mathcal{P}_F(x)\} \tag{5.1.10}$$

where G is the Cauchy operator (see Definition 4.2.2) and $x_0 \in E$ is a given point.

Taking into account Lemma 4.2.1 and applying Corollary 5.1.2 and Theorem 5.1.3 we have the following important fact.

Theorem 5.1.4. *The multioperator Γ is u.s.c. and ν-condensing on every bounded subset of $C([0,d];E)$.*

In the same way we can describe the properties of the family of integral multioperators $\Gamma(\lambda,\cdot)$ when the multivalued nonlinearity F depends on a parameter λ belonging to a metric space Λ.

Assume that the multimap $F : [0,d] \times E \times \Lambda \to Kv(E)$ satisfies the following conditions.

($F1_\lambda$) the multifunction $F(\cdot,x,\lambda) : [0,d] \to Kv(E)$ has a strongly measurable selection for every $(x,\lambda) \in E \times \Lambda$;

($F2_\lambda$) the multimap $F(t,\cdot,\cdot) : E \times \Lambda \to Kv(E)$ is u.s.c. for a.e. $t \in [0,d]$;

($F3_\lambda$) for any nonempty bounded subset $\Omega \subset E$ there exists a function $\mu_\Omega \in L^1_+([0,d])$ such that

$$\| F(t,\Omega,\Lambda) \| \le \mu_\Omega(t)$$

for a.e. $t \in [0,d]$;

($F4_\lambda$) there exists a function $k(\cdot) \in L^1_+([0,d])$ such that

$$\chi(F(t,\Omega,\Lambda)) \le k(t) \cdot \chi(\Omega)$$

for every nonempty bounded subset $\Omega \subset E$.

For every fixed $\lambda \in \Lambda$, by formula (5.1.10) we can construct the integral multioperator $\Gamma(\lambda,\cdot)$ of the form

$$\Gamma(\lambda,x) = \{y : y(t) = \exp\{At\}x_0 + z(t),\ z \in G \circ \mathcal{P}_{F_\lambda}(x)\} \tag{5.1.11}$$

where $F_\lambda(t,x) = F(t,x,\lambda)$.

Following the main lines of the above arguments the reader can easily verify the following statement.

Theorem 5.1.5. *The family of integral multioperators*

$$\Gamma : \Lambda \times C([0,d];E) \to Kv(C([0,d];E))$$

defined by formula (5.1.11) *is u.s.c. and ν-condensing on every bounded subset.*

5.2 Existence and continuous dependence theorems

In this section, using the topological ideas developed in Chapter 3, we consider local and global existence results for semilinear differential inclusions in a Banach space. Moreover, we prove that the solutions set is compact. This fact implies, as a consequence, the existence of optimal trajectories for control systems governed by differential equations in Banach spaces. We conclude, using results of Section 3.5, showing that the solutions set depends continuously on the initial data and parameters.

5.2.1 Local and global existence results. Regularity properties of solutions. Some examples

We study the Cauchy problem for the following *semilinear differential inclusion* in a Banach space E:

$$x'(t) \in Ax(t) + F(t, x(t)) \tag{5.2.1}$$

$$x(0) = x_0 \in E \tag{5.2.2}$$

The following assumptions hold.

(A) The linear operator $A : D(A) \subseteq E \to E$ is the infinitesimal generator of a C_0-semigroup $\exp\{At\}$.

The multivalued nonlinearity $F : [0, d] \times E \to Kv(E)$ satisfies conditions (F1)–(F4) of Section 5.1.

From Lemma 1.3.3 we know that condition (F1) is satisfied if the multifunction $F(\cdot, x)$ is strongly measurable for every $x \in E$. Recall also that the superposition multioperator $\mathcal{P}_F : C([0, d]; E) \to L^1_+([0, d]; E)$ generated by the multimap F is well-defined.

Definition 5.2.1. A function $x \in C([0, h]; E)$, $0 < h \leq d$ is said to be a *mild solution of the problem* (5.2.1), (5.2.2) on an interval $[0, h]$ provided

$$x(0) = x_0$$

$$x(t) = \exp\{At\}x_0 + \int_0^t \exp\{A(t-s)\}f(s)\,ds, \ t \in [0, h]$$

where $f \in \mathcal{P}_F(x)$.

The set of all mild solutions of the problem (5.2.1), (5.2.2) will be denoted as $\Sigma^F_{x_0}[0, h]$ or simply $\Sigma^F_{x_0}$ if the interval of existence is clear.

We can now prove the following local existence theorem.

Theorem 5.2.1. *Under conditions* (A) *and* (F1)–(F4) *there exists* $h > 0$ *such that the set* $\Sigma_{x_0}^F[0, h]$ *is a nonempty subset of the space* $C([0, h]; E)$.

Proof. Let $r > 0$ be any number. As $\exp\{At\}$ is a C_0-semigroup it follows that we can choose h_1, $0 < h_1 \leq d$ so that

$$\| (\exp\{At\} - I)x_0 \| \leq r/2 \quad \text{for all } t \in [0, h_1]. \tag{5.2.3}$$

Let $M > 0$ be a constant such that

$$\| \exp\{At\} \| \leq M \quad \text{for all } t \in [0, d]. \tag{5.2.4}$$

Denote by B the closed ball $\overline{B}_r(x_0) \subset E$ and choose h_2, $0 < h_2 \leq d$ so that

$$M \int_0^{h_2} \mu_B(s)\, ds \leq r/2 \tag{5.2.5}$$

where μ_B is the function from the boundedness condition (F3).

Now take $h = \min\{h_1, h_2\}$ and consider the integral multioperator

$$\Gamma : C([0, h]; E) \to Kv(C([0, h]; E))$$

given by formula (5.1.10). It is clear that Γ can be written as

$$\Gamma(x) = \Big\{ y \in C([0, h]; E) : \; y(t) = \exp\{At\}x_0 + \int_0^t \exp\{A(t - s)\} f(s)\, ds, \quad f \in \mathcal{P}_F(x) \Big\} \tag{5.2.6}$$

and that $\operatorname{Fix} \Gamma = \Sigma_{x_0}^F[0, h]$.

From Theorem 5.1.4 we know that the multioperator Γ is u.s.c. and ν-condensing on bounded sets. Consider the ball $\overline{B}_r(x^0) \subset C([0, h]; E)$ where $x^0(\cdot) \in C([0, h]; E)$ denotes the function identically equal to x_0. We show that the multioperator Γ maps the ball $\overline{B}_r(x^0)$ into itself. In fact, if $x \in \overline{B}_r(x^0)$ then $x(t) \in B$ for all $t \in [0, h]$ and applying the condition (F3) we obtain

$$\| f(t) \| \leq \mu_B(t) \quad \text{a.e. } t \in [0, h]$$

for any $f \in \mathcal{P}_F(x)$.

Therefore for $y \in \Gamma(x)$ expressed as

$$y(t) = \exp\{At\}x_0 + \int_0^t \exp\{A(t - s)\} f(s)\, ds, \quad f \in \mathcal{P}_F(x)$$

using (5.2.3)–(5.2.5) we have the following estimates:

$$\| y(t) - x_0 \| \leq \| (\exp\{At\} - I)x_0 \| + \int_0^t \| \exp\{A(t-s)\} \| \cdot \| f(s) \| \, ds$$
$$\leq r/2 + M \int_0^{h_2} \mu_B(s) \, ds \leq r,$$

implying $y \in \overline{B}_r(x^0)$. To complete the proof we apply Corollary 3.3.1. □

Strengthening boundedness condition (F3) we obtain the following global existence result.

Theorem 5.2.2. *Suppose that conditions* (A), (F1), (F2), (F4) *are satisfied and that*

(F3′) *there exists a function* $\alpha \in L^1_+([0, d])$ *such that*

$$\| F(t, x) \| \leq \alpha(t)(1 + \| x \|) \quad a.e.\ t \in [0, d].$$

Then $\Sigma^F_{x_0}[0, d]$ *is a nonempty compact subset of the space* $C([0, d]; E)$.

Proof. The integral multioperator $\Gamma : C([0, d]; E) \to Kv(C([0, d]; E))$ on the interval $[0, d]$, given by formula (5.1.10) or (5.2.6), is as before, u.s.c. and ν-condensing on bounded sets, and again $\Sigma^F_{x_0}[0, d] = \mathrm{Fix}\,\Gamma$.

We introduce the equivalent norm in the space $C([0, d]; E)$ given by

$$\| x \|_* = \max_{t \in [0,d]} e^{-Lt} \| x(t) \|, \tag{5.2.7}$$

where L is chosen so that

$$\max_{t \in [0,d]} M \int_0^t e^{-L(t-s)} \alpha(s) \, ds \leq q < 1, \tag{5.2.8}$$

where M is the constant from the estimate (5.2.4). In the space $C([0, d]; E)$ with the norm (5.2.7) consider the ball

$$\overline{B}_r(0) = \{x \in C([0, d]; E) : \| x \|_* \leq r\}$$

where $r > 0$ is chosen so that

$$r \geq M(\| x_0 \| + \| \alpha \|_{L^1_+([0,d])})(1 - q)^{-1}.$$

Note that the last inequality implies

$$M(\| x_0 \| + \| \alpha \|_{L^1_+([0,d])}) + qr \leq r. \tag{5.2.9}$$

We show now that the multioperator Γ maps the ball $\overline{B}_r(0)$ into itself. In fact, for $x \in \overline{B}_r(0)$ and $y \in \Gamma(x)$ we have

$$e^{-Lt}\| y(t) \| \leq e^{-Lt}\| \exp\{At\}x_0 \| + e^{-Lt}\int_0^t \| \exp\{A(t-s)\} \| \cdot \| f(s) \| \, ds$$

where $f \in \mathcal{P}_F(x)$ and so, keeping in mind (5.2.8) and (5.2.9), we obtain

$$\begin{aligned} e^{-Lt}\| y(t) \| &\leq M\| x_0 \| + e^{-Lt}M \cdot \int_0^t \alpha(s)(1 + \| x(s) \|) \, ds \\ &\leq M(\| x_0 \| + \| \alpha \|_{L^1_+([0,d])}) + e^{-Lt}M\int_0^t e^{Ls}\alpha(s)e^{-Ls}\| x(s) \| \, ds \\ &\leq M(\| x_0 \| + \| \alpha \|_{L^1_+([0,d])}) + M\| x \|_* \int_0^t e^{-L(t-s)}\alpha(s) \, ds \\ &\leq M(\| x_0 \| + \| \alpha \|_{L^1_+([0,d])}) + qr \leq r \end{aligned}$$

and therefore $\| y \|_* \leq r$.

According to Corollary 3.3.1 the set $\Sigma^F_{x_0}[0, d]$ is nonempty.

It is also apriori bounded. In fact, if $x \in \Sigma^F_{x_0}[0, d]$ then for $t \in [0, d]$ we have that

$$\| x(t) \| \leq \| \exp\{At\} \| \cdot \| x_0 \| + \int_0^t \| \exp\{A(t-s)\} \| \cdot \| f(s) \| \, ds$$

where $f \in \mathcal{P}_F(x)$. So we obtain

$$\begin{aligned} \| x(t) \| &\leq M\| x_0 \| + M\int_0^t \alpha(s)(1 + \| x(s) \|) \, ds \\ &\leq M\left(\| x_0 \| + \| a \|_{L^1} + \int_0^t \alpha(s)\| x(s) \| \, ds\right). \end{aligned}$$

We can use a Gronwall type inequality to get

$$\| x(t) \| \leq M(\| x_0 \| + \| \alpha \|_{L^1})e^{M\| \alpha \|_{L^1}} = N.$$

To complete the proof it remains only to apply Proposition 3.5.1. □

We want to clarify now the question of differentiability properties of mild solutions of the Cauchy problem (5.2.1), (5.2.2). The following two propositions are simple corollaries of Theorems 4.1.3 and 4.1.4 respectively.

Proposition 5.2.1. *Suppose that the space E is reflexive, $x_0 \in D(A)$ and the following hypotheses hold on an interval $[0, h]$ where $0 < h \le d$:*

(AF1) $F(t, E) \subseteq D(A)$ *for a.e.* $t \in [0, h]$;

(AF2) *for any function* $y(\cdot) \in C([0, h]; E)$ *and* $f \in \mathcal{P}_F(y)$ *the function* $t \to Af(t)$ *is summable.*

Then every mild solution $x(\cdot)$ of the problem (5.2.1), (5.2.2) *on an interval* $[0, h]$ *is a Carathéodory solution, i.e., it is absolutely continuous and satisfies the inclusion* (5.2.1) *almost everywhere on* $[0, h]$.

Proposition 5.2.2. *A function $x(\cdot) \in C([0, h]; E)$, $0 < h \le d$, is a mild solution of the problem* (5.2.1), (5.2.2) *if and only if $x(\cdot)$ satisfies the initial condition* (5.2.2) *and for every $v \in D(A^*)$ the function $\langle x(t), v\rangle$ is absolutely continuous on* $[0, h]$ *and satisfies the differential equation*

$$\frac{d}{dt}\langle x(t), v\rangle = \langle x(t), A^* v\rangle + \langle f(t), v\rangle$$

for a.e. $t \in [0, h]$ *with* $f \in \mathcal{P}_F(x)$.

To conclude this section we give two examples of semilinear differential inclusions satisfying our basic assumptions.

The first example is a mathematical model of the transmission line with non linear parameters and it is motivated by a work of S. S. Ceron and O. Lopes [59] (see also [114]). The second example deals with a hybrid system with dry friction in anisotropic environment; it is based on a paper of D. Bothe [49].

Example 5.2.1. Let us consider the transmission lines whose electric schemes have the following forms:

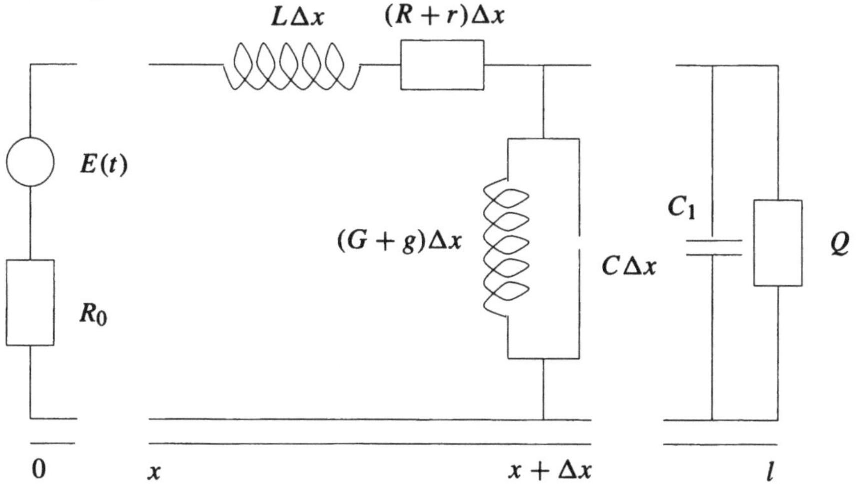

Figure 1a

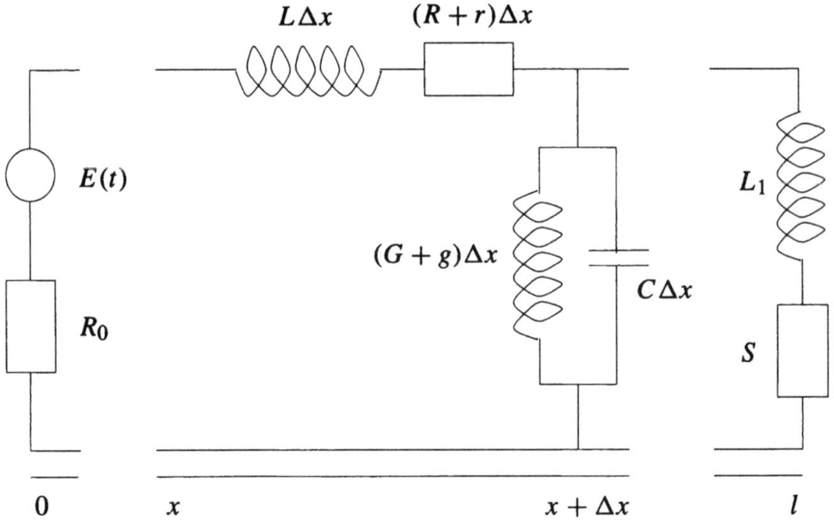

Figure 1b

Voltamper characteristics of the systems can be presented on the following diagrams:

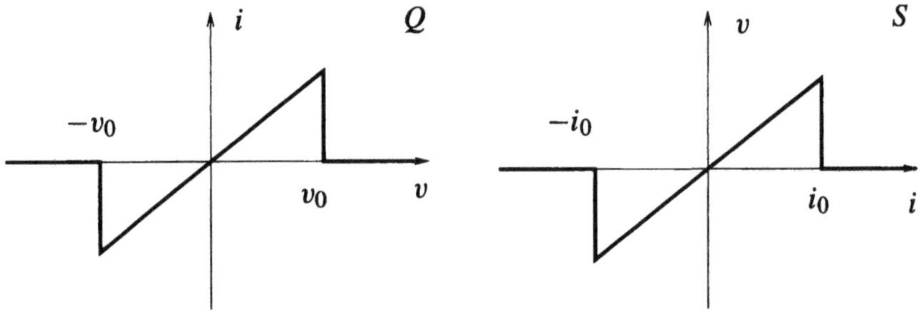

Figure 2

It is clear that both the functions Q and S can be treated as u.s.c. multimaps. Here $\pm v_0$ and $\pm i_0$ are threshold values for a voltage and for a current, respectively, for which the line should be disconnected.

Denoting current and voltage by $i(t,x)$ and $v(t,x)$ respectively, we can describe the mathematical model of the transmission line by the system

$$L\frac{\partial i(t,x)}{\partial t} = -\frac{\partial v(t,x)}{\partial x} - \left(R + r\left(\int_0^l i^2(t,y)\,dy\right)\right) i(t,x)$$

$$C\frac{\partial v(t,x)}{\partial t} = -\frac{\partial i(t,x)}{\partial t} - \left(G + g\left(\int_0^l i^2(t,y)\,dy\right)\right) v(t,x).$$

The line is characterized by the parameters (all per unit length): inductance L, resistance R, capacitance C, and leakage conductance G. The function $r : \mathbb{R}_+ \to \mathbb{R}_+$ expresses the dependence of the resistance (for unit length) of the line on the power of the signal; the function $g : \mathbb{R}_+ \to \mathbb{R}_+$ describes the dependence of the leakage conductance. Both functions r and g are assumed continuous and bounded.

In the first case the boundary conditions are given by the relations

$$E(t) = v(t,0) + R_0 i(t,0) \quad \text{at } x = 0$$
$$v(t,l) \in C_1 \frac{\partial}{\partial t} v(t,l) + Q(v((t,l)) \quad \text{at } x = l.$$

In the second case the last inclusion has the form

$$i(t,l) \in L_1 \frac{\partial}{\partial t} i(t,l) + S(i((t,l)) \quad \text{at } x = l.$$

We suppose that the function $E(t)$ is smooth. The change of variables

$$\begin{cases} I = i, \\ V - v - E, \end{cases}$$

makes the boundary conditions at $x = 0$ homogeneous. In these new variables we have the system

$$\frac{\partial I(t,x)}{\partial t} = -\frac{1}{L}\frac{\partial V(t,x)}{\partial x} - \frac{1}{L}\left(R + r\left(\int_0^l I^2(t,y)\,dy\right) I(t,x)\right.$$
$$\frac{\partial V(t,x)}{\partial t} = -\frac{1}{C}\frac{\partial I(t,x)}{\partial x} - \frac{G}{C}V(t,x)$$
$$- \frac{1}{C} g\left(\int_0^l I^2(t,y)\,dy\right)[V(t,x) + E(t)] - \frac{G}{C}E(t) - E'(t)$$

with boundary conditions:

$$V(t,0) + R_0 I(t,0) = 0 \quad \text{at } x = 0$$

and, for the first case,

$$I(t,l) \in C_1 \frac{\partial}{\partial t} V(t,l) + Q(V(t,l) + E(t)) + C_1 E'(t) \quad \text{at } x = l;$$

while for the second case, we have

$$V(t,l) \in L_1 \frac{\partial}{\partial t} I(t,l) + S(I(t,l)) + E(t) \quad \text{at } x = l.$$

We can interpret both systems as a semilinear differential inclusion

$$X'(t) \in AX(t) + F(t, X(t))$$

in a Hilbert space $E = L^2[0, l] \times L^2[0, l] \times \mathbb{R}$ with a scalar product given as

$$\langle(\varphi_1, \psi_1, b_1), (\varphi_2, \psi_2, b_2)\rangle = \int_0^l \varphi_1\varphi_2\, dx + \int_0^l \psi_1\psi_2\, dx + b_1 b_2.$$

In the first case the operator A is defined by

$$A\begin{pmatrix}\varphi\\ \psi\\ b\end{pmatrix} = \begin{pmatrix} -R/L & (-1/L)D_x & 0\\ (-1/C)D_x & -G/C & 0\\ (1/C_1)\Gamma & 0 & 0\end{pmatrix}\begin{pmatrix}\varphi\\ \psi\\ b\end{pmatrix}$$

where $D_x = \frac{d}{dx}$ and Γ is the evaluation operator $\Gamma(w) = w(l)$. The domain of A can be given as

$$D(A) = \{(\varphi, \psi, b) \in E : \varphi, \psi \in H^1[0, l],\ \psi(0) + R_0\varphi(0) = 0,\ b = \psi(l)\}$$

(cf. Example 4.3.1).

In the second case the operator A is defined by

$$A\begin{pmatrix}\varphi\\ \psi\\ b\end{pmatrix} = \begin{pmatrix} -R/L & (-1/L)D_x & 0\\ (-1/C)D_x & -G/C & 0\\ 0 & (1/L_1)\Gamma & 0\end{pmatrix}\begin{pmatrix}\varphi\\ \psi\\ b\end{pmatrix}$$

with domain

$$D(A) = \{(\varphi, \psi, b) \in E : \varphi, \psi \in H^1[0, l],\ \psi(0) + R_0\varphi(0) = 0,\ b = \varphi(l)\}.$$

The multivalued nonlinearity $F(t, X)$, $X = (\varphi, \psi, b)$ in the first case has a form

$$\begin{aligned} F(t, X) = \Bigg[&-\frac{1}{L} r\Big(\int_0^l \varphi^2(y)\, dy\Big)\varphi, \\ &-\frac{1}{C} g\Big(\int_0^l \varphi^2(y)\, dy\Big)[\psi + E(t)] - \frac{G}{C}E(t) - E'(t), \\ &-\frac{1}{C_1} Q(b + E(t)) - E'(t)\Bigg]. \end{aligned}$$

In the second case the first two components of $F(t, X)$ are the same, and the third is equal to

$$-\frac{1}{L_1} S(b) - E(t)$$

It is not difficult to show that A is the infinitesimal generator of a C_0-semigroup $\exp\{At\}$. Moreover, as we know from Example 4.3.1, this semigroup is sequentially condensing in the equivalent norm.

Let us verify that in the both cases the multioperator $F(t, X)$ satisfies the conditions (F1)–(F4). The hypotheses (F1)–(F3) follow easily from the properties of continuity and boundedness of the corresponding functions and multifunctions.

Since the third component of $F(x, t)$ is one-dimensional and bounded, to prove property (F4), it is sufficient to note that the map $G : L^2[0, l] \times L^2[0, l] \to L^2[0, l]$ defined as

$$\sigma(\varphi, \psi) = h\left(\int_0^l \varphi^2(y)\, dy\right)\psi(x)$$

where $h : \mathbb{R}_+ \to \mathbb{R}_+$ is a continuous bounded function, satisfy the relation

$$\chi_{L^2}(\sigma(\Omega_1, \Omega_2)) \le k\chi_{L^2}(\Omega_2)$$

for any bounded sets $\Omega_1, \Omega_2 \subset L^2[0, l]$ where χ_{L^2} is the Hausdorff MNC in $L^2[0, l]$ and $k = \sup_{t\in\mathbb{R}_+} h(t)$.

In fact for, $\varepsilon > 0$, let the points $y_1, \dots, y_n$ form an ε-net of the set Ω_2. Note that for each y_i, $i = 1, \dots, n$, the set $\sigma(\Omega_1, y_i)$ is relatively compact since it is one-dimensional and bounded. Further, for any $\varphi \in \Omega_1$, $\psi \in \Omega_2$ and y_i such that $\| \psi - y_i \|_{L^2} \le \varepsilon$ the estimate

$$\| \sigma(\varphi, \psi) - \sigma(\varphi, y_i) \|_{L^2} \le k\| \psi - y_i \|_{L^2} \le k\varepsilon$$

can be easily verified. Hence the relatively compact set $\bigcup_{i=1}^n \sigma(\Omega_1, y_i)$ forms a $k\varepsilon$-net of the set $\sigma(\Omega_1, \Omega_2)$

Example 5.2.2. Consider the model illustrated by the figure below.

A block of mass m lies on a rough surface and is connected to a fixed support by a spring.

We assume that the spring moves in an anisotropic environment in presence of dry friction on the block.

To get a semilinear inclusion as a model, consider the Banach space

$$E = \{X = (\varphi, \psi, b_1, b_2) \in H^1[0, 1] \times L^2[0, 1] \times \mathbb{R}^2 : \varphi(0) = 0,\ \varphi(1) = b_1\}$$

with norm given by

$$\| X \| = \left(\alpha \int_0^l \varphi' \, dx + \alpha \int_0^l \psi^2 \, dx + \beta b_1^2 + b_2^2\right)^{1/2}$$

where $\alpha > 0$, $\beta > 0$.

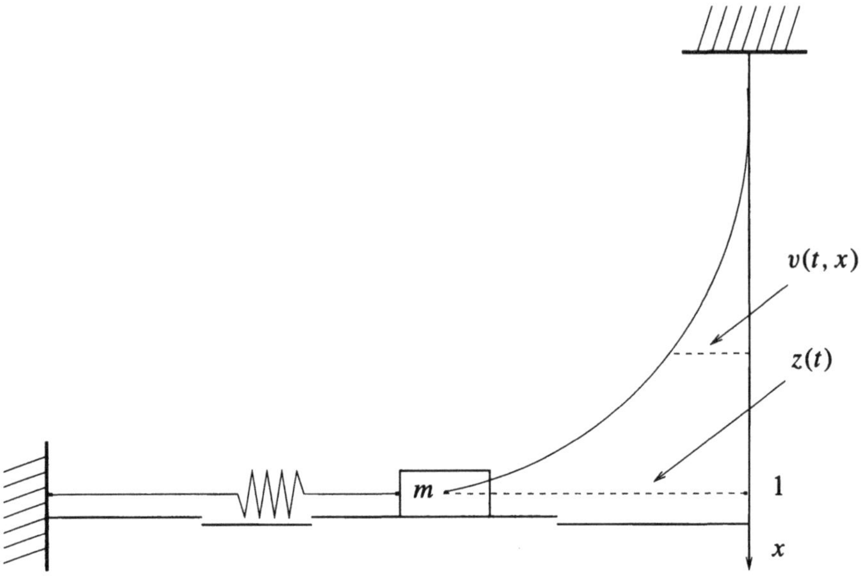

Figure 3

Now set $X(t) = (v(t, \cdot), v_t(t, \cdot), z(t), z'(t))$ where $v(t, x)$ is the elongation of the spring and $z(t)$ denotes the position of the block. The system can be interpreted as a semilinear differential inclusion of the form

$$X'(t) \in A(t)X(t) + F(t, X(t))$$

where

$$A \begin{pmatrix} \varphi \\ \psi \\ b_1 \\ b_2 \end{pmatrix} = \begin{pmatrix} 0 & 1 & 0 & 0 \\ D_x & 0 & 0 & 0 \\ 0 & 0 & 0 & 1 \\ -\alpha \Gamma D_x & 0 & -\beta & 0 \end{pmatrix} \begin{pmatrix} \varphi \\ \psi \\ b_1 \\ b_2 \end{pmatrix}$$

with $D_x = \frac{d}{dx}$ and $\Gamma(w) = w(1)$.

The domain of A is

$$D(A) = \{(\varphi, \psi, b_1, b_2) \in E : \varphi \in H^2[0,1], \psi \in H^1[0,1], \ \psi(0) = 0, \ \psi(1) = b_2\}.$$

As in Example 5.2.1, it is not difficult to see that A generates a C_0-semigroup.

The multivalued nonlinear term F is defined as

$$F(t, X) = (0, \psi W(\varphi), 0, e(t) + G(b_2)).$$

Here $W : C[0,1] \to Kv(C[0,1])$ is a bounded u.s.c. multioperator modelling the effect of anisotropy of the environment as well as the hysteresis phenomena.

Note that multioperators of the same form were considered by J. Appell, E. De Pascale, H. T. Nguyen, and P. P. Zabreiko [14] in connection with the problems of modelling hysteresis introduced by M. A. Krasnoselskii and A. V. Pokrovskii [147].

The function $e(t)$ represent an external force acting on the mass, and the multifunction G is defined as

$$G(s) = \begin{cases} [-\mu_0, \mu_0], & \text{if } s = 0, \\ -\mu(s)\,\mathrm{sign}(s), & \text{if } s \neq 0, \end{cases}$$

where μ_0 and $\mu(\cdot)$ are the coefficients of static and sliding friction, respectively. We assume that functions $e(\cdot)$ and $\mu(\cdot)$ are continuous and bounded and that $\mu(0) \leq \mu_0$.

The hypotheses (F1)–(F3) for F are satisfied due to the continuity and boundedness properties of the functions and multifunctions involved.

To verify (F4), note first that by the Sobolev Imbedding Theorem the space $H^1[0, 1]$ is compactly imbedded into $C[0, 1]$. Therefore the set $W(\Omega)$ is compact in $C[0, 1]$ for any bounded set $\Omega \subset H^1[0, 1]$. Recalling the general scheme of the proof of Proposition 2.2.2 and the previous example we can conclude that for any bounded sets $\Omega_1 \subset H^1[0, 1]$, $\Omega_2 \subset L^2[0, 1]$:

$$\chi_{L^2}(\{\psi W(\varphi) : \varphi \in \Omega_1,\ \psi \in \Omega_2\}) \leq k\chi_{L^2}(\Omega_2)$$

where χ_{L^2} denotes the Hausdorff MNC in $L^2[0, 1]$; $k = \sup_{\varphi \in C[0,1]} \| W(\varphi) \|$. The property (F4) easily follows from the above estimate.

5.2.2 Optimization of semilinear feedback control systems

We consider a semilinear feedback control system of the form

$$x'(t) = Ax(t) + f(t, x(t), u(t)), \quad t \in [0, d]; \tag{5.2.10}$$

$$u(t) \in U(t, x(t)), \quad t \in [0, d], \tag{5.2.11}$$

in a separable Banach space E . The initial condition is

$$x(0) = x_0 \in E. \tag{5.2.12}$$

We study problem (5.2.10)–(5.2.12) under the following assumptions.

The linear operator $A : D(A) \subseteq E \to E$ satisfies the condition (A).

Let E_1 be the separable Banach space of controls. It will be assumed that the nonlinear term

$$f : [0, d] \times E \times E_1 \to E$$

satisfies the following conditions:

(f1) the function

$$f(\cdot, x, u) : [0, d] \to E$$

is measurable for any $(x, u) \in E \times E_1$;

(f2) $\| f(t, x_1, u) - f(t, x_0, u) \| \leq k(t) \cdot \| x_1 - x_0 \|$ for any $x_0, x_1 \in E$; $u \in E_1$ where $k \in L^1_+([0, d])$;

(f3) the map

$$f(t,\cdot,\cdot): E \times E_1 \to E$$

is continuous for a.e. $t \in [0,d]$.

We assume that the feedback multimap

$$U: [0,d] \times E \multimap K(E_1)$$

satisfies the following conditions:

(U1) the multifunction

$$U'(\cdot,x): [0,d] \multimap K(E_1)$$

is measurable for every $x \in E$;

(U2) the multimap

$$U(t,\cdot): E \multimap K(E_1)$$

is u.s.c. for a.e. $t \in [0,d]$;

(U3) the multimap U is superpositionally measurable (see Theorem 1.3.4 and Proposition 1.3.1);

(U4) the set

$$F(t,x) = f(t,x,U(t,x))$$

is convex for all $(t,x) \in [0,d] \times E$;

(U5) the multimap F satisfies the boundedness condition (F3′) of Theorem 5.2.2;

(U6) for every $(t,x) \in [0,d] \times E$ the set

$$f(t,x,U(t,D))$$

is relatively compact for any bounded set $D \subset E$.

Definition 5.2.2. A pair $(x(\cdot),u(\cdot))$ where $x \in C([0,d];E)$, $u: [0,d] \to E_1$ is measurable, is said to be a *mild solution of the problem* (5.2.10)–(5.2.12) if

$$x(0) = x_0$$

$$x(t) = \exp\{At\}x_0 + \int_0^t \exp\{A(t-s)\}f(s,x(s),u(s))\,ds, \quad t \in [0,d],$$

where

$$u(s) \in U(s,x(s)), \quad s \in [0,d].$$

The function $x(\cdot)$ is called a *mild trajectory of the system* and the function $u(\cdot)$ is the *control function* or simply the *control.*

We can prove the following optimization result.

Theorem 5.2.3. *Let* $j : C([0,d]; E) \to \mathbb{R}$ *be a lower semicontinuous functional. Then under conditions* (A), (f1)–(f3) *and* (U1)–(U6) *there exists a mild solution* $(x_*(\cdot), u_*(\cdot))$ *of the problem* (5.2.10)–(5.2.12) *such that*

$$j(x_*) = \min_{x \in \Sigma_{x_0}} j(x),$$

where Σ_{x_0} *is the set of all mild trajectories of the system with the initial value* $x(0) = x_0$.

Proof. We prove that besides condition (F3′) the multimap $F(t,x) = f(t,x,U(t,x))$ defined on $[0,d] \times E$ satisfies conditions (F1), (F2) and (F4). First, we note that from conditions (f3) and (U4) it follows that the multimap F takes its values in $Kv(E)$.

Moreover, from conditions (f1), (f3) and (U1) it follows that the multimap $F(\cdot, x)$ is measurable for every $x \in E$ (see Theorem 1.3.4) hence condition (F1) is fulfilled.

Condition (F2) follows from assumptions (f3), (U2) and Theorem 1.2.8.

The multimap F satisfies condition (F4). In fact, fix $t \in [0,d]$ and consider the multimap $B : E \times E \to K(E)$ defined as

$$B(x,y) = f(t,y,U(t,x)).$$

From condition (U6) it follows directly that the set $B(D,y)$ is relatively compact for every $y \in E$ and bounded $D \subset E$. Fix $x \in E$ and let $y', y'' \in E$. If $b' \in B(x,y')$ then $b' = f(t,y',u)$, where $u \in U(t,x)$. Consider $b'' = f(t,y'',u) \in B(x,y'')$. From condition (f2) it follows that

$$\| b'' - b' \| \le k(t) \| y'' - y' \|$$

i.e., the multimap $B(x,\cdot)$ is $k(t)$-Lipschitz with respect to the Hausdorff metric on $K(E)$. Applying Proposition 2.2.2 to the multimap B we obtain that condition (F4) is satisfied for the multimap $F(t,x) = B(x,x)$.

From Theorem 5.2.2 it follows that there exists a nonempty compact set of functions $x \in C([0,d]; E)$ of the form

$$x(0) = x_0$$

$$x(t) = \exp\{At\}x_0 + \int_0^t \exp\{A(t-s)\}g(s)\,ds, \quad t \in [0,d],$$

where $g(s) \in F(s,x(s)) = f(s,x(s),U(s,x(s)))$ is a measurable selection. Using condition (U3) and applying the Filippov Implicit Function Lemma (Theorem 1.3.3) we conclude that every such function is a trajectory of the system (5.2.10)–(5.2.12). Let x_* be the trajectory minimizing the functional j and let u_* be the corresponding control. The pair $(x_*(\cdot), u_*(\cdot))$ is the required optimal solution of the problem (5.2.10)–(5.2.12). □

5.2.3 On the index of the solutions set

Under conditions of Theorems 5.2.1 and 5.2.2 the multioperator Γ maps a ball into itself, hence the index of the set of solutions of the problem (5.2.1), (5.2.2) (see Section 3.3) is equal to 1. We prove that this is a generic property when all solutions are extendable to the whole interval $[0, d]$ and the solutions set is bounded on $[0, d]$. To get the result we need to prove that it is possible to modify a given inclusion with an arbitrary multimap F to an inclusion with a bounded multimap $\widetilde{F}$ preserving the solutions set of the initial problem.

It is convenient to write the condition for the extendability of all solutions to the interval $[0, d]$ in the following form

(P) $$\Sigma_{x_0}^F[0, \tau] = \Sigma_{x_0}^F[0, d]\big|_{[0,\tau]} \quad \text{for every } \tau \in (0, d].$$

Under conditions (A), (F1)–(F4) if the set $\Sigma_{x_0}^F[0, d]$ is bounded in the space $C([0, d]; E)$, then it is compact by Proposition 3.5.1. Let us consider the compact set

$$Z = \{(t, y) : \exists\, x \in \Sigma_{x_0}^F[0, d],\ x(t) = y\} \tag{5.2.13}$$

in the metric space $[0, d] \times E$. For $r > 0$ let $\kappa : [0, d] \times E \to [0, 1]$,

$$\kappa(t, y) = \begin{cases} 1, & \text{if } (t, y) \in \overline{W_r}(Z), \\ 0, & \text{if } (t, y) \notin W_{2r}(Z), \end{cases}$$

be an Urysohn function (here $W_a(Z)$ denotes an a-neighborhood of the set Z). Consider the new multimap

$$\widetilde{F}(t, x) = \kappa(t, x) F(t, x). \tag{5.2.14}$$

Lemma 5.2.1. *Let us suppose that F satisfies conditions* (A), (F1)–(F4) *and* (P) *and the set $\Sigma_{x_0}^F[0, d]$ is bounded in the space $C([0, d]; E)$. Then the multimap $\widetilde{F} : [0, d] \times E \to Kv(E)$ satisfies conditions* (F1)–(F4) *and*

$$\Sigma_{x_0}^{\widetilde{F}}[0, d] = \Sigma_{x_0}^F[0, d].$$

Proof. The multimap $\widetilde{F}$ satisfies conditions (F1)–(F2) as the function κ is continuous. The boundedness of κ implies property (F3). To verify (F4) we note that

$$\kappa(t, \Omega) F(t, \Omega) \subseteq \overline{\text{co}}(F(t, \Omega) \cup \{0\})$$

for any bounded set $\Omega \subset E$ and hence

$$\chi(\widetilde{F}(t, \Omega)) = \chi(\kappa(t, \Omega) F(t, \Omega)) \le \chi(F(t, \Omega)) \le k(t)\chi(\Omega)$$

for a.e. $t \in [0, d]$.

Further, note that the inclusion

$$\Sigma_{x_0}^{\widetilde{F}}[0,d] \supseteq \Sigma_{x_0}^{F}[0,d]$$

is obvious. We will show that

$$\Sigma_{x_0}^{\widetilde{F}}[0,d] \subseteq \Sigma_{x_0}^{F}[0,d].$$

Supposing the contrary, there exists $y \in \Sigma_{x_0}^{\widetilde{F}}[0,d]$ such that $y \notin \Sigma_{x_0}^{F}[0,d]$. Then we can find $r_0 < r$ and $t_* \in [0,d]$ such that $\operatorname{dist}(y(t), \Sigma_{x_0}^{F}[0,d](t)) < r_0$ for $t \in [0,t_*)$ and

$$\operatorname{dist}(y(t_*), \Sigma_{x_0}^{F}[0,d](t_*)) = r_0. \tag{5.2.15}$$

Note that $\kappa(t, y(t)) \equiv 1$ for $t \in [0,t_*]$ and therefore $y \in \Sigma_{x_0}^{F}[0,t_*]$, contradicting to condition (P) since by virtue of (5.2.15) we have $y \notin \Sigma_{x_0}^{F}[0,d]\big|_{[0,t_*]}$. □

Theorem 5.2.4. *Suppose that conditions of Lemma* 5.2.1 *hold. Then*

$$\operatorname{ind}(\Sigma_{x_0}^{F}[0,d]; \Gamma) = 1.$$

Proof. Consider an arbitrary open neighborhood U of the set $\Sigma_{x_0}^{F}[0,d]$ in the space $C([0,d]; E)$. From the compactness of the set $\Sigma_{x_0}^{F}[0,d]$ it follows that there exists $r > 0$ such that $W_r(\Sigma_{x_0}^{F}[0,d]) \subset U$. For this r, we construct the multimap $\widetilde{F}$ by formula (5.2.14). By Lemma 5.2.1 the multimap $\widetilde{F}$ satisfies conditions (F1)–(F4) and hence the corresponding integral multioperator $\widetilde{\Gamma}$ will be ν-condensing (see Theorem 5.1.4). As it was shown in Theorem 5.2.1 the multioperator $\widetilde{\Gamma}$ maps a ball $B_R(x_0)$ into itself for R large enough. Without loss of generality we can suppose that $W_r(\Sigma_{x_0}^{F}[0,d]) \subset B_R(x_0)$. Note that

$$\widetilde{\Gamma}\big|_{\overline{W_r(\Sigma_{x_0}^{F}[0,d])}} = \Gamma\big|_{\overline{W_r(\Sigma_{x_0}^{F}[0,d])}}.$$

In fact, if $x \in \overline{W_r}(\Sigma_{x_0}^{F}[0,d])$ then $(t, x(t)) \in \overline{W_r}(Z)$ where Z is given by the formula (5.2.13). Therefore $\kappa(t, x(t)) \equiv 1$ and hence $F(t, x(t)) \equiv \widetilde{F}(t, x(t))$. Using the properties of the topological degree (see Section 3.2) we obtain

$$\deg(i - \Gamma, \overline{U}) = \deg(i - \Gamma, \overline{W_r}) = \deg(i - \widetilde{\Gamma}, \overline{W_r}) = \deg(i - \widetilde{\Gamma}, \overline{B_R}(x_0)) = 1. \ \square$$

5.2.4 Continuous dependence of the solutions set on parameters and initial data

We investigate the dependence of the solutions of the parameterized problem on the parameter λ belonging to a metric space (Λ, ρ) :

$$x'(t) \in Ax(t) + F(t, x(t), \lambda), \tag{5.2.16}$$

$$x(0) = x_0. \tag{5.2.17}$$

As usual, we assume that the condition (A) holds and that the multivalued nonlinearity $F : [0, d] \times E \times \Lambda \to Kv(E)$ satisfies conditions (F1$_\lambda$)–(F4$_\lambda$) of Section 5.1.

Theorem 5.2.5. *Assume conditions* (A) *and* (F1$_\lambda$)–(F4$_\lambda$). *Suppose that for some value* $\lambda_0 \in \Lambda$ *of the parameter, the solutions set* $\Sigma_{x_0}^{F(\cdot,\cdot,\lambda_0)}[0, d]$ *is bounded and condition* (P) *is satisfied for* $\lambda = \lambda_0$*:*

$$\Sigma_{x_0}^{F(\cdot,\cdot,\lambda_0)}[0, \tau] = \Sigma_{x_0}^{F(\cdot,\cdot,\lambda_0)}[0, d]\,\big|_{[0,\tau]} \qquad \forall\, \tau \in (0, d]. \tag{5.2.18}$$

Then for every $r > 0$ *there exists* $\delta > 0$ *such that for all* $\lambda \in \Lambda$ *satisfying* $\rho(\lambda_0, \lambda) < \delta$ *the set* $\Sigma_{x_0}^{F(\cdot,\cdot,\lambda)}[0, d]$ *is nonempty and*

$$\Sigma_{x_0}^{F(\cdot,\cdot,\lambda)}[0, d] \subset W_r(\Sigma_{x_0}^{F(\cdot,\cdot,\lambda_0)}[0, d]),$$

i.e., the multimap $\lambda \multimap \Sigma_{x_0}^{F(\cdot,\cdot,\lambda)}[0, d]$ *is u.s.c. at* λ_0.

Proof. Let $C > 0$ be a constant such that

$$\left\| \Sigma_{x_0}^{F(\cdot,\cdot,\lambda_0)}[0, d] \right\| \le C, \tag{5.2.19}$$

and $M = \max_{t\in[0,d]} \| \exp\{At\} \|$. We show that there exists $\delta_0 > 0$ such that for all $\lambda \in \Lambda$ satisfying $\rho(\lambda_0, \lambda) < \delta_0$ we have

$$\left\| \Sigma_{x_0}^{F(\cdot,\cdot,\lambda)}(t) \right\| \le 3C, \quad t \in [0, d]. \tag{5.2.20}$$

Here $\Sigma_{x_0}^{F(\cdot,\cdot,\lambda)}$ denotes the set of all solutions of the problem (5.2.16)–(5.2.17), i.e., $x \in \Sigma_{x_0}^{F(\cdot,\cdot,\lambda)}$ if there exists $\Delta > 0$ such that for any $\tau < \Delta$ and $y^\tau = x|_{[0,\tau]} \in C([0, \tau]; E)$ we have $y^\tau \in \Gamma(\lambda, y^\tau)$ where $\Gamma(\lambda, \cdot)$ is the integral multioperator in the space $C([0, \tau]; E)$ generated by $F(\cdot, \cdot, \lambda)$ (see (5.1.11)).

In fact, assuming that inequality (5.2.20) fails, we can find sequences $\lambda_n \in \Lambda$, $\lambda_n \to \lambda_0$, $t_n \in [0, d]$, $x_n \in C([0, t_n]; E)$ such that

$$x_n(t) \in \Gamma(\lambda_n, x_n)(t) \quad \text{for } t \in [0, t_n]$$

and

$$\operatorname{dist}(x_n(t_n), \Sigma_{x_0}^{F(\cdot,\cdot,\lambda_0)}(t_n)) = 2C, \tag{5.2.21}$$

$$\operatorname{dist}(x_n(t), \Sigma_{x_0}^{F(\cdot,\cdot,\lambda_0)}(t)) < 2C, \quad t \in [0, t_n). \tag{5.2.22}$$

where dist denotes the distance of a point from a set.

Let $t_* = \underline{\lim}_n\, t_n$. Note that $t_* > 0$ by (F3$_\lambda$). In fact, supposing the contrary, there exist a subsequence of $\{t_n\}$ (the whole sequence $\{t_n\}$, without loss of generality) which tends to zero, then from condition (F3$_\lambda$) and the (5.2.19), (5.2.21), (5.2.22) it follows that selectors $f_n \in S^1_{F(\cdot,x_n(\cdot),\lambda_n)}$ satisfy

$$\| f_n(t) \| \le \mu_{B_{3C}}(t) \quad \text{for a.e. } t \in [0, t_n].$$

Condition (5.2.18) implies that the set $\Sigma_{x_0}^{F(\cdot,\cdot,\lambda_0)}$ is equal $\Sigma_{x_0}^{F(\cdot,\cdot,\lambda_0}[0,d]$ and hence is compact, so we obtain that

$$\operatorname{dist}(x_0, \Sigma_{x_0}^{F(\cdot,\cdot,\lambda_0)}(t_n)) \to 0. \tag{5.2.23}$$

Taking into account (5.2.21) we have that

$$\begin{aligned} 0 < 2C &= \operatorname{dist}(x_n(t_n), \Sigma_{x_0}^{F(\cdot,\cdot,\lambda_0)}(t_n)) \le \| x_n(t_n) - x_0 \| + \operatorname{dist}(x_0, \Sigma_{x_0}^{F(\cdot,\cdot,\lambda_0)}(t_n)) \\ &\le \| (\exp\{At_n\} - I)x_0 \| + \int_0^{t_n} \| \exp\{A(t_n - s)\} \| \cdot \| f_n(s) \| \, ds \qquad (5.2.24) \\ &\quad + \operatorname{dist}(x_0, \Sigma_{x_0}^{F(\cdot,\cdot,\lambda_0)}(t_n)). \end{aligned}$$

Note that the first term on the right-hand side of the last inequality tends to zero, while $n \to \infty$, since $\exp\{At\}$ is a C_0-semigroup, and the last one vanishes by (5.2.23). The second term can be estimated as follows:

$$\int_0^{t_n} \| \exp\{A(t_n - s)\} \| \cdot \| f_n(s) \| \, ds \le \int_0^{t_n} M \cdot \mu_{B_{3C}}(s) \, ds$$

and also tends to zero. Passing to the limit in (5.2.24) we obtain the desired contradiction.

We prove now that there exists $\alpha > 0$ such that all the solutions x_n are defined on the interval $[0, t_* - \alpha]$ and for every x_n there exists a point $t_n' \in [0, t_* - \alpha]$ at which the inequality

$$\operatorname{dist}((x_n(t_n'), \Sigma_{x_0}^{F(\cdot,\cdot,\lambda_0)}(t_n')) \ge C/2M, \tag{5.2.25}$$

holds. The first part of this claim follows from the definition of t_*. To prove the second part, note that if at any point $t_1 \in [0, t_n)$ for some $x^0 \in \Sigma_{x_0}^{F(\cdot,\cdot,\lambda_0)}$ we have

$$\left\| x_n(t_1) - x^0(t_1) \right\| < C/2M,$$

then for $t_1 + \tau \in [0, t_n]$ the following estimate holds:

$$\begin{aligned} \left\| x_n(t_1 + \tau) - x^0(t_1 + \tau) \right\| &\le \left\| \exp\{A\tau\} \right\| \cdot \left\| x_n(t_1) - x^0(t_1) \right\| \qquad (5.2.26) \\ &\quad + 2 \int_{t_1}^{t_1+\tau} \left\| \exp\{A(t_1 + \tau - s)\} \right\| \cdot \mu_{B_{3C}}(s) \, ds. \end{aligned}$$

From this estimate we have for small τ that

$$\left\| x_n(t_1 + \tau) - x^0(t_1 + \tau) \right\| < C. \tag{5.2.27}$$

If we assume that the required α does not exist then from inequality (5.2.22) and the choice of t_* it follows that

$$\left\|x_n(t_n) - x^0(t_n)\right\| < C,$$

contradicting (5.2.21).

The family of integral multioperators $\Gamma : \Lambda \times C([0, t_* - \alpha]; E) \to Kv(C([0, t_* - \alpha]; E))$ of problem (5.2.16)–(5.2.17) is ν-condensing in the second variable (Theorem 5.1.5) and hence the sequence $\{x_n |_{[0,t_*-\alpha]}\}$ is relatively compact. Taking the limit point x^* of this sequence and noting that $\{t'_n\}$ tends to t^* we have

$$x^*(t) \in \Gamma(\lambda_0, x^*)(t) \quad \text{for } t \in [0, t_* - \alpha]$$

and

$$\operatorname{dist}(x^*(t^*), \Sigma_{x_0}^{F(\cdot,\cdot,\lambda_0)}(t^*)) \geq C/2M. \tag{5.2.28}$$

By condition (5.2.18) the solution x^* of the problem (5.2.16), (5.2.17) can be extended to the interval $[0, d]$, which contradicts inequality (5.2.28).

To conclude the proof we apply the abstract continuous dependence result (Proposition 3.5.2) taking the ball $\overline{B}_{3C}(0)$ as the set X. □

Remark 5.2.1. Note that in the proof of the above theorem we used the condition of upper semicontinuity of the family Γ and the fact that it is ν-condensing in the second argument, rather than condition (F2$_\lambda$).

The used approach allows us to consider the initial value x_0 as a parameter. Thus we can obtain the following statements.

Corollary 5.2.1. *Under conditions* (A), (F1)–(F4) *and* (P) *let the set* $\Sigma_{x_0}^F[0, d]$ *be bounded in the space* $C([0, d]; E)$. *Then the multimap*

$$y \in E \multimap \Sigma_y^F[0, d] \subset C([0, d]; E)$$

is defined in some neighborhood of the initial value x_0 *and is u.s.c.*

Proof. By the above remark it is sufficient to note that for any convergent sequence $\{y^n\}$ of initial conditions the corresponding family of integral multioperators $\Gamma : \{y^n\} \times C([0, d]; E) \to Kv(C([0, d]; E))$ is u.s.c. and ν-condensing since the sequence $\{\exp\{At\}y^n\}$ is convergent uniformly for $t \in [0, d]$. □

Corollary 5.2.2. *If conditions* (A), (F1), (F2), (F4) *and the boundedness condition* (F3′) *of Theorem* 5.2.2 *hold, then the multimap* $\Sigma^F : E \to K(C([0, d]; E))$,

$$y \in E \multimap \Sigma_y^F[0, d]$$

is defined on the whole space E and is u.s.c.

Proof. The fact that $D(\Sigma^F) = E$, condition (P) and the compactness of every set $\Sigma_y^F[0, d]$ follow from the existence Theorem 5.2.2. □

5.3 The structure of the solutions set

In this section we study the topological structure of the set $\Sigma_{x_0}^F[0,d]$ of all mild solutions of the Cauchy problem for a semilinear differential inclusion in a separable Banach space E:

$$x'(t) \in Ax(t) + F(t, x(t)) \tag{5.3.1}$$

$$x(0) = x_0. \tag{5.3.2}$$

In fact we will prove the following.

Theorem 5.3.1. *Suppose that the set $\Sigma_{x_0}^F[0,d]$ is bounded in the space $\mathcal{C} = C([0,d]; E)$, under assumptions* (A), (F1)–(F4) *and* (P). *Then this set is an R_δ-subset of the space $\mathcal{C}$ (see Definition* 3.4.2*).*

To prove this we need a few preliminaries. First of all, we recall that from Proposition 3.5.1 it follows that $\Sigma_{x_0}^F[0,d]$ is a compact subset of $\mathcal{C}$. Further, redefining the multimap $F(t,x)$ by formula (5.2.14) and using Lemma 5.2.1 we can assume, without loss of generality, that F satisfies

(F 3″) $$\| F(t,x) \| \le \gamma(t)$$

for a.e. $t \in [0,d]$ and $x \in E$ where $\gamma(\cdot) \in L_+^1[0,d]$.

Now we can prove the following proposition.

Lemma 5.3.1. *Under the above assumptions there exists a nonempty compact convex subset $X \subset \mathcal{C}$ such that:*

(i) $x(0) = x_0$ *for all* $x \in X$*;*

(ii) $\exp\{At\}x_0 + \int_0^t \exp\{A(t-s)\} \overline{\mathrm{co}}\, F(s, X(s))\, ds \subseteq X(t)$ *for* $t \in [0,d]$ *where* $X(t) = \{x(t) : x \in X\}$.

Proof. First we construct a decreasing sequence of closed convex sets $\{X^n\}_{n=1}^\infty \subset \mathcal{C}$ by the following inductive process. Let

$$X^0 = \{x \in \mathcal{C} : x(0) = x_0,\ \| x \|_{\mathcal{C}} \le N\}$$

where $N = M(\| x_0 \| + \| \gamma \|_{L^1})$, $M = \max_{t\in[0,d]} \| \exp\{At\} \|$.

Then $X^n = \overline{Y^n}$, $n \ge 1$, where Y^n is defined as

$$Y^n = \left\{ y \in \mathcal{C} : y(t) = \exp\{At\}x_0 + \int_0^t \exp\{A(t-s)\} f(s)\, ds,\ f \in S^1_{\overline{\mathrm{co}}\, F(\cdot, X^{n-1}(\cdot))} \right\}.$$

Let us note that all X^n, $n \geq 1$ are nonempty since $\Sigma^F_{x_0}[0,d] \subset X^n$ for all $n \geq 0$.

In the space $\mathcal{C}$ we consider the following measure of noncompactness

$$\psi(\Omega) = \sup_{t\in[0,d]} \left[\chi(\Omega(t)) \cdot \exp\Big(- R \int_0^t k(s)\,ds\Big)\right]$$

where χ is the Hausdorff MNC in E, $R > M$ is an arbitrary number and $k(\cdot)$ is the function from (F4). It is easy to see that the MNC ψ is monotone and nonsingular. Using formula (2.1.2) and condition (F4) we have the following estimate for $0 \leq s \leq t \leq d$:

$$\begin{aligned}\chi(\exp\{A(t-s)\}\,\overline{\mathrm{co}}\,F(s, X^{n-1}(s))) &\leq \|\exp\{A(t-s)\}\| \cdot \chi(\overline{\mathrm{co}}\,F(s, X^{n-1}(s))) \\ &\leq M \cdot \chi(F(s, X^{n-1}(s))) \\ &\leq Mk(s) \cdot \chi(X^{n-1}(s)) \\ &\leq Mk(s) \cdot \psi(X^{n-1}) \cdot \exp\Big(R\int_0^s k(\theta)\,d\theta\Big)\end{aligned}$$

for all $s \in [0,t]$.

Conditions (A) and (F3″) imply that for every $t \in [0,d]$ the multifunction

$$s \multimap \exp\{A(t-s)\}\,\overline{\mathrm{co}}\,F(s, X^{n-1}(s)), \quad s \in [0,t]$$

is bounded and therefore, using Theorem 4.2.3, we have for every $t \in [0,d]$

$$\begin{aligned}\chi(Y^n(t)) &= \chi\Big(\exp\{At\}x_0 + \int_0^t \exp\{A(t-s)\}\,\overline{\mathrm{co}}\,F(s, X^{n-1}(s))\,ds\Big) \\ &= \chi\Big(\int_0^t \exp\{A(t-s)\}\,\overline{\mathrm{co}}\,F(s, X^{n-1}(s))\,ds\Big) \\ &\leq M\psi(X^{n-1}) \cdot \int_0^t k(s) \cdot \exp\Big(R\int_0^s k(\theta)\,d\theta\Big)\,ds \\ &\leq (M/R)\psi(X^{n-1}) \cdot \exp\Big(R\int_0^t k(s)\,ds\Big).\end{aligned}$$

Thus

$$\chi(Y^n(t)) \exp\Big(- R\int_0^t k(s)\,ds\Big) \leq (M/R)\psi(X^{n-1})$$

and

$$\psi(Y^n) \le (M/R)\psi(X^{n-1}).$$

Finally, we have $\psi(X^n) \le (M/R)\psi(X^{n-1})$, $n \ge 1$ and therefore

$$\psi(X^n) \xrightarrow[n\to\infty]{} 0.$$

If we consider the nonempty closed set $\widetilde{X} = \bigcap_{n\ge 1} X^n$, from the monotonicity of the MNC ψ we obtain that $\psi(\widetilde{X}) = 0$, hence $\chi(\widetilde{X}(t)) = 0$. Moreover, condition (F4) implies that

$$\chi(F(t, \widetilde{X}(t))) = 0$$

for all $t \in [0, d]$. The property of the Cauchy operator described in Corollary 5.1.1 implies that the set $X \subseteq \widetilde{X}$, $X = \overline{Y}$,

$$Y = \{y \in C : y(t) = \exp\{At\}x_0 + \int_0^t \exp\{A(t-s)\}f(s)\,ds,\ f \in S^1_{\overline{\mathrm{co}}\,F(\cdot,\widetilde{X}(\cdot))}\}$$

is compact. The set X is the required one. □

Consider the metric projection $P : [0, d] \times E \to Kv(E)$,

$$P(t, x) = \{y \in X(t),\ \| x - y \| = \operatorname{dist}(x, X(t))\}$$

and the multimap $\widehat{F} : [0, d] \times E \to Kv(E)$ defined by

$$\widehat{F}(t, x) = \overline{\mathrm{co}}\, F(t, P(t, x)).$$

Lemma 5.3.2. *The multimap $\widehat{F}$ satisfies conditions* (F1), (F2), (F3″) *and* (F4).

Proof. First of all we note that the multimap $X(\cdot) : [0, d] \to Kv(E)$, $t \multimap X(t)$ is continuous (see Example 1.1.6) hence the multimap P is closed. In fact, let $t_n \to t_0 \in [0, d]$; $x_n \to x_0 \in E$; $y_n \in P(t_n, x_n)$, $y_n \to y_0 \in E$. From the inequality

$$|\operatorname{dist}(x_n, X(t_n)) - \operatorname{dist}(x_0, X(t_0))| \le \| x_n - x_0 \| + h(X(t_n), X(t_0))$$

where h is the Hausdorff metric it follows that

$$\lim_{n\to\infty} \operatorname{dist}(x_n, X(t_n)) = \operatorname{dist}(x_0, X(t_0)).$$

On the other hand,

$$\lim_{n\to\infty} \operatorname{dist}(x_n, X(t_n)) = \lim_{n\to\infty} \| x_n - y_n \| = \| x_0 - y_0 \|$$

so $y_0 \in P(t_0, x_0)$.

From Theorem 1.1.7 it follows that the set $X([0, d])$ is compact so Theorem 1.1.5 implies that the multimap P is u.s.c. Applying Theorems 1.2.8 and 1.2.17 we see that the multimap $\widehat{F}$ satisfies the condition (F2).

For any fixed $x \in E$ the multifunction $P(\cdot, x) : [0, d] \to Kv(E)$ is clearly measurable and hence it has a measurable selection $p(\cdot)$. From Theorem 1.3.5 it follows that the multifunction $F(t, p(t))$ has a measurable selection which implies that the multimap $\widehat{F}$ satisfies the condition (F1).

The multimap $\widehat{F}$ clearly satisfies condition (F3$''$).To verify property (F4) it is sufficient to note that due to Theorem 1.1.7 the multimap $\widehat{F}(t, \cdot)$ is compact for a.e. $t \in [0, d]$. □

The above result implies that the set $\Sigma_{x_0}^{\widehat{F}}[0, d]$ of all mild solutions of the problem

$$\begin{aligned} x'(t) &\in Ax(t) + \widehat{F}(t, x(t)), \quad t \in [0, d] \\ x(0) &= x_0 \end{aligned}$$

is nonempty. Moreover,

Lemma 5.3.3. $\Sigma_{x_0}^{\widehat{F}}[0, d] = \Sigma_{x_0}^{F}[0, d]$.

Proof. In fact, let $x \in \Sigma_{x_0}^{\widehat{F}}[0, d]$. Then

$$\begin{aligned} x(t) &\in \exp\{At\}x_0 + \int_0^t \exp\{A(t-s)\}\widehat{F}(s, x(s))\, ds \\ &= \exp\{At\}x_0 + \int_0^t \exp\{A(t-s)\}\,\overline{\mathrm{co}}\, F(s, P(s, x(s)))\, ds \\ &\subseteq \exp\{At\}x_0 + \int_0^t \exp\{A(t-s)\}\,\overline{\mathrm{co}}\, F(s, X(s))\, ds \subset X(t), \end{aligned}$$

hence $P(t, x(t)) = \{x(t)\}$. Then

$$x(t) = \exp\{At\}x_0 + \int_0^t \exp\{A(t-s)\}f(s)\, ds$$

where $f \in S^1_{\widehat{F}(\cdot, x(\cdot))} = S^1_{F(\cdot, x(\cdot))}$, and so $x \in \Sigma_{x_0}^{F}[0, d]$.

The inclusion $\Sigma_{x_0}^{F}[0, d] \subseteq \Sigma_{x_0}^{\widehat{F}}[0, d]$ follows easily from the observation that $\Sigma_{x_0}^{F}[0, d] \subset X$. □

Now, following [71], [73] and [213] we can prove the following approximation result.

Lemma 5.3.4. *There exists a sequence of multimaps* $\{F_n\}_{n=1}^{\infty}$, $F_n : [0, d] \times E \to Kv(E)$ *such that:*

(i) *each multimap* $F_n(t, \cdot) : E \to Kv(E)$, $n \geq 1$ *is continuous for a.e.* $t \in [0, d]$;

(ii) $\widehat{F}(t, x) \subset \cdots \subset F_{n+1}(t, x) \subset F_n(t, x) \subset \overline{\text{co}}\, F(t, X(t))$, $n \geq 1$;

(iii) $\widehat{F}(t, x) = \bigcap_{n \geq 1} F_n(t, x)$;

(iv) *for each* $n \geq 1$ *there exists a selection* $g_n(t, x) \in F_n(t, x)$ *such that* $g_n(\cdot, x)$ *is measurable and* $g_n(t, \cdot)$ *is locally Lipschitz.*

Proof. Consider the sequence $r_n = 3^{-n}$, $n \geq 1$ and let $\{V_j\}_{j \in J}$ be a locally finite refinement of the cover $\{B_{r_n}(x)\}_{x \in E}$, $\{p_j\}_{j \in J}$ a locally Lipschitz partition of unity subordinated to $\{V_j\}_{j \in J}$. For every $j \in J$ let x_j be such that $V_j \subset B_{r_n}(x_j)$ and define

$$F_n(t, x) = \sum_{j \in J} p_j(x) \cdot \overline{\text{co}}\, \widehat{F}(t, B_{2r_n}(x_j)).$$

Property (i) follows easily from the local finiteness of the cover $\{V_j\}_{j \in J}$. To prove (ii) and (iii) note that $p_j(x) > 0$ implies that $x \in V_j \subset B_{r_n}(x_j)$, hence $B_{2r_n}(x_j) \subset B_{3r_n}(x)$ and therefore

$$\widehat{F}(t, x) \subset F_n(t, x) \subseteq \overline{\text{co}}\, \widehat{F}(t, B_{3r_n}(x)).$$

Let

$$F_{n+1}(t, x) = \sum_{i \in I} \widehat{p}_i(x) \cdot \overline{\text{co}}\, \widehat{F}(t, B_{2r_{n+1}}(\widehat{x}_i))$$

and suppose that $p_j(x) > 0$ and $\widehat{p}_i(x) > 0$. Then $\| x_j - \widehat{x}_i \| < r_n + r_{n+1}$, hence $\overline{\text{co}}\, \widehat{F}(t, B_{2r_{n+1}}(\widehat{x}_i)) \subset \overline{\text{co}}\, \widehat{F}(t, B_{2r_n}(x_j))$ and therefore $F_{n+1}(t, x) \subset F_n(t, x)$.

Finally, to prove (iv) we take, for every x_j, $j \in J$, a measurable selection g_j of the multifunction $\widehat{F}(\cdot, x_j)$ and define $g_n : [0, d] \times E \to E$ as

$$g_n(t, x) = \sum_{j \in J} p_j(x) \cdot g_j(t). \qquad \square$$

Proof of Theorem 5.3.1. From (i), (ii) and (iv) of Lemma 5.3.4 it follows that each F_n satisfies the conditions (F1), (F2), (F3″) and (F4) and hence each set $\Sigma_{x_0}^{F_n}[0, d]$, $n \geq 1$ is nonempty and compact.

We show that the set $\Sigma_{x_0}^{F_n}[0, d]$ is contractible for $n \geq 1$. In fact, let $x_n \in \Sigma_{x_0}^{F_n}[0, d]$ and for any $\tau \in [0, 1)$ the function z_n^{τ} be a unique solution on $[\tau d, d]$ of the integral equation

$$z_n(t) = \exp\{A(t - \tau d)\} x_n(\tau d) + \int_{\tau d}^{t} \exp\{A(t - s)\} g_n(s, z_n(s))\, ds. \tag{5.3.3}$$

The functions

$$w_n^\tau(t) = \begin{cases} x_n(t), & 0 \le t \le \tau d, \\ z_n^\tau(t), & \tau d \le t \le d \end{cases}$$

belong to $\Sigma_{x_0}^{F_n}[0, d]$. Define the deformation $h : [0, 1] \times \Sigma_{x_0}^{F_n}[0, d] \to \Sigma_{x_0}^{F_n}[0, d]$ by the formula

$$h(\tau, x_n) = \begin{cases} w_n^\tau & \text{for } 0 \le \tau < 1, \\ x_n, & \tau = 1. \end{cases}$$

Since $\exp\{At\}$ is a C_0-semigroup and the function g_n is locally Lipschitz in the second argument, the solutions of the problem (5.3.3) depend continuously on (τ, x_n), therefore the deformation h is continuous. But $h(0, \cdot) = z_n^0$ and $h(1, \cdot)$ is the identity, hence $\Sigma_{x_0}^{F_n}[0, d]$ is contractible.

We prove that

$$\Sigma_{x_0}^{\widehat{F}}[0, d] = \bigcap_{n \ge 1} \Sigma_{x_0}^{F_n}[0, d].$$

It is clear that

$$\Sigma_{x_0}^{\widehat{F}}[0, d] \subset \bigcap_{n \ge 1} \Sigma_{x_0}^{F_n}[0, d].$$

Let $x \in \bigcap_{n \ge 1} \Sigma_{x_0}^{F_n}[0, d]$, then for each $n \ge 1$ we have

$$x(t) = \exp\{At\}x_0 + \int_0^t \exp\{A(t-s)\} f_n(s)\, ds$$

where $f_n \in \mathcal{P}_{F_n}(x)$. From (ii) of Lemma 5.3.4 it follows that the sequence $\{f_n\}_{n=1}^\infty$ is semicompact and by Remark 4.2.2 we can assume, without loss of generality, that

$$f_n \underset{w}{\to} f_0 \in L^1([0, d]; E).$$

Condition (iii) of Lemma 5.3.4 implies that $f_0 \in \mathcal{P}_{\widehat{F}}(x)$. Applying Corollary 5.1.1 we see that the sequence of functions

$$\left\{ \exp\{At\}x_0 + \int_0^t \exp\{A(t-s)\} f_n(s)\, ds \right\}_{n=1}^\infty$$

converges to the function $\exp\{At\}x_0 + \int_0^t \exp\{A(t-s)\} f_0(s)\, ds$ from the solutions set $\Sigma_{x_0}^{\widehat{F}}[0, d]$. Since each member of this sequence is equal to x we have $x \in \Sigma_{x_0}^{\widehat{F}}[0, d]$, so

$$\Sigma_{x_0}^{\widehat{F}}[0, d] = \bigcap_{n \ge 1} \Sigma_{x_0}^{F_n}[0, d].$$

Therefore $\Sigma_{x_0}^{F}[0, d] = \Sigma_{x_0}^{\widehat{F}}[0, d]$ is an R_δ-set. □

Corollary 5.3.1. *Under conditions* (A), (F1), (F2), (F4) *and the boundedness condition* (F3′) *of Theorem* 5.2.2 *the set* $\Sigma_{x_0}^F$ *is an* R_δ*-subset of the space* $\mathcal{C}$.

Proof. See the proof of Corollary 5.2.2. □

5.4 The averaging principle

We consider the Cauchy problem for a semilinear differential inclusion in a Banach space E of the form

$$x'(t) \in Ax(t) + F(t/\varepsilon, x(t)) \tag{5.4.1}$$

$$x(0) = x_0 \tag{5.4.2}$$

where ε is a small positive parameter and the linear operator $A : D(A) \subseteq E \to E$ satisfies (A). We assume that, besides condition (F1), the multimap $F : \mathbb{R} \times E \to Kv(E)$ satisfies the following additional assumptions.

(F_T) The multioperator F is T-periodic in the first variable, i.e., for a.e. $t \in \mathbb{R}$ we have that

$$F(t + T, x) = F(t, x)$$

for all $x \in E$ where $T > 0$.

It is clear that this condition implies the existence of a T-periodic measurable selection from the multifunction $F(\cdot, x)$ for every $x \in E$.

(F2′) The multioperator F is u.s.c. in the second variable, uniformly with respect to the first one, i.e., for every $x \in E$ and any $\varepsilon > 0$ there exists $\delta > 0$ such that

$$F(t, B_\delta(x)) \subset W_\varepsilon(F(t, x))$$

for a.e. $t \in \mathbb{R}$.

(F4′) For every bounded set $\Omega \subset E$ we have

$$\chi(F([0, T] \times \Omega)) \leq k \cdot \chi(\Omega)$$

where χ is the Hausdorff MNC in E and $k \geq 0$.

Condition (F4′) implies that the multioperator F is bounded on bounded subsets and then condition (F3) is fulfilled.

Parallel to the inclusion (5.4.1) we consider the averaged inclusion

$$x'(t) \in Ax(t) + F_0(x(t)) \tag{5.4.3}$$

where

$$F_0(x) = \frac{1}{T}\int_0^T F(s, x)\, ds.$$

Applying the properties (F1) and (F4′) we can easily see that every value $F_0(x)$ is a nonempty compact convex set.

Lemma 5.4.1. *The multioperator $F_0 : E \to Kv(E)$ is u.s.c.*

Proof. Let $\{x_n\}_{n=1}^{\infty} \subset E$ be a sequence such that $x_n \to x_0$. If we take arbitrary $y_n \in F_0(x_n)$, $n \geq 1$ then there exist selections $f_n \in \mathcal{P}_F(\overline{x}_n)$ where $\overline{x}_n(t) \equiv x_n$ is such that

$$y_n = \frac{1}{T}\int_0^T f_n(s)\, ds, \quad n \geq 1.$$

The sequence $\{f_n\}_{n=1}^{\infty}$ is semicompact and from Remark 4.2.2 it follows that we can assume without loss of generality that $f_n \underset{w}{\to} f_0 \in \mathcal{P}_F(\overline{x}_0)$ where $\overline{x}_0(t) \equiv x_0$ (see Lemma 5.1.1).

Taking into consideration Remark 4.2.3 and Corollary 5.1.1 we conclude that

$$\frac{1}{T}\int_0^T f_n(s)\, ds \to \frac{1}{T}\int_0^T f_0(s)\, ds,$$

i.e., $y_n \to y_0 \in F_0(x_0)$. □

For every bounded set $\Omega \subset E$ the following estimate holds:

$$\chi(F_0(\Omega)) \leq k \cdot \chi(\Omega) \tag{5.4.4}$$

where k is the constant from the condition (F4′). In fact, for any $x \in \Omega$ and $f \in \mathcal{P}_F(x)$ we have that $f(t) \in F([0, T] \times \Omega)$ for a.e. $t \in [0, T]$. Hence

$$\frac{1}{T}\int_0^T f(s)\, ds \in \overline{\text{co}}\, F([0, T] \times \Omega))$$

and therefore $F_0(\Omega) \subseteq \overline{\text{co}}\, F([0, T] \times \Omega))$ and inequality (5.4.4) now follows from the properties of the Hausdorff MNC χ and the estimate (F4′).

So we can consider the integral multioperator Γ described in Section 5.1 for the inclusion (5.4.3). This multioperator is ν-condensing (see Theorem 5.1.4).

Let us give the following "multivalued" analogue of the Krasnoselskii–Krein lemma ([146]).

Lemma 5.4.2. *Let us consider the sequences* $\{x_n\}_{n=1}^{\infty} \subset C([0,d];E)$ *and* $g_n \in S^1_{\mathcal{F}_n}$, *where* $\mathcal{F}_n(s) = F(\frac{s}{\varepsilon_n}, x_n(s))$, *for a.e.* $s \in [0,d]$. *Suppose that* $\varepsilon_n \to 0$, $x_n \to x^0$, *and* $g_n \underset{w}{\to} g_\infty$. *Then* $g_\infty \in S^1_{\mathcal{F}_0}$, *where* $\mathcal{F}_0(s) = F_0(x^0(s))$ *for a.e.* $s \in [0,d]$.

Proof. First we show that there exists $g^0 \in L^1([0,d];E)$ such that $g^0 \in S^1_{\mathcal{F}_0}$, where $\mathcal{F}_0(s) = F_0(x^0(s))$ and for every linear functional $\phi \in [L^1([0,d];E)]^*$ of the form

$$\phi = \kappa_{[\tau,t]}\varphi, \tag{5.4.5}$$

where $\kappa_{[\tau,t]}$ is the characteristic function of the interval $[\tau,t]$ and $\varphi \in E^*$ we have

$$\phi(g_n) \to \phi(g^0).$$

Given a sequence of partitions of the interval $[0,d]$

$$0 = t_0^m < t_1^m < \cdots < t_{k_m}^m = d \tag{5.4.6}$$

with prescribed points $\bar{t}_i^m \in [t_i^m, t_{i+1}^m)$ set

$$\lambda_m = \max_{0 \le i \le k_m - 1} |t_{i+1}^m - t_i^m|.$$

Let $\{\gamma_m\}$, $\gamma_m \to 0$ be a monotone decreasing sequence. There exists a sequence of partitions of the form (5.4.6), with prescribed points $\bar{t}_i^m$, and a sequence of indices $\{n_m\}$ such that $\lambda_m < \gamma_m$ and

$$g_n(s) \in F\Big(\frac{s}{\varepsilon_n}, x_m^0(s)\Big) + \gamma_m B_E, \quad s \in [0,d],\ n \ge n_m \tag{5.4.7}$$

where $x_m^0(s) \equiv x^0(\bar{t}_i^m)$, $s \in [t_i^m, t_{i+1}^m)$, $i = 0, \ldots, k_m$ and B_E denotes the unit ball in E.

By (F2′) for every $s \in [0,d]$ there exists $\eta(s,m)$ such that for each x satisfying the inequality $\| x - x^0(s) \| < \eta(s,m)$ we have

$$F(t,x) \subset F(t, x^0(s)) + \gamma_m B_E, \quad \text{for a.e. } t \in \mathbb{R}.$$

Then there exists $\delta(s)$ and $n(s)$ such that the inequalities

$$|\tau - s| < \delta(s) \text{ and } n \ge n(s,m)$$

imply for a.e. $t \in \mathbb{R}$

$$F(t, x_n(\tau)) \subset F(t, x^0(s)) + \gamma_m B_E, \quad s \in [0,d].$$

Therefore the interval $[0,d]$ is covered by intervals

$$(s - \delta(s), s + \delta(s)).$$

Choosing a finite subsystem of these intervals and taking the finest partition of the interval $[0, d]$ generated by this system we obtain (5.4.7).

Denote by K the set $\overline{\mathrm{co}}\, F([0,T] \times \overline{\bigcup_{n=1}^{\infty} x_n([0,T])})$. K is a compact set. Set $Q = K - K$ and $Q_m = Q \cap \gamma_m B_E$. Therefore Q_m is a convex compact set. We consider the multioperators $F_m(t,x) = F(t,x) + Q_m$. It is easy to see that for every fixed m the multioperator F_m satisfies (F_T), (F1), (F2′) and (F4′). From (5.4.7) we have that

$$g_n(s) \in F_m\left(\frac{s}{\varepsilon_n}, x_m^0(s)\right), \quad s \in [0,d],\ n \geq n_m. \tag{5.4.8}$$

We study now the properties of the sequence of multioperators

$$F_m^0(x) = \frac{1}{T} \int_0^T F_m(t,x)\, dt.$$

Note that the inclusions $F_{m+1}(t,x) \subset F_m(t,x)$, $m \geq 1$ imply that

$$F_{m+1}^0(x) \subset F_m^0(x) \subset \cdots \subset F_1^0(x), \quad m \geq 1.$$

As for F_0, taking F_1^0 we have

$$\chi(F_1^0(\Omega)) \leq k\chi(\Omega)$$

for any bounded set $\Omega \subset E$. Therefore

$$\chi\Big(\bigcup_{m \geq 1} F_m^0(\Omega)\Big) \leq k\chi(\Omega).$$

In particular, if $\{x_m\}_{m=1}^{\infty}$ is a convergent sequence then the set $\bigcup_m F_m^0(x_m)$ is relatively compact.

The convergence $x_m \to x^0$ implies that for every $\delta > 0$ there exists m_0 such that

$$F_m^0(x_m) \subset F_0(x^0) + \delta B_E \tag{5.4.9}$$

for all $m \geq m_0$.

In fact, if we suppose the contrary there exist δ_0 and $y_{m_k} \in F_{m_k}^0(x_{m_k})$ such that $y_{m_k} \notin F_0(x^0) + \delta_0 B_E$. Without loss of generality we can assume that $y_{m_k} \to y^0$. Hence $\mathrm{dist}(y^0, F_0(x^0)) \geq \delta_0$. Let $y_{m_k} = \frac{1}{T}\int_0^T s_{m_k}(t)\, dt$ where $s_{m_k} \in \mathcal{P}_{F_{m_k}}(x_{m_k})$. The sequence $\{s_{m_k}\}$ is weakly compact so we can assume without loss of generality that $s_{m_k} \underset{w}{\to} s^0$. As before (see Section 5.1) using Mazur's Theorem we get that $s^0(t) \in F(t, x^0)$ for a.e. $t \in [0,d]$. Then

$$y^0 = \frac{1}{T}\int_0^T s^0(t)\, dt \in F_0(x^0)$$

which is a contradiction.

We can show now that if $\{x_m^0\}_{m=1}^{\infty}$ is a sequence of step functions converging uniformly to the function x^0 and $g_m^0(\cdot) \in S_{F_m^0(x_m(\cdot))}$ then the sequence $\{g_m^0\}_{m=1}^{\infty}$ is weakly compact. Moreover if $g_m^0 \underset{w}{\to} g^0$ then

$$g^0 \in S^1_{F_0(x^0(\cdot))}. \tag{5.4.10}$$

In fact by Mazur's Theorem there exists a double sequence $\{\lambda_{ik}\}$ such that $\lambda_{ik} \geq 0$, $\sum_{k=i}^{\infty} \lambda_{ik} = 1$, $\lambda_{ik} = 0$ for $k \geq k_0(i)$ and

$$\sum_{k=i}^{\infty} \lambda_{ik} g_k \underset{L^1}{\to} g^0.$$

Starting from m_0 all functions $g_m^0(s)$ lie in a δ-neighborhood of the set $F_0(x^0(s))$ (see (5.4.9). Therefore inclusion (5.4.10) holds.

Return now to the sequence $\{g_n\}_{n=1}^{\infty}$.

We take a linear functional $\phi \in [L^1([0, d]; E)]^*$ which has the form of (5.4.5). Then

$$\begin{aligned}\phi(g_n) &= \int_{\tau}^{t} \langle \varphi, g_n(s) \rangle \, ds \\ &= \int_{\tau}^{t^m_{i_m(\tau)+1}} \langle \varphi, g_n(s) \rangle \, ds + \sum_{k=i_m(\tau)+1}^{i_m(t)-1} \int_{t_k^m}^{t^m_{k+1}} \langle \varphi, g_n(s) \rangle \, ds + \int_{t_{i_m(t)}}^{t} \langle \varphi, g_n(s) \rangle \, ds\end{aligned}$$

where $i_m(\tau) = \max\{i : t_i^m < \tau\}$. Note that

$$\left| \int_{\tau}^{t^m_{i_m(\tau)+1}} \langle \varphi, g_n(s) \rangle \, ds \right| \leq C\lambda_m \leq C\gamma_m \,, \quad n \geq n_m$$

and

$$\left| \int_{t_{i_m}(t)}^{t} \langle \varphi, g_n(s) \rangle \, ds \right| \leq C\lambda_m \leq C\gamma_m \,, \quad n \geq n_m.$$

For every fixed m we have

$$\int\limits_{t_k^m}^{t_{k+1}^m} \langle \varphi, g_n(s)\rangle\, ds = \varepsilon_n \int\limits_{(t_k^m/\varepsilon_n)}^{(t_{k+1}^m/\varepsilon_n)} \langle \varphi, g_n(\varepsilon_n\tau)\rangle\, d\tau$$

$$= \varepsilon_n \int\limits_{(t_k^m/\varepsilon_n)}^{(p(t_k^m/\varepsilon_n)+1)T} \langle \varphi, g_n(\varepsilon_n\tau)\rangle\, d\tau + \varepsilon_n \sum_{j=p(t_k^m/\varepsilon_n)+1}^{p(t_{k+1}^m/\varepsilon_n)-1} \int\limits_{jT}^{(j+1)T} \langle \varphi, g_n(\varepsilon_n\tau)\rangle\, d\tau$$

$$+ \varepsilon_n \int\limits_{p(t_{k+1}^m/\varepsilon_n)T}^{t_{k+1}^m/\varepsilon_n} \langle \varphi, g_n(\varepsilon_n\tau)\rangle\, d\tau$$

where $p(\tau) = \max\{i : iT < \tau\}$.

Note that

$$\left|\varepsilon_n \int\limits_{(t_k^m/\varepsilon_n)}^{(p(t_k^m/\varepsilon_n)+1)T} \langle \varphi, g_n(\varepsilon_n\tau)\rangle\, d\tau\right| \leq C\varepsilon_n$$

and

$$\left|\varepsilon_n \int\limits_{p(t_{k+1}^m/\varepsilon_n)T}^{t_{k+1}^m/\varepsilon_n} \langle \varphi, g_n(\varepsilon_n\tau)\rangle\, d\tau\right| \leq C\varepsilon_n.$$

Since from (5.4.7) it follows that the function $g_n(\varepsilon_n\tau)$ is the selector of $F_m(\tau, x^0(\overline{t}_k^m))$ when $\tau \in [jT, (j+1)T]$ we have that

$$\widetilde{g}_m(s) = \frac{1}{(p(t_{k+1}^m/\varepsilon_n) - 1) - (p(t_k^m/\varepsilon_n) + 1)} \sum_{j=p(t_k^m/\varepsilon_n)+1}^{p(t_{k+1}^m/\varepsilon_n)-1} g_n(\varepsilon_n(s + jT))$$

is the measurable selection of the multifunction

$$F_m(s, x^0(\overline{t}_k^m)), \quad s \in [0, T].$$

Denote $\dfrac{1}{T}\displaystyle\int_0^T \widetilde{g}_m(s)\, ds$ by $\widetilde{x}_m$. Then

$$\varepsilon_n \sum_{j=p(t_k^m/\varepsilon_n)+1}^{p(t_{k+1}^m/\varepsilon_n)-1} \int\limits_{jT}^{(j+1)T} \langle \varphi, g_n(\varepsilon_n\tau)\rangle\, d\tau$$

$$= \langle \varphi, \widetilde{x}_m\rangle \varepsilon_n (p(t_{k+1}^m/\varepsilon_n) - 1 - (p(t_k^m/\varepsilon_n) + 1))T$$
$$= \langle \varphi, \widetilde{x}_m\rangle (t_{k+1}^m - t_k^m) + \varepsilon_n \nu(n, m, k)$$

where $\| \nu(n, m, k) \| \leq C$.

Therefore if we set $x_m^0(s) = \widetilde{x}_m$, $s \in [t_k^m, t_{k+1}^m)$, then $g_m^0 \in S_{F_m^0(x_m^0(\cdot))}$ and for every fixed m the inequality

$$|\phi(g_n) - \phi(g_m^0)| < 2C\varepsilon_n + 2C\gamma_m, \quad n \geq n_m \tag{5.4.11}$$

holds. As was proved earlier,

$$g_{m_k} \underset{w}{\to} g^0 \in S_{F_0(x^0(\cdot))}.$$

From (5.4.11) it follows that there exists a subsequence $\{g_{n_k}\}$ such that $\phi(g_{n_k}) \to \phi(g^0)$, hence

$$\phi(g_n) \to \phi(g^0).$$

We prove that $g^0 = g_\infty$. Note that $g^0(s), g_\infty(s) \in K$ for $s \in [0, d]$. Let $\{z_k\}_{k=1}^\infty$ be a countable dense subset of Q and $\{\varphi_k\}_{k=1}^\infty$ a sequence of functionals from E^* such that $\| \varphi_k \| = 1$ and $\langle \varphi_k, z_k \rangle = \| z_k \|$. Consider the linear functionals

$$\phi_k^{\tau,t} = \kappa_{[\tau,t]}\varphi_k \in [L^1([0, d]; E)]^*.$$

Set $\xi(\tau) = g_\infty(\tau) - g^0(\tau)$. Then $0 = \phi_k^{\tau,t}(\xi) = \dfrac{1}{\tau - t}\displaystyle\int_\tau^t \langle \varphi_k, \xi(\tau)\rangle\, d\tau$, so $\langle \varphi_k, \xi(\tau)\rangle \underset{\text{a.e.}}{=} 0$ for all $k \geq 1$. Take $z_{k_p} \to \xi(\tau)$ then $\left\| z_{k_p} \right\| \to \| \xi(\tau) \|$. Since $\langle \varphi_{k_p}, z_{k_p} - \xi(\tau)\rangle \to 0$ we have

$$\langle \varphi_{k_p}, \xi(\tau)\rangle \to \| \xi(\tau) \|$$

and hence $\| \xi(\tau) \| = 0$ a.e. $\tau \in [0, d]$. □

For $\varepsilon > 0$ we set

$$F_\varepsilon(t, x) = F(t/\varepsilon, x).$$

Recall that by the symbol $\Sigma_{x0}^F[0, d]$ we denote the set of all mild solutions of the Cauchy problem

$$x'(t) \in Ax(t) + F(t, x(t))$$
$$x(0) = x_0.$$

Theorem 5.4.1. *Suppose that the multimap F satisfies conditions* (F_T), (F1), (F2′) *and* (F4′), *and the solutions set $\Sigma_{x0}^{F_0}[0, d]$ satisfies condition* (P) *(see Section 5.2.3). Then for any $r > 0$ there exists $\varepsilon_0 > 0$ such that $\Sigma_{x0}^{F_\varepsilon}[0, d] \neq \emptyset$ and*

$$\Sigma_{x0}^{F_\varepsilon}[0, d] \subset W_r(\Sigma_{x0}^{F_0}[0, d])$$

for every $\varepsilon \in (0, \varepsilon_0]$.

Proof. Let us consider the family of multioperators $\Gamma : [0,1] \times C([0,d];E) \to Kv(C([0,d];E))$ defined as

$$\Gamma(\varepsilon, x) = \{y : y(t) = \exp\{At\}x_0 + \int_0^t \exp\{A(t-s)\}g_\varepsilon(s)\,ds,\ g_\varepsilon \in \mathcal{P}_{F_\varepsilon}(x)\}.$$

We prove that Γ is u.s.c. in each point $(0, x)$. In fact, assume that $\varepsilon_n \to 0$, $x_n \to x$ for some sequences $\{\varepsilon_n\}_{n=1}^\infty \subset [0,1]$, $\{x_n\}_{n=1}^\infty \subset C([0,d];E)$. Then for any sequence of selections $g_n \in \mathcal{P}_{F_{\varepsilon_n}}(x_n)$, $n \geq 1$ for a.e. $t \in [0,d]$ the set $\{g_n(t)\}_{n=1}^\infty$ lies in the relatively compact set $F([0,T] \times \{x_n(t)\}_{n=1}^\infty)$ (see condition (F4$'$)). Therefore by Remark 4.2.2 we can assume, without loss of generality, that $g_n \underset{w}{\to} g_\infty$. Applying Lemma 5.4.2 we obtain that $g_\infty \in \mathcal{P}_{F_0}(x)$ proving the upper semicontinuity.

Let us show that the family Γ is ν-condensing with respect to the MNC ν given by formula (2.1.7). In fact, assume that for a bounded set $\Omega \subset E$

$$\nu(\Omega) \leq \nu(\Gamma([0,1] \times \Omega)). \tag{5.4.12}$$

Recall that

$$\nu(\Gamma([0,1]\times\Omega)) = \max_{\mathcal{D}\in\Delta(\Gamma([0,1]\times\Omega))} (\gamma(\mathcal{D}), \mathrm{mod}_C(\mathcal{D}))$$

where the max on the right-hand side is taken over all denumerable subsets $\mathcal{D}$ of the set $\Gamma([0,1]\times\Omega)$, $\gamma(\mathcal{D}) = \sup_{t\in[0,d]} e^{-Lt}\chi(\mathcal{D}(t))$, and mod_C denotes the modulus of equicontinuity (see (2.1.5)). Suppose that the maximum is attained for a subset $\mathcal{D}_0 = \{y_n\}_{n=1}^\infty$. Then there exist sequences $\{\varepsilon_n\}_{n=1}^\infty \subset [0,1]$ and $\{x_n\}_{n=1}^\infty \subset \Omega$ such that

$$y_n(t) = \exp\{At\}x_0 + \int_0^t \exp\{A(t-s)\}g_{\varepsilon_n}(s)\,ds, \quad g_{\varepsilon_n} \in \mathcal{P}_{F_{\varepsilon_n}}(x_n). \tag{5.4.13}$$

From inequality (5.4.12) we have that

$$\gamma(\{x_n\}_{n=1}^\infty) \geq \gamma(\{y_n\}_{n=1}^\infty). \tag{5.4.14}$$

Note that

$$e^{-Lt}\chi(\{y_n(t)\}_{n=1}^\infty) \leq e^{-Lt}\chi\left(\left\{\int_0^t \exp\{A(t-s)\}g_{\varepsilon_n}(s)\,ds\right\}_{n=1}^\infty\right). \tag{5.4.15}$$

From the monotonicity property of the Hausdorff MNC and property (F4$'$) it follows that

$$\chi(\{g_{\varepsilon_n}(s)\}_{n=1}^\infty) \leq \chi(F([0,T]\times\{x_n(s)\}_{n=1}^\infty) \leq k\cdot\chi(\{x_n(s)\}_{n=1}^\infty),$$

then we have

$$\chi(\{\exp\{A(t-s)\}g_{\varepsilon_n}(s)\}_{n=1}^{\infty}) \le kMe^{Lt} \sup_{\theta\in[0,d]} e^{-L\theta}\chi(\{x_n(\theta)\}_{n=1}^{\infty})$$
$$= kMe^{Lt}\gamma(\{x_n\}_{n=1}^{\infty})$$

where M is the constant from the estimate (5.2.4) for the semigroup $\exp\{At\}$.

The function on the right-hand side of the last inequality is continuous and therefore, applying Theorem 4.2.2 we have

$$e^{-Lt}\chi\left(\left\{\int_0^t \exp\{A(t-s)\}g_{\varepsilon_n}(s)\,ds\right\}_{n=1}^{\infty}\right) \le \left(2kMe^{-Lt}\int_0^t e^{Ls}\,ds\right)\gamma\left(\{x_n\}_{n=1}^{\infty}\right)$$

Now from (5.4.14), (5.4.15) and the last inequality we get

$$\gamma(\{x_n(t)\}_{n=1}^{\infty}) \le (2kM/L)\gamma(\{x_n\}_{n=1}^{\infty})$$

implying

$$\chi(\{x_n(t)\}_{n=1}^{\infty}) \equiv 0. \qquad (5.4.16)$$

The boundedness of the multioperator F on bounded sets implies that the values of the selections g_{ε_n} are uniformly bounded. From the condition (F4′) and the relation (5.4.16) it follows that the set $F([0,T]\times\{x_n(t)\}_{n=1}^{\infty})$ is relatively compact and hence the sequence $\{g_{\varepsilon_n}(t)\}_{n=1}^{\infty}$ is also relatively compact for a.e. $t\in[0,d]$. Then the sequence $\{g_{\varepsilon_n}\}_{n=1}^{\infty}$ is semicompact and from Corollary 5.1.1 we know that the sequence of functions $\left\{\int_0^t \exp\{A(t-s)\}g_{\varepsilon_n}(s)\,ds\right\}_{n=1}^{\infty}$ is relatively compact in the space $C([0,d];E)$. From the inequality (5.4.12) it follows that

$$\mathrm{mod}_C\left(\{x_n\}_{n=1}^{\infty}\right) \le \mathrm{mod}_C\left(\{y_n\}_{n=1}^{\infty}\right) = \mathrm{mod}_C\left(\{Gg_{\varepsilon_n}\}_{n=1}^{\infty}\right) = 0$$

where $Gg_{\varepsilon_n}(t) = \int_0^t \exp\{A(t-s)\}g_{\varepsilon_n}(s)\,ds$. So $\nu(\{x_n\}_{n=1}^{\infty}) = 0$ and hence the sequence $\{x_n\}_{n=1}^{\infty}$ is relatively compact. This implies that the sequence $\{y_n\}_{n=1}^{\infty}$ defined by (5.4.13) is also relatively compact. Therefore the left-hand side of the inequality (5.4.12) is equal to zero, the set Ω is relatively compact and the family Γ is ν-condensing.

Recalling Remark 5.2.1 we can complete the proof following the scheme of Theorem 5.2.5. □

5.5 On inclusions with almost lower semicontinuous nonlinearities

In this section, using the continuous selection theorem for multimaps with decomposable values, we discuss briefly the existence of solutions for a semilinear differential inclusion whose linear part generate a noncompact semigroup, and whose multivalued nonlinearity is almost lower semicontinuous and nonconvex-valued. We prove also a result on the topological structure of the solutions set.

Consider the Cauchy problem for a semilinear differential inclusion in a separable Banach space E

$$x'(t) \in Ax(t) + F(t, x(t)), \quad t \in [0, d] \tag{5.5.1}$$

$$x(0) = x_0 \tag{5.5.2}$$

under the assumption that the linear part A satisfies the condition (A) and the multimap $F : [0, d] \times E \to K(E)$ satisfies the following condition of *almost lower semicontinuity* (a.l.s.c.):

(F_L) there exists a sequence of disjoint compact sets $\{I_n\}$, $I_n \subseteq [0, d]$ such that: (i) $\text{meas}([0, d] \setminus I) = 0$ where $I = \bigcup_n I_n$; (ii) the restriction of F on each set $J_n = I_n \times E$ is l.s.c. (cf. [212], [73], [52]).

We assume also that F satisfies the condition of boundedness (F3) and χ-regularity condition (F4).

Proposition 1.3.1 states that for every function $x(\cdot) \in C([0, h]; E)$, $0 < h \leq d$ the multifunction $F(t, x(t))$ is measurable (and integrable, as follows from property (F3)), therefore the superposition multioperator

$$\mathcal{P}_F : C([0, h]; E) \to P(L^1([0, h]; E))$$

is correctly defined. We conclude that the concepts of mild solution and integral multioperator for the problem (5.5.1)–(5.5.2) have the same sense as above.

We can prove the following local existence theorem for the problem (5.5.1)–(5.5.2).

Theorem 5.5.1. *Under conditions* (A), (F_L), (F3), (F4) *there exists a mild solution* $x(\cdot) \in C([0, h]; E)$ *of the problem* (5.5.1)–(5.5.2) *for some* h, $0 < h \leq d$.

Proof. Using the strong continuity of the semigroup $\exp\{At\}$ and condition (F3) we can, analogously to what was done in the proof of Theorem 5.2.1, find h, $0 < h \leq d$ and a ball $\mathcal{B} = \overline{B}_\delta(x_0(\cdot))$ where $x_0(\cdot)$ is the function identically equal to x_0, such that the integral multioperator Γ_F of the problem (5.5.1)–(5.5.2) maps $\mathcal{B}$ into itself.

Consider now the MNC ψ in $C([0, h]; E)$

$$\psi(\Omega) = \sup_{t \in [0,h]} \left\{ \chi(\Omega(t)) \exp\left(-R \int_0^t k(s)\, ds \right) \right\}$$

where $R > M = \sup_{t \in [0,h]} \| \exp\{At\} \|$ and $k(\cdot)$ is the function from condition (F4) (cf. Section 5.3). As in the proof of Lemma 5.3.1, we obtain the following estimate for every bounded $\Omega \subset B$ and $0 \leq s \leq t \leq h$:

$$\chi(\exp\{A(t-s)\}F(s, \Omega(s))) \leq Mk(s)\psi(\Omega)\exp\left(R\int_0^s k(\theta)\,d\theta\right)$$

then, using Theorem 4.2.3, we get that for every t, $0 \leq t \leq h$:

$$\chi(\Gamma_F(\Omega(t))) \leq (M/R)\psi(\Omega)\exp\left(R\int_0^t k(s)\,ds\right)$$

hence

$$\psi(\Gamma_F(\Omega)) \leq (M/R)\psi(\Omega),$$

i.e., the multioperator Γ_F is $(M/R, \psi)$-condensing on $\mathcal{B}$.

Taking into account that, in general, the MNC ψ is not regular we conclude from Remark 2.2.2 that there exists a non-empty fundamental set $T \subset B$ of Γ_F such that

$$\psi(T) = 0$$

and

$$\chi(T(t)) = 0 \quad \text{for all } t \in [0, h].$$

Condition (F4) implies

$$\chi(F(t, T(t)) = 0 \quad \text{for all } t \in [0, h]$$

and using the property of the Cauchy operator proved in Corollary 5.1.1 we obtain that $\Gamma_F(T)$ is a relatively compact subset of $\mathcal{B}$. Now

$$X = \overline{\text{co}}\,\Gamma_F(T)$$

is the compact convex subset of $\mathcal{B}$ invariant with respect to the action of Γ_F.

To continue the proof we need the following notion.

Definition 5.5.1. A nonempty subset $\mathcal{M} \subset L^1([0, h]; E)$ is said to be *decomposable* if for every $f, g \in \mathcal{M}$ and each measurable subset m in $[0, h]$,

$$f \cdot \kappa_m + g \cdot \kappa_{[0,h]\setminus m} \in \mathcal{M}.$$

It is clear that the superposition multioperator $\mathcal{P}_F$ has closed and decomposable values. We want to show that it is l.s.c. To this aim we need some auxiliary results.

Lemma 5.5.1. *For every function $x^0(\cdot) \in C([0, h]; E)$ and every $\varepsilon > 0$ there exists $\delta > 0$ such that*

$$F(t, x^0(t)) \subset W_\varepsilon(F(t, x(t)) \quad \textit{for a.e. } t \in [0, h]$$

for any $x(\cdot) \in C([0, h]; E)$, $\|x(\cdot) - x^0\| < \delta$.

Proof. Since the multifunction $F(t, x^0(t))$ is measurable, using Theorems 1.3.1 (f) and 1.1.9, we get that for any $\rho > 0$ there exists a closed subset $m_\rho \subset I$, meas$\{I \setminus m_\rho) < \rho$ and a number $\delta > 0$ such that

$$h\big(F(t_1, x^0(t_1)), F(t_2, x^0(t_2))\big) < \varepsilon/2 \tag{5.5.3}$$

for all $|t_1 - t_2| < \delta$, $t_1, t_2 \in m_\rho$ where h is the Hausdorff metric.

Since the multimap F is a.l.s.c., we can cover the compact set $D = \{(t, x^0(t)) : t \in [0, h]\} \subset [0, h] \times E$ by a finite number of balls $B_i = \{(t, x) : |t_i - t| < \delta_i, \| x^0(t_i) - x \|_E < \delta_i\}$, $t_i \in I$, $i = 1, \dots, m$; $0 < \delta_i \leq \delta$ so that

$$F(t_i, x^0(t_i)) \subset W_{\varepsilon/2} F(t, x) \tag{5.5.4}$$

for all $(t, x) \in B_i \cap (\bigcup_n J_n)$. Then the inclusions (5.5.3) and (5.5.4) imply that

$$F(t, x^0(t)) \subset W_{\varepsilon/2}\big(F(t_i, x^0(t_i))\big) \subset W_{\varepsilon/2} F(t, x) \tag{5.5.5}$$

for $(t, x) \in B_i \cap (\bigcup_n I_n)$, $t \in m_\rho$. From the arbitrariness of ρ it follows that the inclusion (5.5.5) holds for a.e. $t \in [0, h]$, $(t, x) \in B_i$. To find the required number $\delta > 0$ it is sufficient to choose a δ-neighborhood of a compact D which is contained in $\bigcup_{i=1}^m B_i$. □

Lemma 5.5.2. *If $x(\cdot) \in C([0, h]; E)$ and $z : [0, h] \to E$ is measurable, then there exists $y(\cdot) \in \mathcal{P}_F(x)$ such that*

$$\| z(t) - y(t) \|_E = \operatorname{dist}(z(t), F(t, x(t))) \quad \textit{for a.e. } t \in [0, h].$$

Proof. The proof of this statement is not complicated and is based on the property that the intersection of measurable multifunctions is measurable (cf. Corollary 1.3.1 (a)). Since in Section 1.3 we considered only compact-valued multifunctions we refer the reader to [210], Lemma 2.1.4 or to [73], Proposition 3.5 (b). □

Note that from the previous lemma it follows that for any measurable $z : [0, h] \to E$ we have

$$\operatorname{dist}(z, \mathcal{P}_F(x)) = \int_0^h \operatorname{dist}(z(s), F(s, x(s)))\, ds. \tag{5.5.6}$$

Let $x^0(\cdot) \in C([0, h]; E)$ be an arbitrary function and $\varepsilon > 0$. In accordance with Lemma 5.5.1, let $\delta > 0$ be such that

$$F(t, x^0(t)) \subset W_{\varepsilon/h}(F(t, x(t)) \quad \text{for a.e. } t \in [0, h]$$

for any $x(\cdot) \in C([0, h]; E)$, $\| x(\cdot) - x^0(\cdot) \| < \delta$. If we take an arbitrary $z \in \mathcal{P}_F(x^0)$ from Lemma 5.5.2 and equality (5.5.6) it follows that $\operatorname{dist}(z, \mathcal{P}_F(x)) < \varepsilon$ and hence

$$\mathcal{P}_F(x^0) \subset W_\varepsilon(\mathcal{P}_F(x)).$$

To conclude that the multimap $\mathcal{P}_F$ is l.s.c. it remains only to apply Theorem 1.1.8 (ii) (b). (Note that in the last statement the compactness of the values of the multimap is not essential).

We need now the following continuous selection result which is due to A. Fryszkowski [90] (a proof can be found in [73], Lemma 9.1).

Lemma 5.5.3. *Let X be a compact metric space, Y a Banach space, $Z = L^1([0, d]; Y)$. Then every l.s.c. multimap $\mathcal{F} : X \to C(Z)$ with decomposable values has a continuous selection.*

Applying this result to the restriction of the superposition multimap $\mathcal{P}_F$ to X we see that it has a continuous selection $p : X \to L^1([0, h]; E)$. Consider the map $\gamma : X \to X$ defined by

$$\gamma(x)(t) = \exp\{At\}x_0 + \int_0^t \exp\{A(t-s)\}p(x)(s)\,ds.$$

Since the Cauchy operator is continuous, the map γ is also continuous and therefore it is a continuous selection for the integral multioperator Γ_F. From the Schauder fixed point theorem it follows that Fix $\gamma \neq \emptyset$ and hence Fix $\Gamma_F \neq \emptyset$ proving Theorem 5.5.1.

□

Remark 5.5.1. It is clear that changing the boundedness condition (F3) to condition (F3′) of Theorem 5.2.2 we can obtain in the same manner a result on the existence of a mild solution of the problem (5.5.1)–(5.5.2) on the whole interval $[0, d]$.

Let us consider the question of the topological structure of the solutions set for the problem (5.5.1)–(5.5.2). We prove that this set has the classical Kneser connectedness property.

Suppose that a linear part of the inclusion (5.5.1) satisfies condition (A) and the multivalued nonlinearity F satisfies condition (F_L) of the almost lower semicontinuity. Moreover, assume for F the following slightly more strict conditions of boundedness and χ-regularity:

($F3_L$) there exists a constant $K > 0$ such that

$$\| F(t, x) \| \leq K(1 + \| x \|)$$

for a.e. $t \in [0, d]$;

($F4_L$) there exists a function $k(\cdot) \in L^1_+[0, d]$ such that $\lim_{\tau \to +0} \chi(F(J_{t,\tau} \times \Omega) \leq k(t) \cdot \chi(\Omega)$ for a.e. $t \in [0, d]$ for all bounded $\Omega \subset E$, where χ is the Hausdorff MNC in E and $J_{t,\tau} = [t - \tau, t + \tau] \cap [0, d]$.

From Remark 5.5.1 it follows that under conditions (A), (F_L), ($F3_L$) and ($F4_L$) the set $\Sigma_{x_0}^F = \Sigma_{x_0}^F[0,d]$ of all mild solutions of the problem (5.5.1)–(5.5.2) is nonempty. Our aim is to prove the following statement concerning the topological structure of this set.

Theorem 5.5.2. *Under conditions* (A), (F_L), ($F3_L$) *and* ($F4_L$) *the set* $\Sigma_{x_0}^F$ *is connected. In particular each set* $P(t) = \{x(t) : x \in \Sigma_{x_0}^F\}$, $t \in [0,d]$ *is also connected.*

The proof can be divided into the following main stages.

Step 1. Let us consider the following multimap
$\widehat{F} : [0,d] \times E \to Cv(E)$,

$$\widehat{F}(t,x) = \bigcap_{\varepsilon > 0} F^{\varepsilon}(t,x)$$

where $F^{\varepsilon}(t,x) = \overline{\text{co}}\{F(s,y) : |s-t| < \varepsilon,\ \| y - x \| < \varepsilon\}$.

We state that the multimap $\widehat{F}$ satisfies our usual χ-regularity condition:

$$\chi(\widehat{F}(t,\Omega)) \leq k(t) \cdot \chi(\Omega) \quad \text{for a.e. } t \in [0,d] \tag{5.5.7}$$

for any bounded set $\Omega \subset E$. In fact for any $t \in [0,d]$ for which the estimation ($F4_L$) holds, take arbitrary $\delta > 0$ and choose τ, $0 < \tau \leq \delta$ such that

$$\begin{aligned} \chi(F(J_{t,\tau} \times W_{\delta}(\Omega)) &\leq k(t) \cdot \chi(W_{\delta}(\Omega)) + \delta \\ &\leq k(t) \cdot (\chi(\Omega) + \delta) + \delta. \end{aligned}$$

Now

$$\begin{aligned} \chi(\widehat{F}(t,\Omega)) &\leq \chi(F^{\tau}(t,\Omega)) \leq \chi(F(J_{t,\tau} \times W_{\delta}(\Omega)) \\ &\leq k(t) \cdot (\chi(\Omega) + \delta) + \delta \end{aligned}$$

and the estimate (5.5.7) follows from the arbitrariness of δ.

As in Lemma 5.3.1 we can construct a nonempty compact convex subset $X \subset C([0,d]); E)$ with the following properties:

$$x(0) = x_0 \quad \text{for all } x \in X; \tag{5.5.8}$$

$$\exp\{At\} \cdot x_0 + \int_0^t \exp\{A(t-s)\} \cdot \overline{\text{co}}\, \widehat{F}(s, X(s))\, ds \subseteq X(t) \quad \text{for } t \in [0,d]; \tag{5.5.9}$$

$$\Sigma_{x_0}^F \subset X. \tag{5.5.10}$$

To verify (5.5.10) it is sufficient to remember that by construction,

$$F(t,x) \subseteq \widehat{F}(t,x) \quad \text{for all } t \in [0,d] \times E. \tag{5.5.11}$$

Step 2. An associated differential inclusion with an u.s.c. nonlinearity. Consider now a compact set $D \subset [0,d] \times E$,

$$D = \{(s,y) : s \in [0,d],\ y = x(s),\ x \in X\}.$$

Let K be the constant from the condition (F3$_L$), as usual

$$M = \max_{t \in [0,d]} \| \exp\{At\} \|,$$

and $r > 0$ is the radius of the ball X^0 in the inductive process of the construction of the set X (see Lemma 5.3.1).

For $(t,x) \in D$, $\varepsilon > 0$ and $N > \mu = MK(1+r)$ consider the set

$$V(t,x,\varepsilon) = \Big\{(s,y) \in D : t \le s < t+\varepsilon,\ \big\|y - e^{A\tau}x\,\big\| \le N(s-t) \\ \text{for some } \tau \in [0, s-t]\Big\}.$$

The following properties are known from [51], [50] (see also [114]).

Lemma 5.5.4. *The family of sets $\{V(t,x,\varepsilon) : (t,x) \in D,\ \varepsilon > 0\}$ form a basis of closed-open neighborhoods for a topology $\mathcal{T}^+$ on D, stronger than usual metric topology.*

Lemma 5.5.5. *The multifunction $F : D \to K(E)$ admits an almost $\mathcal{T}^+$-continuous selection $\gamma : D \to E$ in the sense that γ is $\mathcal{T}^+$-continuous on every set $D_n = J_n \cap D$.*

Now for $(t,x) \in D_n$ set

$$G_n(t,x) = \bigcap_{\varepsilon>0} G_n^{\varepsilon}(t,x) = \bigcap_{\varepsilon>0} \overline{\text{co}}\{\gamma(s,y) : (s,y) \in D_j,\ |s-t| < \varepsilon,\ \| y - x \| < \varepsilon\},$$

and, for $(t,x) \in D$ define

$$G(t,x) = \begin{cases} G_n(t,x), & \text{if } (t,x) \in D_n, \\ \{\gamma(t,x)\}, & \text{if } (t,x) \notin \bigcup_n D_n. \end{cases}$$

Note that from the construction it clearly follows that

$$\gamma(t,x) \in G(t,x) \tag{5.5.12}$$

and

$$G(t,x) \subseteq \widehat{F}(t,x) \tag{5.5.13}$$

for all $(t,x) \in D$.

Observe that from condition (F4$_L$) and compactness of the sets D_n it follows that for every sequence $\varepsilon_k \to +0$ the corresponding sequence $\chi(F_n^{\varepsilon_k}(t,x))$ tends to zero and hence every $G(t,x)$ is a nonempty compact convex set. Further, we can verify the following statement.

Lemma 5.5.6. *The multimap G is u.s.c. on $\bigcup_n D_n$.*

Proof. To the contrary, suppose that there exist a point $(t, x) \in D_n$, a number $\delta > 0$ and the sequences $(t_k, x_k) \in D_n$, $(t_k, x_k) \to (t, x)$, $y_k \in G_n(t_k, x_k)$ with the property that $\operatorname{dist}(y_k, G_n(t, x)) > \delta$. Without loss of generality we can assume that the sequence

$$\varepsilon_k = 2 \cdot \max\{|t_k - t|, \ |x_k - x|\}$$

is nonincreasing. Then we obtain that $y_k \in G_n^{\varepsilon_k}(t, x)$ and we can consider the decreasing sequence of closed sets

$$L_k = G_n^{\varepsilon_k}(t, x) \cap (E \setminus W_\delta(G_n(t, x))).$$

Every L_k is nonempty since it contains y_k and, as mentioned above, $\chi(L_k) \to 0$. The nonsingularity property of the Hausdorff MNC gives that $\chi(\{y_k\}) = 0$, so the sequence $\{y_k\}$ is relatively compact and has a limit point y. From one side $y \in G_n(t, x)$ and from the other $\operatorname{dist}(y, G_n(t, x)) \geq \delta$ giving the desired contradiction. □

Consider now the following problem on the compact set D

$$x'(t) \in Ax(t) + G(t, x(t)), t \in [0, d] \tag{5.5.14}$$

$$x(0) = x_0. \tag{5.5.15}$$

It is clear that the multimap G satisfies upper Carathéodory conditions. Since the multimap G is also obviously integrably bounded, the integral multioperator Γ_G is defined and u.s.c. on X. Let us study its action on X in more detail.

Choose an arbitrary function $x \in X$. From the construction of X (see Lemma 5.3.1) it is easy to see that for any $\eta > 0$ x can be uniformly η-approximated by the function

$$\widetilde{x}(t) = \exp\{At\} \cdot x_0 + \int_0^t \exp\{A(t-s)\} \cdot v(s)\, ds,$$

where $v(\cdot) \in L^1([0, d]; E)$ and $\| v(t) \| \leq K(1 + r)$ for a.e. $t \in [0, d]$.

Lemma 5.5.7. *The function x satisfies the following condition:*

$$\{(s, y) \in D : t \leq s < t + \varepsilon; \ y \in \overline{B}_{(s-t)(N-\mu)}(x(s))\} \subset V(t, x(t), \varepsilon).$$

In other words, for points s close enough to t, the set $V(t, x(t), \varepsilon)$ is a metric neighborhood of the point $(s, x(s))$ (in the relative topology of the space D).

Proof of Lemma 5.5.7. Let $(s, y) \in D$ and $y \in \overline{B}_{(s-t)(N-\mu)}(x(s))$. Then

$$\begin{aligned}
&\| y - \exp\{A(s-t)\} \cdot x(t) \| \\
&\quad \le \| y - \exp\{A(s-t)\} \cdot \widetilde{x}(t) \| + \| \exp\{A(s-t)\} \cdot \widetilde{x}(t) - \exp\{A(s-t)\} \cdot x(t) \| \\
&\quad \le \| y - \widetilde{x}(s) \| + \| \widetilde{x}(s) - \exp\{A(s-t)\} \cdot \widetilde{x}(t) \| + M\eta \\
&\quad \le (s-t)(N-\mu) + \eta + \Big\| \exp\{A(s-t)\} \cdot \widetilde{x}(t) \\
&\qquad + \int_t^s \exp\{A(s-\theta)\} \cdot v(\theta)\, d\theta - \exp\{A(s-t)\} \cdot \widetilde{x}(t) \Big\| + M\eta
\end{aligned}$$

$$\begin{aligned}
&\le (s-t)(N-\mu) + \int_t^s \| \exp\{A(s-\theta)\} \| \cdot \| v(\theta) \| \, d\theta \\
&\le (s-t)(N-\mu) + (s-t) \cdot \mu + \eta(1+M) \\
&= N(s-t) + \eta(1+M).
\end{aligned}$$

Since η is arbitrary it follows that

$$\| y - \exp\{A(s-t)\} \cdot x(t) \| \le N(s-t). \qquad \square$$

Consider now an arbitrary function $z \in \Gamma_G(x)$. It has the form

$$z(t) = \exp\{At\} \cdot x_0 + \int_0^t \exp\{A(t-s)\} \cdot g(s)\, ds,$$

where $g \in \mathcal{P}_G(x)$.

Lemma 5.5.8. *The following equality holds:*

$$g(t) = \gamma(t, x(t)) \quad \textit{for a.e. } t \in [0, d].$$

Proof. For each index value n, consider the set $I_n^* \subseteq I_n$ consisting of all $t \in I_n$ with the properties that:

(i) $g(t) \in G_n(t, x(t))$,

(ii) there exists a sequence $\{t_k\} \subset I_n$ strictly decreasing to t such that $g(t_k) \in G_n(t, x(t_k))$ for all k and $g(t_k) \to g(t)$.

It is known (see Lemma 2.3 in [51]) that $\operatorname{meas}(I_n \setminus I_n^*) = 0$. Following [50], we prove that $g(t) = \gamma(t, x(t))$ for all $t \in I_n^*$. Using a contradiction argument, let us suppose that there exists a point $t \in I_n^*$ and $\varepsilon > 0$ such that

$$\| g(t) - \gamma(t, x(t)) \| = \varepsilon. \tag{5.5.16}$$

Since γ is $\mathcal{T}^*$-continuous on D_n there exists $\delta > 0$ such that

$$\| \gamma(s, y) - \gamma(t, x(t)) \| < \varepsilon/2$$

for all $(s, y) \in V(t, x(t), \delta)$, $s \in I_n$.

Now let $\{t_k\} \subset I_n$ be a sequence with the properties described in (ii). Choose k_0 such that for $k \geq k_0$,

$$0 < t_k - t < \delta$$

and

$$\| g(t_k) - g(t) \| < \varepsilon/2. \tag{5.5.17}$$

From Lemma 5.5.7 we know that the set $V(t, x(t), \delta)$ is a neighborhood of $(t_k, x(t_k))$ in the usual relative metric topology of D for all $k \geq k_0$. So, let the ω-neighborhood W_ω of $(t_k, x(t_k))$ in D contained in $V(t, x(t), \delta)$.

Then we have for $k \geq k_0$ that

$$\begin{aligned} g(t_k) \in G_n(t_k, x(t_k)) &\subseteq \overline{\text{co}}\{\gamma(s, y) : (s, y) \in W_\omega,\ s \in I_n\} \\ &\subset \overline{\text{co}}\{\gamma(s, y) : (s, y) \in V(t, x(t), \delta),\ s \in I_n\} \\ &\subset B_{\varepsilon/2}(\gamma(t, x(t))), \end{aligned}$$

i.e.,

$$\| \gamma(t, x(t)) - g(t_k) \| < \varepsilon/2. \tag{5.5.18}$$

It is clear that (5.5.17) and (5.5.18) give the contradiction to (5.5.16) proving the lemma. □

From the above lemma it follows that the integral multioperator Γ_G is single-valued on X, and moreover, since it has the form

$$\Gamma_G(x) = \exp\{At\} \cdot x_0 + \int_0^t \exp\{A(t-s)\} \cdot \gamma(s, x(s))\, ds,$$

it is a continuous selection of the integral multioperator Γ_F. Consequently, Γ_G transforms the set X into itself since Γ_F has the same property (see (5.5.9) and (5.5.11)). Moreover, every mild solution of the problem (5.5.14), (5.5.15) is a mild solution of (5.5.1), (5.5.2).

Now we pass from problem (5.5.14), (5.5.15) to the Cauchy problem for a differential inclusion whose right-hand side nonlinearity is defined on the whole $[0, d] \times E$.

To this aim we consider a metric projection $P : [0, d] \times E \to Kv(E)$,

$$P(t, x) = \{y \in X(t),\ \| x - y \| = \text{dist}(x, X(t))\}$$

and the multimap $\widetilde{G} : [0, d] \times E \to Kv(E)$ defined by

$$\widetilde{G}(t, x) = \overline{\text{co}}\, G(t, P(t, x)).$$

Following the lines of Lemma 5.3.2 the next lemma can be proved.

Lemma 5.5.9. *The multimap $\widetilde{G}$ satisfies the following conditions:*

(i) *the multifunction $\widetilde{G}(\cdot, x) : [0, d] \to Kv(E)$ admits a measurable selection for every $x \in E$;*

(ii) *the multimap $\widetilde{G}(t, \cdot) : E \to Kv(E)$ is u.s.c. for a.e. $t \in [0, d]$;*

(iii) *there exists a constant $R > 0$ such that $\| \widetilde{G}(t, x) \| \leq R$ for a.e. $t \in [0, d]$ and $x \in E$;*

(iv) *the multimap $\widetilde{G}(t, \cdot) : E \to Kv(E)$ is compact for a.e. $t \in [0, d]$.*

In particular, the validity of condition (iii) for the multimap $\widetilde{G}$ easily follows from the property $(F3_L)$, the definition of the multimap G and the boundedness of the set X. To justify property (iv) it is sufficient to note that the image of a compact set under the action of a compact-valued u.s.c. multimap is compact (see Theorems 1.1.7 and 1.2.17). □

We can consider now the Cauchy problem on $[0, d] \times E$

$$x'(t) \in Ax(t) + \widetilde{G}(t, x(t)), \ t \in [0, d] \tag{5.5.19}$$

$$x(0) = x_0. \tag{5.5.20}$$

From Lemma 5.5.9 and corresponding existence result (Theorem 5.2.2) we conclude that the set $\Sigma_{x_0}^{\widetilde{G}}$ of all mild solutions of (5.5.19), (5.5.20) is a nonempty compact subset of $C([0, d]; E)$. Moreover, from the result on the topological structure of the solutions set (see Corollary 5.3.1) it follows that the set $\Sigma_{x_0}^{\widetilde{G}}$ is connected.

Keeping in mind relations (5.5.9) and (5.5.13) and following the lines of Lemma 5.3.3, we can prove that the solutions sets of problems (5.5.14), (5.5.15) and (5.5.19), (5.5.20) coincide:

$$\Sigma_{x_0}^{\widetilde{G}} = \Sigma_{x_0}^{G}.$$

Problem (5.5.19), (5.5.20) with u.s.c. nonlinear part is said to be associated with the Cauchy problem (5.5.1), (5.5.2).

Step 3. Connectedness of the integral funnel for (5.5.1), (5.5.2). We pass directly to the proof of Theorem 5.5.2 and to show the connectedness of the set $\Sigma_{x_0}^{F}$.

In fact, let $x^1, x^2 \in \Sigma_{x_0}^{F}$ be mild solutions. They have the form

$$x^i(t) = \exp\{At\} \cdot x_0 + \int_0^t \exp\{A(t-s)\} \cdot f^i(s)\, ds$$

where $f^i \in \mathcal{P}_F(x^i)$, $i = 1, 2$. Consider the multimaps $F^i(t, x) \subseteq F(t, x)$, $i = 1, 2$ defined as

$$F^i(t, x) = \begin{cases} \{f^i(t)\}, & \text{if } x = x^i(t), \\ F(t, x), & \text{if } x \notin x^i(t). \end{cases}$$

Since the function f^i is measurable, there exists a sequence of disjoint compact sets $\{I_k\}$, $I_k \subset [0, d]$ such that $\text{meas}([0, d] \setminus \bigcup_k I_k) = 0$ and the restriction of f^i on each I_k is continuous. Hence each multimap F^i, $i = 1, 2$ is a.l.s.c. and satisfies properties (F3$_L$), (F4$_L$). In accordance to Step 1, we construct a nonempty compact convex subset $X \subset C([0, d]; E)$ containing mild solutions x^1, x^2 and invariant with respect to the action of Γ_F and, consequently Γ_{F^i}, $i = 1, 2$.

Following Step 2, for each of semilinear inclusions with nonlinearities F^1, F^2 let us pass to the associated differential inclusions with u.s.c. nonlinear parts $\widetilde{G}^i$, $i = 1, 2$. Note that from construction (remind (5.5.12)) it follows that

$$f^i(t) = F^i(t, x^i(t)) = \gamma^i(t, x^i(t)) \in \widetilde{G}^i(t, x^i(t))$$

for a.e. $t \in [0, d]$ and $i = 1, 2$, hence each of x^i is the solution of the associated problem with $\widetilde{G}^i$, $i = 1, 2$.

Consider now the parametrized family of semilinear differential inclusions

$$x'(t) \in Ax(t) + \widetilde{G}_\lambda(t, x(t)), \quad t \in [0, d],\ \lambda \in [0, 1] \tag{5.5.21}$$

$$x(0) = x_0 \tag{5.5.22}$$

where the one-parameter family $\widetilde{G}_\lambda$ is defined as

$$\widetilde{G}_\lambda(t, x) = \begin{cases} \widetilde{G}^1(t, x), & \text{if } t \in [0, \lambda d]; \\ \overline{\text{co}}(\widetilde{G}^1(t, x) \cup \widetilde{G}^2(t, x)), & \text{if } t = \lambda d; \\ \widetilde{G}^2(t, x), & \text{if } t \in [\lambda d, d]. \end{cases}$$

From results of Sections 5.2.1 and 5.3 it follows that for every $\lambda \in [0, 1]$ the set $\Sigma_{x_0}^{\widetilde{G}_\lambda}$ of all mild solutions of (5.5.21), (5.5.22) is a nonempty compact connected subset of $C([0, d]; E)$ and it is easy to see that

$$\Sigma_{x_0}^{\widetilde{G}_\lambda} \subset \Sigma_{x_0}^F$$

for all $\lambda \in [0, 1]$.

Moreover, the family (5.5.21), (5.5.22) satisfies the conditions of the theorem on continuous dependence of the solutions set on a parameter (Theorem 5.2.5) so the multimap

$$\lambda \multimap \Sigma_{x_0}^{\widetilde{G}_\lambda}, \quad \lambda \in [0, 1]$$

is u.s.c. and the set $\bigcup_{\lambda \in [0,1]} \Sigma_{x_0}^{\widetilde{G}_\lambda}$ is connected (Proposition 1.1.1). It remains to observe only that $x^1 \in \Sigma_{x_0}^{\widetilde{G}_1}$ and $x^2 \in \Sigma_{x_0}^{\widetilde{G}_0}$.

Bibliographic notes

Semilinear differential inclusions in a Banach space (SDIBS's) have been the object of intensive study by many researchers in recent years. Inclusions of such form appear, in a very natural way, in the general theory of control systems (see, e.g., [1], [2], [3], [6], [12], [46], [131], [133], [135], [180], [182] [190], [192], [210]), in the problem of controlled heat transfer ([12], [46], [64], [180]), in obstacle problems ([60]), in the study of hybrid systems with dry friction ([49]), in the control of a transmission line ([59]), in the general theory of partial differential equations ([194]) and in other fields.

Since the pioneering work of A. Lasota and Z. Opial [158] the method of integral multioperators has been often applied to existence problems for differential inclusions. In the case of inclusions in infinite-dimensional spaces its direct application is complicated by the fact that the integral multioperator is noncompact. Conditions under which the integral multioperator for the Cauchy problem of SDIBS is condensing were considered by N. S. Papageorgiou [187], [189] and V. V. Obukhovskii [180]. Section 5.1 follows the paper of J. F. Couchouron and M. I. Kamenskii [67].

Existence results for SDIBS's arise in the works of C. Castaing and M. Valadier [56] and G. Anichini and P. Zecca [12]. Existence theorems with a χ-regularity condition for the multivalued nonlinearity were obtained in the works of N. S. Papageorgiou [187] (the case of a compact semigroup and Carathéodory nonlinearity) and [189] (the case of an (implicitly supposed) analytic semigroup and measurable nonlinearity). A general existence theorem was given by V. V. Obukhovskii [180] for a semilinear functional differential inclusion with an analytic semigroup and an upper Carathéodory type nonlinearity. The local and global existence results given in Section 5.2 (Theorems 5.2.1 and 5.2.2) are new. They consider a situation where the Banach space is not necessarily separable, the linear part generates a strongly continuous semigroup, and the multivalued nonlinearity satisfies simple and general conditions of boundedness and χ-regularity. These results are close to those in the paper of J. P. Couchouron and M. I. Kamenskii [67]. Note that in the work of M. I. Kamenskii, V. V. Obukhovskii and P. Zecca [136] the method of solutions sets and fixed point theorem for condensing acyclic multimaps were used to prove a local existence theorem for a quasilinear functional differential inclusion in a Banach space. We also mention results on the existence of solutions of SDIBS's with a χ-regularity condition on closed subsets (viability theorems) given in the papers of L. Górniewicz, P. Nistri and V. V. Obukhovskii [108], D. Bothe [49], and R. Bader [21]. Concerning viability results for SDIBS's with a compact semigroup we mention to the works of Shi Shuzhong [205]–[207], H. Frankowska [89], Ya. I. Umanskii [215], [216] and the references therein.

Section 5.5.2 is based on the papers of V. V. Obukhovskii [180] and Yu. G. Borisovich and V. V. Obukhovskii [46].

Continuous dependence of the solutions set on initial data and parameters for SDIBS's with χ-regularity conditions was studied in the works of N. S. Papageorgiou [189], V. V. Obukhovskii [180], A. Andreini, M. Kamenskii and P. Nistri [7]. The results of Section 5.2.4 are new and rather general.

The topological properties of the solutions sets of differential inclusions were studied in a large number of works. We refer the reader to the book [80] and the papers [64] and [214] where one can find detailed references. For SDIBS's this problem was studied in the works of N. S. Papageorgiou [187], M. I. Kamenskii and

V. V. Obukhovskii [135] and G. Conti, V. V. Obukhovskii and P. Zecca [64]. The treatment of this question in Section 5.3 is a generalization (for the case of SDIBS's) of the main result in [64].

The results of Section 5.4 are presented in the work of J. C. Couchouron and M. Kamenskii [68]. The method of proof of the averaging principle for the Cauchy problem based on topological degree theory was suggested by M. I. Kamenskii [128], [129].

The existence results of Section 5.5 are generalizations from the nonlinear case ($A = 0$) of a theorem of A. A. Tolstonogov and I. A. Finogenko [212] (see also Theorem 9.3 of [73]). We note that existence results for differential inclusions in a Banach space with m-accretive or maximal monotone operators generating compact semigroups and perturbed by lower semicontinuous multimaps were given in papers of E. Mitidieri and I. I. Vrabie [169], and A. Bressan and V. Staicu [52].

Theorem 5.5.2 is an extension of the results of A. Bressan [50], [51] and A. Bressan and V. Staicu [52]. We point out that we do not assume that the semigroup $\exp\{At\}$ is compact or contractive and the nonlinearity is not supposed to be bounded. Moreover, we do not pose any additional assumptions on the geometry of a Banach space E and do not suppose reflexivity.

The results of Section 5.5 are based on the papers of M. Kamenskii, V. Obukhovskii and P. Zecca [138], N. S. Papageorgiou [191], and V. Obukhovskii and P. Zecca [184].

Chapter 6
Semilinear inclusions: periodic problems

In this chapter we develop an operator method for the study of periodic problems for semilinear differential inclusions in a Banach space. This method is based on an application of the topological degree theory for condensing multimaps, described in Chapter 3, to the corresponding operator inclusions. The first of these inclusions is studied in the space of continuous periodic functions and the multioperator is of integral type with convex values. The second inclusion will be studied in a Banach phase space. In the latter case we deal with a nonconvex-valued multioperator: the translation multioperator along the trajectories of the differential inclusion. The theoretical foundation for the study of these inclusions by topological degree methods leads to sufficient conditions for the condensivity of these multioperators with respect to convenient measures of noncompactness.

We use also topological methods to justify the averaging principle in periodic problem.

6.1 The integral multioperator for a periodic problem

Our aim in this section is to develop the method of integral multioperators for the problem of finding periodic solutions of semilinear differential inclusions. The idea of this method consists in the introduction of a special integral multioperator in the space of periodic functions whose fixed points coincide with the desired periodic solutions. As a result we obtain some existence principles for periodic solutions.

Let E be a separable Banach space. For $T > 0$, let $C_T(E)$ denote the space of all continuous T-periodic functions from $\mathbb{R}_+$ into E with the usual sup-norm. When we need to consider the function $y \in C_T(E)$ defined on the whole of $\mathbb{R}$ we denote its T-periodic extension with the same symbol.

Suppose that

(A′) $A : D(A) \subseteq E \to E$ is a linear operator generating a C_0-semigroup $\exp\{At\}$ with the property that 1 does not belong to the spectrum $\mathrm{sp}(\exp\{AT\})$.

Note that from the above condition it follows that the operator $[I - \exp\{AT\}]^{-1}$ is well defined.

Throughout all of this chapter it will be supposed that

(F_T) the multioperator $F : \mathbb{R}_+ \times E \to Kv(E)$ is T-periodic in the first argument: for a.e. $t \in \mathbb{R}_+$ we have that

$$F(t + T, x) = F(t, x)$$

for all $x \in E$.

We study the problem of the existence of T-periodic mild solutions for differential inclusions of the form

$$y'(t) \in Ay(t) + F(t, y(t)). \tag{6.1.1}$$

It is assumed that the restriction $F : [0, T] \times E \to Kv(E)$ (which is denoted by the same symbol), possesses properties (F1)–(F4) of the Section 5.1 (with the change of d to T). Therefore the superposition multioperator $\mathcal{P}_F : C([0, T]; E) \to P(L^1([0, T]; E)$ generated by F is well-defined (see Sections 1.3.3 and 5.1). Note that for $y \in C_T(E)$, every $f \in \mathcal{P}_F(y)$ is considered as T-periodically extended on $\mathbb{R}_+$.

We introduce the multioperator $\mathcal{C} : C_T(E) \to P(C(\mathbb{R}_+; E))$ defined in the following way

$$\mathcal{C}y = \Big\{z : z(t) = \exp\{At\}[I - \exp\{AT\}]^{-1} \int_0^T \exp\{A(T - s)\} f(s)\, ds + \int_0^t \exp\{A(t - s)\} f(s)\, ds,\ f \in \mathcal{P}_F(y)\Big\}.$$

Proposition 6.1.1. *If $y \in C_T(E)$ then $\mathcal{C}y \subset C_T(E)$.*

Proof. Let $z \in \mathcal{C}y$, then

$$z(t) = \exp\{At\}[I - \exp\{AT\}]^{-1} \int_0^T \exp\{A(T - s)\} f(s)\, ds + \int_0^t \exp A(t - s) f(s)\, ds,$$

where $f \in \mathcal{P}_F(y)$ and f is T-periodic. Therefore

$$z(t + T) = \exp\{At\} \exp\{AT\}[I - \exp\{AT\}]^{-1} \int_0^T \exp\{A(T - s)\} f(s)\, ds + \int_0^{t+T} \exp\{A(t + T - s)\} f(s)\, ds$$

$$= \exp\{At\}\exp\{AT\}[I - \exp\{AT\}]^{-1} \int_0^T \exp\{A(T-s)\} f(s)\,ds$$

$$+ \exp\{At\} \int_0^T \exp\{A(T-s)\} f(s)\,ds + \int_T^{t+T} \exp\{A(t+T-s)\} f(s)\,ds$$

$$= \exp\{At\}\{\exp\{AT\}[I - \exp\{AT\}]^{-1} + I\} \int_0^T \exp\{A(T-s)\} f(s)\,ds$$

$$+ \int_T^{t+T} \exp\{A(t+T-s)\} f(s)\,ds$$

$$= \exp\{At\}[I - \exp\{AT\}]^{-1} \int_0^T \exp\{A(T-s)\} f(s)\,ds$$

$$+ \int_0^t \exp\{A(t-\tau)\} f(\tau)\,d\tau,$$

where $\tau = s - T$, i.e., $z(t+T) \equiv z(t)$. □

Proposition 6.1.2. *The set of all mild T-periodic solutions of* (6.1.1) *coincides with the fixed points set* Fix $\mathcal{C}$.

Proof. Let $y \in C_T(E)$ be a mild T-periodic solution of (6.6.1). Then

$$y(0) = \exp\{AT\}y(0) + \int_0^T \exp\{A(T-s)\} f(s)\,ds,$$

where $f \in \mathcal{P}_F(y)$. Hence

$$y(0) = [I - \exp\{AT\}]^{-1} \int_0^T \exp\{A(T-s)\} f(s)\,ds,$$

and

$$y(t) = \exp\{At\}[I - \exp\{AT\}]^{-1} \int_0^T \exp\{A(T-s)\} f(s)\,ds + \int_0^t \exp\{A(t-s)\} f(s)\,ds,$$

i.e., $y \in \mathcal{C}y$. The converse is evident. □

As the multioperator F has convex values the multioperator $\mathcal{C}$ is convex-valued.

Theorem 6.1.1. *The multioperator $\mathcal{C}$ is u.s.c. and has compact values.*

Proof. It is sufficient to prove that the multioperator

$$\widetilde{\mathcal{C}} : C([0,T]; E) \to P(C([0,T]; E))$$

given by the formula

$$\widetilde{\mathcal{C}}(y) = \Big\{ z : z \in C([0,T]; E),$$
$$z(t) = \exp\{At\}[I - \exp\{AT\}]^{-1} \int_0^T \exp\{A(T-s)\} f(s)\,ds$$
$$+ \int_0^t \exp\{A(t-s)\} f(s)\,ds, \quad f \in \mathcal{P}_F(y) \Big\}$$

is u.s.c. and has compact values.

Note that the multioperator $\widetilde{\mathcal{C}}$ can be written as the composition

$$\widetilde{\mathcal{C}} = S \circ \mathcal{P}_F$$

where the operator $S : L^1([0,T]; E) \to C([0,T]; E)$ is given by the formula

$$Sf(t) = \exp\{At\}[I - \exp\{AT\}]^{-1} \int_0^T \exp\{A(T-s)\} f(s)\,ds$$
$$+ \int_0^t \exp\{A(t-s)\} f(s)\,ds. \tag{6.1.2}$$

The operator S can be represented as the composition of the following maps:

(a) the Cauchy operator $G : L^1([0,T]; E) \to C([0,T]; E)$ (see Definition 4.2.2);

(b) $\overline{g} : C([0,T]; E) \to C([0,T]; E) \times C([0,T]; E)$,

$$\overline{g}(u) = (u, u)$$

(c) $\overline{e} : C([0,T]; E) \times C([0,T]; E) \to C([0,T]; E) \times C([0,T]; E)$,

$$\overline{e}(u, v) = (w, v), \quad w(t) = \exp\{At\}[I - \exp\{AT\}]^{-1} u(T);$$

and

(d) $\sigma : C([0,T];E) \times C([0,T];E) \to C([0,T];E)$,

$$\sigma(u, v) = u + v.$$

Therefore the operator S is the composition of the Cauchy operator G and the bounded linear operator $\sigma \circ \overline{e} \circ \overline{g} : C([0,T];E) \to C([0,T];E)$. Since the Cauchy operator G satisfies (S1) and (S2) (see Lemma 4.2.1) and hence properties (S1′) and (S2), the operator S has these last two properties.

To conclude the proof it is sufficient to apply Theorem 5.1.2 and Corollary 5.1.2. □

In order to obtain sharper results we investigate the condensivity of the multioperator $\mathcal{C}$ in the case when the space E is a Cartesian product of separable Banach spaces

$$E = E_1 \times \cdots \times E_N.$$

In this situation we assume that

(A_N) the linear operator A in E has a diagonal form

$$Ax = (A_1x_1, \ldots, A_Nx_N),$$

and for each $i = 1, \ldots, N$ the closed linear operator $A_i : D(A_i) \subseteq E_i \to E_i$ generates the C_0-semigroup $\exp\{A_it\}$ with the property that $1 \notin \mathrm{sp}(\exp\{A_iT\})$. We suppose also that semigroups $\exp\{A_it\}$ are χ-strongly contractive:

$$\| \exp\{A_it\} \|^{(\chi)} \le e^{-\gamma_i t} \tag{6.1.3}$$

where $\gamma_i > 0,\ i = 1, \ldots, N$.

Condition (A_N) implies that $1 \notin \mathrm{sp}(\exp\{AT\})$ and therefore $0 \notin \mathrm{sp}(A)$.

We assume that the multivalued nonlinearity $F : \mathbb{R}_+ \times E \to Kv(E)$ has the form

$$F(t,x) = F_1(t,x) \times \cdots \times F_N(t,x) \tag{6.1.4}$$

where each component $F_i : \mathbb{R}_+ \times E \to Kv(E_i)$, $i = 1, \ldots, N$ satisfies conditions (F1)–(F3). We assume also that the χ-regularity property holds in the following form:

($\widetilde{\mathrm{F}}4_N$) there exists a $(N \times N)$-matrix composed of functions

$$\{m_{ij}(\cdot)\}_{i,j=1}^{N} \in L^1_+([0,T])$$

such that:

(a) for any bounded sets $D_1 \subset E_1, \dots, D_n \subset E_N$ we have

$$\chi_i(F_i(t, D_1 \times \cdots \times D_N)) \leq \sum_{j=1}^{N} m_{ij}(t)\chi_j(D_j) \quad \text{for a.e. } t \in [0, T]$$

where χ_i denotes the Hausdorff measure of noncompactness in the space $E_i, i = 1, \dots, N$;

(b) the zero solution of the system of ordinary differential equations

$$z_i'(t) = -\gamma_i z_i(t) + \sum_{j=1}^{N} m_{ij}(t)z_j(t), \quad i = 1, \dots, n, \tag{6.1.5}$$

is exponentially stable. (Here γ_i, $i = 1, \dots, N$ are the coefficients from the estimates (6.1.3) and solutions of (6.1.5) are understood in the Carathéodory sense).

The above-listed conditions on the multioperator F will be denoted as (F_N).

So under conditions (A_N) and (F_N) the inclusion (6.1.1) takes the form of a system

$$y_i'(t) \in A_i y_i(t) + F_i(t, y_1(t), \dots, y_N(t)), \quad i = 1, \dots, N \tag{6.1.6}$$

whose mild solution is the collection of functions of the form

$$y_i(t) = \exp\{A_i t\} y_i(0) + \int_0^t \exp\{A_i(t-s)\} f_i(s)\, ds$$

where $f_i \in S^1_{F_i(\cdot, y_1(\cdot), \dots, y_N(\cdot))}$, $i = 1, \dots, N$.

When $N = 1$ condition (A_N) takes the form of condition (A′) with semigroup $\exp\{At\}$ χ-strongly contractive:

$$\| \exp\{At\} \|^{(\chi)} \leq e^{-\gamma t}, \gamma > 0. \tag{6.1.7}$$

We denote such a combination as condition (A″). Condition ($\widetilde{\mathrm{F}4}_N$) can then be written as

($\widetilde{\mathrm{F}4}$) there exists a function $m(\cdot) \in L^1_+([0, T])$ such that:

(a) for any bounded set $D \subset E$ we have that

$$\chi(F(t, D) \leq m(t) \cdot \chi(D) \quad \text{for a.e. } t \in [0, T]$$

where χ denotes the Hausdorff measure of noncompactness in the space E;

(b) for the coefficient γ from the estimate (6.1.7) the following relation holds:

$$\gamma > \frac{1}{T}\int_0^T m(s)\,ds$$

implying that the zero solution of the ordinary differential equation

$$z'(t) = -\gamma z(t) + m(t)\cdot z(t) \qquad (6.1.8)$$

is exponentially stable.

Remark 6.1.1. Without loss of generality we can assume that the semigroup $\exp\{At\} = \exp\{A_1 t\} \times \cdots \times \exp\{A_N t\}$ is decreasing in the space E:

$$\|\exp\{At\}\| \le Ce^{-at}, \qquad (6.1.9)$$

with $a > 0,\ C \ge 1$. In order to obtain this estimate it is sufficient to subtract and add to the right side of every inclusion (6.1.6) the term Lx_i with constant L sufficiently large and to consider the operator $A_i - L$ instead of A_i and the multimap $F_i + Lx_i$ instead of F_i. It is easy to see that for these new operators all previous conditions and estimate (6.1.3) are valid. From (6.1.9) it follows that the spectral radius of $\exp\{AT\}$ is less or equal to e^{-aT}. Therefore $1 \notin \mathrm{sp}(\exp\{AT\})$ and so condition (A_N) can be omitted without loss of generality.

Remark 6.1.2. Note that the semilinear differential inclusion considered in Example 5.2.1 satisfies conditions (A″) and $\widetilde{\mathrm{F}}4$ if the function $E(t)$ is T-periodic and with an appropriate choice of the constants.

To establish the condensivity of the multioperator $\mathfrak{C}$ in the periodic problem we will need the following statement.

Lemma 6.1.1. *For $v \in L_T^\infty(\mathbb{R}^N)$, let every component $v_i(t)$, $i = 1, \dots, N$ satisfy the integral inequality*

$$v_i(t) \le e^{-\gamma_i t}(I - e^{-\gamma_i T})^{-1}\int_0^T e^{-\gamma_i (T-s)}\sum_{j=1}^N m_{ij}(s)v_j(s)\,ds$$
$$+ \int_0^t e^{-\gamma_i (t-s)}\sum_{j=1}^N m_{ij}(s)v_j(s)\,ds, \quad i = 1, \dots, N,$$

where γ_i and m_{ij} are taken from (6.1.3) *and the condition* ($\widetilde{\mathrm{F}}4_N$). *Then $v(t) = 0$ a.e. on $\mathbb{R}$.*

Proof. Let us take $z^0 = (z_1^0, \dots, z_N^0) \in \mathbb{R}^N$, where

$$z_i^0 = (I - e^{-\gamma_i T})^{-1} \int_0^T e^{-\gamma_i (T-s)} \sum_{j=1}^N m_{ij}(s) v_j(s)\, ds, \quad i = 1, \dots, N.$$

Denote by $w^0(t) = (w_1^0(t), \dots, w_N^0(t))$ the continuous vector-function given by the formula

$$w_i^0(t) = e^{-\gamma_i t} z_i^0 + \int_0^t e^{-\gamma_i (t-s)} \sum_{j=1}^N m_{ij}(s) v_j(s)\, ds, \quad i = 1, \dots, N.$$

On every bounded interval we have for a.e. $t \in \mathbb{R}$

$$v_i(t) \le w_i^0(t),$$

$i = 1, \dots, N$. Therefore for $w_i^0(t)$ we have

$$w_i^0(t) \le e^{-\gamma_i t} z_i^0 + \int_0^t e^{-\gamma_i (t-s)} \sum_{j=1}^N m_{ij}(s) w_j^0(s)\, ds, \quad i = 1, \dots, N.$$

We define the sequence of continuous vector-functions by induction:

$$w_i^{n+1}(t) = e^{-\gamma_i t} z_i^0 + \int_0^t e^{-\gamma_i (t-s)} \sum_{j=1}^N m_{ij}(s) w_j^n(s)\, ds, \quad i = 1, \dots, N.$$

Then $w_i^n(t) \le w_i^{n+1}(t)$ for all $t \in \mathbb{R}$ and $n = 0, 1, \dots$. Therefore $v_i(t) \le w_i^n(t)$ for a.e. $t \in \mathbb{R}$ and $n = 0, 1, \dots$. The sequence $\{w^n\}$ converges on every bounded interval to the solution $z(t) = (z_1(t), \dots, z_N(t))$ of the system (6.1.5) with the initial condition

$$z(0) = z^0.$$

So $v_i^0(t) \le z_i(t)$ for a.e. $t \in \mathbb{R}$, $i = 1, \dots, N$.

All the solutions of the system (6.1.5) tend to zero when $t \to +\infty$. Therefore by the periodicity of v we obtain what was required. □

Following Example 2.1.5 and Remark 2.1.2 we consider the measure of noncompactness $\varphi_N(\Omega) = (\chi_1(\Omega_1(\cdot)), \dots, \chi_N(\Omega_N(\cdot)), \mathrm{mod}_C(\Omega))$ on the space $C_T(E)$ with values in $L_T^\infty(\mathbb{R}^N) \times \mathbb{R}$ ordered by the cone $\hat{K} = \hat{K}_T^+ \times [0, +\infty)$, $\hat{K}_T^+ = \{y : y \in L_T^\infty(\mathbb{R}^N), y(t) \in \mathbb{R}_+^N \text{ a.e.}\}$ (see (2.1.10)).

Recall that the MNC φ_N is monotone, regular and invariant with respect to union with a compact set.

Theorem 6.1.2. *Under conditions* (A_N) *and* (F_N) *the integral multioperator* $\mathcal{C}$ *is* φ_N*-condensing on bounded subsets of* $C_T(E)$.

Proof. Let Ω be a bounded subset of $C_T(E)$ such that

$$\varphi_N(\Omega) \leq \varphi_N(\mathcal{C}\Omega) \tag{6.1.10}$$

in the sense of the order generated by the cone $\boldsymbol{K}$. Therefore we have

$$\chi_i(\Omega_i(t)) \leq \chi_i\Big(\Big\{ \exp\{A_i t\}[I - \exp\{A_i T\}]^{-1} \int_0^T \exp\{A_i(T-s)\} f_i(s)\, ds + \int_0^t \exp\{A_i(t-s)\} f_i(s)\, ds, \ f_i \in S^1_{F_i(\cdot,\Omega(\cdot))},\ f_i \text{ is } T\text{-periodic}\Big\}\Big)$$

Using now the estimate (6.1.3), properties of the Hausdorff MNC (see Section 2.1.1) and Theorems 4.2.3 and 4.2.4 we can continue the above inequality as follows

$$\chi_i\Big(\Big\{ \exp\{A_i t\}[I - \exp\{A_i T\}]^{-1} \int_0^T \exp\{A_i(T-s)\} f_i(s)\, ds + \int_0^t \exp\{A_i(t-s)\} f_i(s)\, ds, \ f_i \in S^1_{F_i(\cdot,\Omega(\cdot))},\ f_i \text{ is } T\text{-periodic}\Big\}\Big)$$
$$\leq e^{-\gamma_i t}[I - e^{-\gamma_i T}]^{-1} \int_0^T e^{-\gamma_i(T-s)} \sum_{j=1}^N m_{ij}(s)\chi_j(\Omega_j(s))\, ds + \int_0^t e^{-\gamma_i(t-s)} \sum_{j=1}^N m_{ij}(s)\chi_j(\Omega_j(s))\, ds, \quad i = 1, \ldots, N.$$

The functions $v_i(t) = \chi_i(\Omega_i(t))$ satisfy the conditions of Lemma 6.1.1, hence $\chi_i(\Omega_i(t)) = 0$ almost everywhere.

Now we show that the set Ω is equicontinuous. Note that from

$$\operatorname{mod}_C(\Omega) \leq \operatorname{mod}_C(\mathcal{C}(\Omega))$$

it follows that it is sufficient to show that the set $\mathcal{C}(\Omega)$ is equicontinuous. This is true if we can prove that every sequence $\{g_n\} \subset \mathcal{C}(\Omega)$ is equicontinuous.

For a sequence $\{g_n\}$ there exist a sequence $\{y_n\} \subset \Omega$ and a sequence of selectors $\{f_n\}$, $f_n \in \mathcal{P}_F(y_n)$ such that $g_n = S(f_n)$, $n = 1, 2, \ldots$ where the operator S is given by formula (6.1.2).

Condition (F3) implies that the sequence of functions $\{f_n\}$ is integrably bounded. The sequence $\{y_n\}$ satisfies the equality

$$\chi(\{y_n(t)\}) = 0 \quad \text{a.e.},$$

hence, by condition $(\widetilde{\mathrm{F}4}_N)$ we have

$$\chi(\{f_n(t)\}) = 0 \quad \text{a.e.}$$

Therefore the sequence $\{f_n\}$ is semicompact.

Since the operator S satisfies conditions (S1′) and (S2) we can use Theorem 5.1.1 to conclude that the sequence $\{g_n\}$ is relatively compact and consequently equicontinuous. Then the relative compactness of Ω follows from the Arzela–Ascoli criterion. □

The following statement is useful in applications to optimal control theory.

Corollary 6.1.1. *Under conditions* (A_N) *and* (F_N), *if the set of all mild T-periodic solutions of the inclusion* (6.1.6) *is bounded, then it is compact.*

Proof. This follows from Proposition 6.1.2, Theorem 6.1.1, the above theorem and Proposition 3.5.1. □

The properties described above for the integral multioperator $\mathcal{C}$ allow the use of the topological degree theory for condensing multimaps developed in Sections 3.2 and 3.3 to find periodic solutions of the problem (6.1.6). The following general principle is valid.

Theorem 6.1.3. *Let $\Omega \subseteq C_T(E)$ be an open bounded set not containing mild T-periodic solutions of the problem* (6.1.6) *on its boundary $\partial\Omega$. Under conditions* (A_N) *and* (F_N), *if* $\deg(i - \mathcal{C}, \partial\Omega) \neq 0$ *(cf. Section* 3.2.2*) then the collection Σ_Ω of all mild T-periodic solutions of* (6.1.6) *contained in Ω is a nonempty compact set.*

Proof. This is a direct consequence of Theorem 3.3.1, Proposition 6.1.2 and Corollary 6.1.1. □

To ensure that conditions of the above general principle hold we can use apriori bounds for mild T-periodic solutions. As an example, we have the following result. (For simplicity we will write system (6.1.6) in a general form (6.1.1)).

Theorem 6.1.4. *Let A and F satisfy the conditions* (A_N) *and* (F_N) *and assume that there exists an apriori bound for mild T-periodic trajectories of the family of semilinear inclusions*

$$y'(t) \in Ay(t) + \lambda F(t, y(t)), \quad t \in \mathbb{R}_+,\ \lambda \in [0, 1]. \tag{6.1.11}$$

Then the collection of all mild T-periodic solutions of the problem (6.1.1) *is a nonempty compact set.*

Proof. If $\mathcal{C}$ is an integral multioperator for the problem (6.1.1) then, as is easy to see, for every $\lambda \in [0, 1]$ the multimap $\lambda\mathcal{C}$ is an integral multioperator for the corresponding inclusion from the family (6.1.11). Then, for $r > 0$ large enough, the family of multifields $i - \lambda\mathcal{C}$, $\lambda \in [0, 1]$ realizes a homotopy joining the identity field i and the multifield $i - \mathcal{C}$ on $\partial B_r(0)$. Using the normalization and homotopy invariance properties of the topological degree we conclude that $\deg(i - \mathcal{C}, \partial B_r(0)) = 1$ and apply Theorem 6.1.3. □

Corollary 6.1.2. *Let A and F satisfy conditions in Theorem* 6.1.4. *Suppose moreover that F satisfies the stronger global estimate*

(F3″) *there exists a function $\gamma \in L^1_+([0, T])$ such that*

$$\| F(t, x) \| \le \gamma(t)$$

for all $x \in E$ and a.e. $t \in [0, T]$ (cf. Section 5.3*).*

Then the conclusion of Theorem 6.1.4 *is valid.*

Proof. It is enough to note that the fixed points y of the multimaps $\lambda\mathcal{C}$, $0 \le \lambda \le 1$ satisfy in this case the apriori bound

$$\| y \| \le M \cdot \| \gamma \|_{L^1} \cdot T \cdot \left(M \cdot \left\|[I - \exp\{AT\}]^{-1}\right\| + 1\right)$$

where $M = \sup_{t \in [0,T]} \| \exp\{At\} \|$. □

This general principle can also be applied when it is possible to join the initial inclusion to an inclusion for which the topological degree of the integral multioperator can be easily evaluated. We discuss briefly an example of such a procedure.

Suppose that

$$y'(t) \in Ay(t) + F(t, y(t), \lambda) \tag{6.1.12}$$

is a family of semilinear differential inclusions depending on a parameter $\lambda \in [0, 1]$. We assume that the linear part A satisfies the condition (A″) and the family $F : \mathbb{R}_+ \times E \times [0, 1] \to Kv(E)$ satisfy the periodicity condition (F_T) for every $\lambda \in [0, 1]$ and conditions ($F1_\lambda$)–($F3_\lambda$) of Section 5.1 (for $d = T$ and $\Lambda = [0, 1]$). Suppose also that condition ($\widetilde{F}4$) holds for the multimap $F(\cdot, \cdot, [0, 1])$.

Then for every $\lambda \in [0, 1]$ the integral multioperator

$$\mathcal{C}(\lambda, \cdot) : C_T(E) \to Kv(C_T(E))$$

is defined as

$$\mathcal{C}(\lambda, y) = \Big\{z \in C_T(E) : z(t) = \exp\{At\}\left[I - \exp\{AT\}\right]^{-1} \tag{6.1.13}$$

$$\int_0^T \exp\{A(T - s)\} f_\lambda(s)ds + \int_0^t \exp\{A(t - s)\} f_\lambda(s)ds, \quad f_\lambda \in \mathcal{P}_{F_\lambda}(y)\Big\}$$

where $F_\lambda(t,x) = F(t,x,\lambda)$. It is not difficult to verify that the family of integral multioperators

$$\mathcal{C} : [0,1] \times C_T(E) \to Kv(C_T(E))$$

is u.s.c. and φ-condensing.

Let $d : E \to E$, $\| d \| = 1$ be a periodic operator of period $p \geq 2$ (see Definition 3.3.1) commuting with the operator A and such that $\operatorname{Fix} d = \{0\}$. It is then clear that d also commutes with the linear operator $\exp\{At\}$ for every $t \in \mathbb{R}_+$ and generates a periodic operator d_1 of the same period in the space $C_T(E)$ by the formula

$$(d_1 f)(t) = d(f(t)). \tag{6.1.14}$$

It is also obvious that $\operatorname{Fix} d_1 = \{0\}$.

Theorem 6.1.5. *Under the above conditions suppose that the boundary S of a ball $B_r(0) \subset C_T(E)$ does not contain mild T-periodic solutions of the family* (6.1.12) *and that the multimap $F(\cdot,\cdot,0)$ from this family is equivariant in the sense that*

$$F(t,dx,0) = dF(t,x,0)$$

for a.e. $t \in [0,1]$ and all $x \in E$. Then the inclusion

$$y'(t) \in Ay(t) + F(t,y(t),1)$$

has a mild T-periodic solution in the ball $B_r(0)$.

Proof. It is easy to verify that the superposition multioperator $\mathcal{P}_{F_0}$ generated by $F(\cdot,\cdot,0)$ is equivariant in the sense that

$$\mathcal{P}_{F_0}(d_1 y) = d_2(\mathcal{P}_{F_0}(y))$$

where d_2 is the periodic operator induced by d in the space $L^1_T(E)$ of T-periodic summable functions by the formula analogous to (6.1.14). Then the multioperator $\mathcal{C}(0,\cdot)$ is equivariant with respect to the operator d_1 and hence by Theorem 3.3.8 we have

$$\deg(\mathcal{C}(0,\cdot), S) \equiv 1 \pmod{p}$$

and we can apply Homotopy Invariance and Fixed Point Properties of the topological degree (see Chapter 3). □

Corollary 6.1.3. *Assume that the boundary S of a ball $B_r(0) \subset C_T(E)$ does not contain mild T-periodic solutions of the family* (6.1.12) *and that the multimap $F(\cdot,\cdot,0)$ from this family is odd in the sense that*

$$F(t,-x,0) = -F(t,x,0)$$

for a.e. $t \in [0,T]$ and all $x \in E$. Then the inclusion

$$y'(t) \in Ay(t) + F(t,y(t),1)$$

has a mild T-periodic solution in the ball $B_r(0)$.

6.2 The averaging principle for a periodic problem

In this section we return to the study of systems of differential inclusions containing high frequency terms. We consider the problem of periodic solutions. Inclusions of this type can be written as

$$y'(\tau) \in Ay(\tau) + F(\tau/\varepsilon, y(\tau))$$

where the multioperator F is T-periodic in the first argument and $\varepsilon > 0$ is a small parameter. After the change of variables

$$\tau = \varepsilon t, \quad y(\varepsilon t) = x(t)$$

the above inclusion takes the form

$$x'(t) \in \varepsilon Ax(t) + \varepsilon F(t, x(t)).$$

Then the problem of finding εT-periodic solutions of the initial inclusion is equivalent to the existence of T-periodic solutions of this one. Conditions for the existence of T-periodic solutions of the last inclusion are formulated in terms of the nontriviality of the topological degree for a certain multifield on the boundary of a neighborhood of the equilibrium state of the system obtained by averaging in t.

Note that in the case of ordinary differential equations statements of this form are generalizations of the classical second averaging theorem of N. N. Bogolyubov. We give a precise formulation.

Let E_i, $i = 1, \ldots, N$ be a collection of separable Banach spaces and χ_i denote the Hausdorff MNC in the space E_i. Suppose that

$$A_i : D(A_i) \subseteq E_i \to E_i, \quad i = 1, \ldots, N$$

are the infinitesimal generators of the C_0-semigroups $\exp\{A_i t\}$.

Moreover, let

$$F_i : \mathbb{R} \times E_1 \times \cdots \times E_N \to Kv(E_i), \quad i = 1, \ldots, N$$

be given multimaps.

We consider a system of inclusions of the form

$$x_i'(t) \in \varepsilon A_i x_i(t) + \varepsilon F_i(t, x_1(t), \ldots, x_N(t)), \quad i = 1, \ldots, N. \tag{6.2.1}$$

Let $E = E_1 \times \cdots \times E_N$. We assume that

(A''_N) The semigroups $\exp\{A_i t\}$ are sequentially condensing (see Definition 4.3.1 and Example 4.3.1) with exponents γ_i, $i = 1, \ldots, N$.

We assume that the multimap $F : \mathbb{R} \times E \to Kv(E)$ is defined as the Cartesian product

$$F(t, x) = F_1(t, x) \times \cdots \times F_N(t, x)$$

and satisfies conditions (F_N) of Section 6.1 with the one exception that the condition ($\widetilde{F4}_N$) have the following slightly different form:

($F4''_N$) there exists an $(N \times N)$-matrix M with nonnegative entries m_{ij} such that for each choice of bounded sets $D_1 \subset E_1, \ldots, D_N \subset E_N$ we have

$$\chi_i(F_i([0,T] \times D_1 \times \cdots \times D_N)) \leq \sum_{j=1}^{N} m_{ij}\chi_j(D_j) \tag{6.2.2}$$

and the zero solution of the system of ordinary differential equations

$$z_i' = -\gamma_i z_i + \sum_{j=1}^{N} m_{ij} z_j, \quad i = 1, \ldots, N \tag{6.2.3}$$

is exponentially stable (where γ_i are the coefficients from condition (A''_N)).

The collection of conditions (F_N) in which the condition ($\widetilde{F4}_N$) is replaced by ($F4''_N$) is denoted as (F''_N).

Consider now the multioperator $\Phi_0 : E \to Cv(E)$ defined as

$$\Phi_0 = F_1^0 \times \cdots \times F_N^0 \tag{6.2.4}$$

where $F_i^0 : E \to Cv(E_i)$,

$$F_i^0(y) = \frac{1}{T}\int_0^T F_i(s, y)\, ds.$$

Since we can assume without loss of generality (see Remark 6.1.1) that the linear operator $A : D(A_1) \times \cdots \times D(A_N) \to E$ defined as

$$Ax = (A_1 x_1, \ldots, A_N x_N)$$

satisfies the condition $1 \notin \mathrm{sp}(\exp\{AT\})$, it follows that $0 \notin \mathrm{sp}(A)$, and the inverse operator $A^{-1} : E \to E$ is well defined.

Consider then, the multioperator $-A^{-1} \circ \Phi_0 : E \to Cv(E)$ where Φ_0 is given by formula (6.2.4).

Lemma 6.2.1. *Under the above conditions the multioperator $-A^{-1} \circ \Phi_0$ is u.s.c. and condensing with respect to the MNC X in E given by*

$$X(\Omega) = (\chi_1(\Omega_1), \ldots, \chi_N(\Omega_N)), \tag{6.2.5}$$

where $\Omega_i = \mathrm{pr}_i(\Omega)$ and $\mathrm{pr}_i : E \to E_i$, $i = 1, \ldots, N$ are the natural projections.

Proof. Note that the multioperator Φ_0 can be represented as the composition

$$\Phi_0 = e_T \circ S \circ \mathcal{P}_F \circ j$$

where

$j : E \to C([0, T]; E)$ is the natural injection as the constant function;

$\mathcal{P}_F : C([0, T]; E) \to Cv(L^1([0, T]; E))$ is the superposition multioperator;

$S : L^1([0, T]; E) \to C([0, T]; E)$ is the integral operator given by the formula (6.1.2); and

$e_T : C([0, T]; E) \to E$ is the evaluation operator:

$$e_T(y) = \frac{1}{T} y(T).$$

By Theorem 6.1.1 the multioperator $S \circ \mathcal{P}_F$ is u.s.c. and an application of Theorem 1.2.8 gives the upper semicontinuity of the multioperator $-A^{-1} \circ \Phi_0$.

By the estimate (6.2.2) and Theorem 4.2.3 we have

$$\chi_i(F_i^0(D_1 \times \cdots \times D_N)) \leq \sum_{j=1}^{N} m_{ij} \chi_j(D_j).$$

We note that

$$-A^{-1} \circ \Phi_0(v) = \Big\{ y : y \in C_T(E),\ y(t) = \exp\{At\}[I - \exp\{AT\}]^{-1} \tag{6.2.6}$$

$$\int_0^T \exp\{A(T-s)\} g(s)\, ds + + \int_0^t \exp\{A(t-s)\} g(s)\, ds, \quad g(s) \equiv g \in \Phi_0(v) \Big\}.$$

Therefore if for any bounded set $\Omega \subset E$ we have

$$X(\Omega) \leq X(-A^{-1} \Phi_0(\Omega)),$$

then for the set $W \subset C_T(E),\ W = \{y(t) \equiv y \in \Omega\}$ the following estimate holds:

$$\chi_i(W_i(t)) \leq \chi\Big(\Big\{ \exp\{A_i t\}[I - \exp\{A_i T\}]^{-1} \int_0^T \exp\{A_i(T-s)\} g_i^0(s)\, ds \tag{6.2.7}$$

$$+ \int_0^t \exp\{A_i(t-s)\} g_i^0(s)\, ds \ :\ g_i^0(s) \equiv g_i^0 \in F_i^0(\Omega) \Big\} \Big),$$

where $W_i(t) = \mathrm{pr}_i\, W(t),\ i = 1, \ldots, N$.

The monotonicity of the Hausdorff MNC and the inequality (6.2.2) imply

$$\begin{aligned} \chi_i(\{g_i^0(s) : g_i^0(s) \equiv g_i^0 \in F_i^0(\Omega)\}) &\leq \chi_i(F_i^0(\Omega_1, \ldots, \Omega_N)) \\ &\leq \sum_{j=1}^{N} m_{ij} \chi_j(\Omega_j) = \sum_{j=1}^{N} m_{ij} \chi_j(W_j(s)). \end{aligned}$$

Inserting this estimate into the inequality (6.2.7) and using (A''_N))we obtain

$$\chi_i(W_i(t)) \leq e^{-\gamma_i t}[I - e^{-\gamma_i T}]^{-1} \int_0^T e^{-\gamma_i(T-s)} \sum_{j=1}^N m_{ij}\chi_j(W_j(s))\,ds$$

$$+ \int_0^t e^{-\gamma_i(t-s)} \sum_{j=1}^N m_{ij}\chi_j(W_j(s))\,ds, \quad i = 1, \dots, N.$$

Therefore the constant periodic functions

$$v_i(t) = \chi_i(W_i(t)) \equiv \chi_i(\Omega_i), \quad i = 1, \dots, N$$

satisfy conditions of Lemma 6.1.1 and hence

$$\chi_i(\Omega_i) = 0, \quad i = 1, \dots, N,$$

i.e., the set Ω is relatively compact. □

Now we are in a position to prove the following main result.

Theorem 6.2.1. *Under conditions* (A''_N) *and* (F''_N) *suppose that the inclusion*

$$u \in -A^{-1} \circ \Phi_0(u)$$

has a solution $u^* \in E$ *and* $\deg(i + A^{-1} \circ \Phi_0, S_E(u^*, r)) \neq 0$ *where* $S_E(u^*, r)$ *is a sphere of radius* $r > 0$ *centered at* u^* *(see Section* 3.2.2*). Then, for* $\varepsilon > 0$ *small enough the inclusion* (6.2.1) *has a* T*-periodic solution* $x^\varepsilon(\cdot)$ *satisfying*

$$\max_{t\in[0,T]} \| x^\varepsilon(t) - u^* \|_E \leq r.$$

Proof. It is clear that the MNC X is monotone and invariant with respect to union with compact sets and therefore from the previous lemma it follows that in accordance with Section 3.2.3 the multioperator $-A^{-1} \circ \Phi_0$ has a compact homotopy approximation $-\widehat{A^{-1} \circ \Phi_0}$ on $S_E(u^*, r)$. Without loss of generality we can assume that the multioperator $-A^{-1} \circ \Phi_0$ is joined with its homotopy approximation by the linear deformation

$$-((1-\lambda)A^{-1} \circ \Phi_0 + \lambda \widehat{A^{-1} \circ \Phi_0}), \quad \lambda \in [0, 1]$$

(see Lemma 3.2.1).

In the space $C_T(E)$, consider the multioperator associating to each function $u \in C_T(E)$ the set of all constant functions with values in the set $\widehat{A^{-1} \circ \Phi_0}(u(0))$. By the principle of map restriction (see Section 3.1) we have

$$\deg_{C_T(E)}(i + \widehat{A^{-1} \circ \Phi_0}, S_{C_T(E)}(u^*, r)) = \deg_E(i + \widehat{A^{-1} \circ \Phi_0}, S_E(u^*, r)) \neq 0. \tag{6.2.8}$$

For $\varepsilon > 0$ the multioperator $\mathcal{C}(\varepsilon)$ in the space $C_T(E)$ given by the formula

$$\mathcal{C}(\varepsilon)u = \Big\{y : y \in C_T(E),\, y(t) = \exp\{\varepsilon At\}[I - \exp\{\varepsilon AT\}]^{-1} \varepsilon \int_0^T \exp\{\varepsilon A(T-s)\}g(s)\,ds + \varepsilon \int_0^t \exp\{\varepsilon A(t-s)\}g(s)\,ds, \\ g \text{ is a } T\text{-periodic measurable selection of } F(\cdot, u(\cdot))\Big\}, \tag{6.2.9}$$

and $\widehat{A^{-1} \circ \Phi_0}$ are linearly homotopic on $S_{C_T(E)}(u^*, r)$.

If we assume the contrary then there exist sequences $\{\varepsilon_m\}$, $\{\lambda_m\}$, $\{u_m\}$ such that $\varepsilon_m \to 0$, $\lambda_m \in [0, 1]$, $\lambda_m \to \lambda_0$, $u_m \in S_{C_T(E)}(u^*, r)$ and

$$u_m \in \lambda_m \mathcal{C}(\varepsilon_m)u_m - (1 - \lambda_m)\widehat{A^{-1} \circ \Phi_0}(u_m(0))$$

or more explicitly

$$u_m(t) = \lambda_m\varepsilon_m \exp\{-\varepsilon_m At\}[I - \exp\{-\varepsilon_m AT\}]^{-1} \int_0^T \exp\{-\varepsilon_m A(T-s)\}g_m(s)\,ds \\ + \lambda_m\varepsilon_m \int_0^t \exp\{-\varepsilon_m A(t-s)\}g_m(s)\,ds + (1 - \lambda_m)w_m \tag{6.2.10}$$

with g_m T-periodic selections of $F(\cdot, u_m(\cdot))$ and $w_m \in -\widehat{A^{-1} \circ \Phi_0}(u_m(0))$. We assume that $\varepsilon_m \in (0, 1/2)$.

Note that $u_m([1/\varepsilon_m]T) = u_m(0)$, where $[p]$ denotes the integer part of p. Therefore we have

$$\lambda_m\varepsilon_m \exp\{-\varepsilon_m A[1/\varepsilon_m]T\}[I - \exp\{\varepsilon_m AT\}]^{-1} \int_0^T \exp\{\varepsilon_m A(T-s)\}g_m(s)\,ds \\ + \lambda_m\varepsilon_m \int_0^{[1/\varepsilon_m]T} \exp\{\varepsilon_m A([1/\varepsilon_m]T - s)\}g_m(s)\,ds + (1 - \lambda_m)w_m \\ = \lambda_m\varepsilon_m[I - \exp\{\varepsilon_m AT\}]^{-1} \int_0^T \exp\{\varepsilon_m A(T-s)\}g_m(s)\,ds + (1 - \lambda_m)w_m.$$

Hence

$$[I - \exp\{\varepsilon_m AT\}]^{-1} \int_0^T \exp\{\varepsilon_m A(T-s)\} g_m(s)\, ds$$

$$= [I - \exp\{\varepsilon_m A[1/\varepsilon_m]T\}]^{-1} \int_0^{[1/\varepsilon_m]T} \exp\{\varepsilon_m A([1/\varepsilon_m](T-s)\} g_m(s)\, ds.$$

Substituting this expression into (6.2.10) we obtain

$$\begin{aligned} u_m(t) = \lambda_m \varepsilon_m \exp\{\varepsilon_m At\}[I - \exp\{\varepsilon_m A[1/\varepsilon_m]T\}]^{-1} \\ \int_0^{[1/\varepsilon_m]T} \exp\{\varepsilon_m A([1/\varepsilon_m]T - s)\} g_m(s)\, ds \\ + \lambda_m \varepsilon_m \int_0^t \exp\{\varepsilon_m A(t-s)\} g_m(s)\, ds + (1-\lambda_m) w_m. \end{aligned} \tag{6.2.11}$$

If we make the change of variables $u_m(\tau[1/\varepsilon_m]) = v_m(\tau)$ in (6.2.11) and set $s = \xi[1/\varepsilon_m]$, we get

$$\begin{aligned} v_m(\tau) = \lambda_m \varepsilon_m [1/\varepsilon_m] \exp\{\varepsilon_m [1/\varepsilon_m] A\tau\}[I - \exp\{\varepsilon_m A[1/\varepsilon_m]T\}]^{-1} \\ \times \int_0^T \exp\{\varepsilon_m [1/\varepsilon_m] A(T-\xi)\} g_m(\xi[1/\varepsilon_m])\, d\xi \\ + \lambda_m \varepsilon_m [1/\varepsilon_m] \int_0^\tau \exp\{\varepsilon_m [1/\varepsilon_m] A(\tau - s)\} g_m(\xi[1/\varepsilon_m])\, d\xi + (1-\lambda_m) w_m. \end{aligned} \tag{6.2.12}$$

The functions $v_m(\cdot)$ are T-periodic. We now estimate $\chi_i(\{(v_m)_i(\tau)\}_{m=1}^{\infty})$.

Note that for every bounded sequence $\{x_m\} \subset E$ we have

$$\chi\big(\big\{ (I - \exp\{\varepsilon_m A_i[1/\varepsilon_m]T\})^{-1} x_m \big\}\big) \le (1 - e^{-\gamma_i T})^{-1} \chi(\{x_m\}). \tag{6.2.13}$$

In fact, by the estimate (6.1.9) (see Remark 6.1.1) we have

$$(I - \exp\{\varepsilon_m A_i[1/\varepsilon_m]T\})^{-1} x_m$$

$$= \sum_{j=0}^{k} \exp\{j\varepsilon_m A_i[1/\varepsilon_m]T\} x_m + \sum_{j=k+1}^{\infty} \exp\{j\varepsilon_m A_i[1/\varepsilon_m]T\} x_m.$$

Further, by (6.1.9) for any given $\delta > 0$ we can find $k > 0$ such that

$$NC^2 e^{-(k+1)aT/2}(1 - e^{-aT/2})^{-1} < \delta/2 \tag{6.2.14}$$

$$N\Big((1 - e^{-\gamma_i T})^{-1} - \sum_{j=0}^{k} e^{-j\gamma_i T}\Big) < \delta/2. \tag{6.2.15}$$

Here $N > 0$ is a constant such that $\| x_m \| \leq N$, C and a are constants from the estimate (6.1.9).

Applying inequality (6.2.14) we obtain

$$\begin{aligned}
&\Big\| \sum_{j=k+1}^{\infty} \exp\{j\varepsilon_m A_i[1/\varepsilon_m]T\}x_m \Big\| \\
&\qquad \leq \| \exp\{-(k+1)\varepsilon_m A_i[1/\varepsilon_m]T\} \| \, \Big\| \sum_{j=0}^{\infty} \exp\{j\varepsilon_m A_i[1/\varepsilon_m]T\}x_m \Big\| \\
&\qquad \leq Ce^{-(k+1)\varepsilon_m[1/\varepsilon_m]aT} \sum_{j=0}^{\infty} Ce^{-j\varepsilon_m[1/\varepsilon_m]aT} N \\
&\qquad \leq NC^2 e^{-(k+1)aT/2} \sum_{j=0}^{\infty} e^{-jaT/2} \\
&\qquad = NC^2 e^{-(k+1)aT/2}(1 - e^{-aT/2})^{-1} < \delta/2.
\end{aligned}$$

From the property of algebraic semiadditivity of the MNC χ and inequality (6.2.15) we obtain

$$\begin{aligned}
&\chi\big(\{(I - \exp\{\varepsilon_m A_i[1/\varepsilon_m]T\})^{-1}x_m\}\big) \leq \sum_{j=0}^{k} \chi\big(\{\exp\{j\varepsilon_m A_i[1/\varepsilon_m]T\}x_m\}\big) \\
&\qquad + \chi\Big(\Big\{ \sum_{j=k+1}^{\infty} \exp\{j\varepsilon_m A_i[1/\varepsilon_m]T\}x_m \Big\}\Big) \leq \Big(\sum_{j=0}^{k} e^{-j\gamma_i T}\Big)\chi(\{x_m\}) + \delta/2 \\
&\quad \leq \big(1 - e^{-\gamma_i T}\big)^{-1}\chi(\{x_m\}) + N\Big(\big(1 - e^{-\gamma_i T}\big)^{-1} - \sum_{j=0}^{k} e^{-j\gamma_i T}\Big) + \delta/2 \\
&\quad = \big(1 - e^{-\gamma_i T}\big)^{-1}\chi(\{x_m\}) + \delta.
\end{aligned}$$

Now inequality (6.2.13) follows from the arbitrariness of δ.

From inequality (6.2.2) it follows that

$$\begin{aligned}
&\chi_i(\{g_m(\xi[1/\varepsilon_m])\}) \leq \chi_i(\{F_i(\xi[1/\varepsilon_m], v_m(\xi))\}) \\
&\qquad \leq \chi_i(\{F_i([0,T] \times \{v_m(\xi)\})\}) \leq \sum_{j=1}^{N} m_{ij}\chi_j(\{(v_m)_j(\xi)\}).
\end{aligned}$$

Applying this fact, inequality (6.2.13) and the compactness of the sequence $\{w_m\}$ we deduce from (6.2.12) that

$$\begin{aligned}
&\chi_i(\{(v_m)_i(\tau)\}_{m=1}^{\infty}) \\
&\quad \le \chi_i\Big(\Big\{\lambda_m\varepsilon_m[1/\varepsilon_m]\exp\{\varepsilon_m[1/\varepsilon_m]A\tau\}(I-\exp\{\varepsilon_m A[1/\varepsilon_m]T\})^{-1} \\
&\quad\quad \times \int_0^T \exp\{\varepsilon_m[1/\varepsilon_m]A(T-\xi)\}g_m(\xi[1/\varepsilon_m])\,d\xi\Big\} \\
&\quad\quad + \chi_i\Big\{\lambda_m\varepsilon_m[1/\varepsilon_m]\int_0^\tau \exp\{\varepsilon_m[1/\varepsilon_m]A(\tau-\xi)\}g_m(\xi[1/\varepsilon_m])\,d\xi\Big\} \\
&\quad \le e^{-\gamma_i\tau}(I-e^{-\gamma_i T})^{-1}\int_0^T e^{-\gamma_i(T-\xi)}\sum_{j=1}^N m_{ij}\chi_j(\{(v_m)_j(\xi)\})\,d\xi \\
&\quad\quad + \int_0^\tau e^{-\gamma_i(\tau-\xi)}\sum_{j=1}^N m_{ij}\chi_j(\{(v_m)_j(\xi)\})\,d\xi, \quad i=1,\dots,N.
\end{aligned}$$

By Lemma 6.1.1

$$\chi_i(\{(v_m)_i(\tau)\}_{m=1}^{\infty}) = 0, \quad i=1,\dots,N,$$

a.e. on $[0,T]$. From Theorem 5.1.1 it follows that functions $v_m(\,\cdot\,)$ are equicontinuous and therefore

$$\chi_i(\{(v_m)_i(\tau)\}_{m=1}^{\infty}) \equiv 0, \quad i=1,\dots,N.$$

In particular,

$$\chi_i((u_m)_i(0)) = \chi_i((v_m)_i(0)) = 0, \quad i=1,\dots,N,$$

i.e., the sequence

$$\Big\{\lambda_m\varepsilon_m(I-\exp\{\varepsilon_m AT\})^{-1}\int_0^T \exp\{\varepsilon_m A(T-s)\}g_m(s)\,ds\Big\}$$

is relatively compact as a difference of relatively compact sequences $\{u_m(0)\}$ and $\{(1-\lambda_m)w_m\}$. Therefore

$$\begin{aligned}
&u_m(t)-u_m(0) \qquad\qquad (6.2.16)\\
&\quad = (\exp\{\varepsilon_m At\}-I)\lambda_m\varepsilon_m(I-\exp\{\varepsilon_m AT\})^{-1}\int_0^T \exp\{\varepsilon_m A(T-s)\}g_m(s)\,ds \\
&\quad\quad + \lambda_m\varepsilon_m\int_0^t \exp\{\varepsilon_m A(t-s)\}g_m(s)\,ds \xrightarrow[m\to\infty]{} 0
\end{aligned}$$

uniformly with respect to t. In fact the first term on the right-hand side tends to zero by the C_0-continuity of the semigroup $\exp\{At\}$ and the second term can be estimated on the interval $[0, T]$ by $\varepsilon_m MCT$ where M is the constant bounding the norms of $F(s, u_m(s))$ and C is the constant from the inequality (6.1.9). Because of the relation (6.2.16) and the relative compactness of the sequence $\{u_m(0)\}$ we can assume, without loss of generality, that the sequence $\{u_m\}$ uniformly converges to the constant function u_0 and moreover $\| u_0 - u^* \|_E = r$. We will keep the notation u_0 for the value of the function u_0, $u_0 \in S_E(u^*, r)$.

Furthermore, the condition ($F4''_N$) implies that the set

$$\mathcal{G} = \{g_m(s) : s \in [0, T],\ m = 1, 2, \dots\} \tag{6.2.17}$$

is relatively compact, and from Remark 4.2.2 we conclude that the sequence $\{g_m\}$ is weakly compact in the space $L^1([0, T]; E)$. Without loss of generality we can assume that this sequence weakly converges to a function g_0.

By Lemma 5.1.1, g_0 is a selection from the multifunction $F(\cdot, u_0(\cdot))$. Since the multioperator $-\widehat{A^{-1} \circ \Phi_0}$ is completely continuous we can assume without loss of generality that

$$w_m \to w_0 \quad \text{for } m \to \infty \tag{6.2.18}$$

and

$$w_0 \in -\widehat{A^{-1} \circ \Phi_0}(u_0).$$

We set $t = 0$ in (6.2.10) to get

$$u_m(0) = \lambda_m \varepsilon_m (I - \exp\{\varepsilon_m AT\})^{-1} \int_0^T \exp\{\varepsilon_m A(T - s)\} g_m(s)\, ds + (1 - \lambda_m) w_m. \tag{6.2.19}$$

Note that since the sequences $\{u_m(0)\}$ and $\{(1 - \lambda_m) w_m\}$ are convergent, the sequence

$$\Big\{\lambda_m \varepsilon_m (I - \exp\{\varepsilon_m AT\})^{-1} \int_0^T \exp\{\varepsilon_m A(T - s)\} g_m(s)\, ds\Big\}$$

is also convergent. We show that

$$\varepsilon_m\, (I - \exp\{\varepsilon_m AT\})^{-1} \int_0^T \exp\{\varepsilon_m A\,(T - s)\}\, g_m\,(s)\, ds \tag{6.2.20}$$

$$\underset{w}{\longrightarrow} -A^{-1} \frac{1}{T} \int_0^T g_0\,(s)\, ds$$

as $m \to \infty$.

From the relative compactness of the set $\mathcal{G}$ (see (6.2.17)) and the fact that $\exp\{At\}$ is a C_0-semigroup we can find, for an arbitrary μ, m_1 such that for $m > m_1$ we have

$$\Big\| \varepsilon_m (I - \exp\{\varepsilon_m AT\})^{-1} \int_0^T \exp\{\varepsilon_m A(T-s)\} g_m(s)\,ds \tag{6.2.21}$$

$$- \varepsilon_m (I - \exp\{\varepsilon_m AT\})^{-1} \int_0^T g_m(s)\,ds \Big\| < \mu/3.$$

By Theorem 4.1.5 and again the relative compactness of the set $\mathcal{G}$ we can find m_2 such that for every $m > m_2$ we have

$$\Big\| \varepsilon_m (I - \exp\{\varepsilon_m AT\})^{-1} \int_0^T g_m(s)\,ds + A^{-1}\frac{1}{T}\int_0^T g_m(s)\,ds \Big\| < \mu/3 \tag{6.2.22}$$

The relative compactness of the set $\mathcal{G}$ implies also that the sequence $\{g_m\}$ is semicompact. Hence, bearing in mind Remark 4.2.2, Lemma 4.2.1, the weak convergence $g_m \underset{w}{\to} g_0$ and the boundedness of the operator A^{-1} we can find m_3 such that for $m > m_3$ we have

$$\Big\| A^{-1}\frac{1}{T}\int_0^T g_m(s)\,ds - A^{-1}\frac{1}{T}\int_0^T g_0(s)\,ds \Big\| < \mu/3. \tag{6.2.23}$$

From (6.2.21)–(6.2.23) it follows that for $m > \max\{m_1, m_2, m_3\}$ the inequality

$$\Big\| \varepsilon_m (I - \exp\{\varepsilon_m AT\})^{-1} \int_0^T \exp\{\varepsilon_m A(T-s)\} g_m(s)\,ds + A^{-1}\frac{1}{T}\int_0^T g_0(s)\,ds \Big\| < \mu$$

holds, proving (6.2.20).

Passing to the limit in (6.2.19) we obtain

$$u_0(0) = -\lambda_0 A^{-1}\frac{1}{T}\int_0^T g_0(s)\,ds + (1-\lambda_0) w_0 \tag{6.2.24}$$

It follows that

$$u_0 \in -\lambda_0 A^{-1} \circ \Phi_0(u_0) - (1-\lambda_0)\widehat{A_0^{-1} \circ \Phi_0(u_0)}$$

in opposition to the homotopy of multioperators $-A^{-1} \circ \Phi_0$ and $-\widehat{A^{-1} \circ \Phi_0}$ on $S_E(u^*, r)$.

So the multioperators $\mathcal{C}(\varepsilon)$ (see (6.2.9)) and $-\widehat{A^{-1} \circ \Phi_0}$ are homotopic on $S_{C_T(E)}(u^*, r)$. To complete the proof we use the relation (6.2.8) and the properties of the degree. □

6.3 The translation multioperator

We consider conditions under which the translation multioperator along mild solutions of semilinear differential inclusions in Banach spaces is condensing. Using Theorem 5.3.1 on the structure of the integral funnel and the topological degree theory for condensing nonconvex-valued multimaps developed in Section 3.4, we give some conditions for the existence of periodic solutions.

In conclusion we prove the existence of a periodic solution and a global attractor for a semilinear differential inclusion satisfying a dissipativity condition.

6.3.1 The translation multioperator along the solutions of semilinear differential inclusions

Consider the Cauchy problem for a semilinear differential inclusion on a separable Banach space E of the form

$$y'(t) \in Ay(t) + F(t, y(t)), \quad t \geq 0 \tag{6.3.1}$$

$$y(0) = x_0 \in E, \tag{6.3.2}$$

under the usual assumption that the linear part $A : D(A) \subseteq E \to E$ satisfies the condition (A). For the multivalued nonlinearity $F : \mathbb{R}_+ \times E \to Kv(E)$ the T-periodicity condition (F_T) is assumed to hold and this obviously implies that it is enough to work with the restriction $F : [0, T] \times E \to Kv(E)$. It will be assumed that conditions (F1)–(F3) of Section 5.1 hold but the χ-regularity condition has the following sharpened form:

$(\mathrm{F4}_g)$ for every nonempty bounded set $\Omega \subset E$, we have

$$\chi(F(t, \Omega) \leq g(t, \chi(\Omega)) \quad \text{for a.e. } t \in [0, T]$$

where $g : [0, T] \times \mathbb{R}_+ \to \mathbb{R}_+$ is a Carathéodory type function such that $g(t, \cdot) : \mathbb{R}_+ \to \mathbb{R}_+$ is nondecreasing for almost every $t \in [0, T]$, $g(t, 0) \equiv 0$ for almost every $t \in [0, T]$ and

$$|g(t, v_1) - g(t, v_2)| \leq k(t) \cdot |v_1 - v_2|$$

for every $v_1, v_2 \in \mathbb{R}_+$ and almost every $t \in [0, T]$ where $k \in L^1_+([0, T])$.

It is clear that from the T-periodicity condition (F_T) that we can extend the functions $g(t, v)$ and $k(t)$ on the whole half-axis $\mathbb{R}_+$ in t.

We say that *the translation multioperator along mild solutions of the inclusion* (6.3.1) *for the time* T is defined at a point $x \in E$ if the set $\Sigma_x^F[0, T] \subset C([0, T]; E)$ of mild solutions of the inclusion (6.3.1) with initial value $y(0) = x$ is nonempty, bounded and satisfies condition (P) of Section 5.2 with $d = T$. (Recall that these assumptions hold if F satisfies condition (F3′) from the existence Theorem 5.2.2). In this case the translation multioperator P_T is given by the formula

$$P_T(x) = \{z = y(T) \ : \ y \in \Sigma_x^F[0, T]\}.$$

The multioperator P_T is defined on a set $D \subseteq E$ if it is defined at every point of D and the set $\bigcup_{x \in \Omega} \Sigma_x^F[0, T]$ is bounded for every bounded $\Omega \subset D$. Note that from Corollary 5.2.1 it follows that without loss of generality D can be considered as an open set. It is clear that the multioperator $P_T : D \multimap E$ can be represented as a composition of the multimap $\Sigma : D \multimap C([0, T]; E)$, $\Sigma(x) = \Sigma_x^F[0, T]$ and the continuous map $\theta : C([0, T]; E) \to E$, $\theta(y) = y(T)$. From Corollary 5.2.1 and Theorem 5.3.1 it follows immediately that Σ is an R_δ-multimap (see Definition 3.4.6) and hence the multimap P_T is quasi R_δ.

Note that Corollaries 5.2.2 and 5.3.1 in combination with apriori estimates (see the proof of Theorem 5.2.2) give sufficient conditions ((A), (F1), (F2), (F3′) and (F4)) for the translation multioperator P_T to be quasi R_δ and defined on the whole space E.

Our aim now is to describe conditions under which this multioperator is condensing.

Theorem 6.3.1. *Let us suppose that the translation multioperator $P_T : D \subseteq E \multimap E$ is defined. In addition assume that A satisfies*

(A1) *the C_0-semigroup* $\exp\{At\}$ *generated by A is χ-decreasing:*

$$\| \exp\{At\} \|^{(\chi)} \leq Ce^{-\gamma t}$$

where $C > 0$ and $\gamma > 0$ (see Section 4.3).

Assume also that:

(H1) *the zero solution of the scalar differential equation*

$$r'(t) = -\gamma r(t) + Cg(t, r(t)), \quad t \geq 0 \tag{6.3.3}$$

is exponentially stable in the sense that $r(t) \leq C_1 e^{-\mu t} r(0)$, $t \geq 0$, for any nonnegative solution $r(\cdot)$ of (6.3.3) where $C_1 > 0$, $\mu > 0$;

(H2) $T > (1/\mu) \ln(C \cdot C_1)$.

Then the translation multioperator P_T is (k, χ)-condensing on bounded subsets of D, with $k = C \cdot C_1 e^{-\mu T} < 1$ (and hence P_T is $(e^{-\mu T}, \chi)$-condensing for every $T > 0$ if $C \cdot C_1 = 1$).

Proof. Let $\Omega \subset D$ be a nonempty bounded set and denote by $\Sigma(\Omega)(t)$, $0 \leq t \leq T$ the subset of E defined as

$$\Sigma(\Omega)(t) = \{y(t) : y \in \Sigma(\Omega)\}.$$

It is clear that $\Sigma(\Omega)(0) = \Omega$ and

$$\Sigma(\Omega)(t) \subseteq \exp\{At\}\Omega + \int_0^t \exp\{A(t-s)\}F(s, \Sigma(\Omega)(s))\, ds, \quad 0 \le t \le T.$$

We have the following estimate for the measure of noncompactness of $\Sigma(\Omega)(t)$, $0 \le t \le T$ (cf. Theorems 4.2.4 and 4.2.3):

$$\begin{aligned}\chi(\Sigma(\Omega)(t)) &\le \chi\Big(\exp\{At\}\Omega + \int_0^t \exp\{A(t-s)\}F(s, \Sigma(\Omega)(s))\, ds\Big) \\ &\le Ce^{-\gamma t}\chi(\Omega) + C\int_0^t e^{-\gamma(t-s)} g(s, \chi(\Sigma(\Omega)(s))\, ds.\end{aligned}$$

Therefore the function $\omega(t) = \chi(\Sigma(\Omega)(t))$ satisfies the integral inequality

$$\omega(t) \le Ce^{-\gamma t}\omega(0) + C\int_0^t e^{-\gamma(t-s)} g(s, \omega(s))\, ds, \quad 0 \le t \le T.$$

Denoting the right-hand side of this inequality by $\rho(t)$ and using the monotonicity of the function g in its second argument we easily obtain

$$\rho'(t) = -\gamma\rho(t) + Cg(t, \omega(t)) \le -\gamma\rho(t) + Cg(t, \rho(t)).$$

Using standard results on differential inequalities we get that $\rho(t)$ is bounded from above by the solution $r(t)$ of the equation (6.3.3) with initial value $r(0) = \rho(0) = C\omega(0)$. From assumption (H1) it follows that

$$\omega(t) \le \rho(t) \le r(t) \le C_1 e^{-\mu t} r(0) = C \cdot C_1 e^{-\mu t}\omega(0), \quad t \ge 0.$$

Therefore

$$\chi(P_T(\Omega)) = \omega(T) \le C \cdot C_1 e^{-\mu T} \cdot \omega(0) = C \cdot C_1 e^{-\mu T} \cdot \chi(\Omega),$$

concluding the proof. □

Remark 6.3.1. It is easy to see that the exponential bound on the solutions of equation (6.3.3) required in condition (H1) is satisfied if the function $k(\cdot)$ from the condition $(F4_g)$ satisfies $\frac{1}{T}\int_0^T k(s)\, ds < \gamma/C$. (Recall that we assume that the functions k and g are T-periodic in t).

As a direct application of the topological degree theory developed in Section 3.4.3, we obtain the following general principle for the existence of periodic solutions of the differential inclusion (6.3.1).

Theorem 6.3.2. *Suppose that conditions of Theorem* 6.3.1 *hold. Let* $B \subset D$ *be an open ball such that* $\operatorname{Fix} P_T \cap \partial B = \emptyset$ *and* $\deg(i - P_T, \overline{B}) \neq 0$. *Then the differential inclusion* (6.3.1) *has a* T*-periodic mild solution* $y(\cdot)$.

As an example of application of this general principle, consider the following statement.

Theorem 6.3.3. *Suppose that conditions* (A), (A1), (F1), (F2), ($F4_g$) *and* (H1) *hold. Furthermore, suppose that:*

(F3″) $\| F(t, y) \| \leq \alpha(t)$ *for every* $y \in E$ *and almost every* $t \in [0, T]$ *where* $\alpha \in L^1_+([0, T])$,

and the multimap F *satisfies the periodicity condition* (F_T) *with*

$$T > \max\{(1/\gamma) \cdot \ln C, (1/\mu)(C \cdot C_1)\}.$$

If $1 \notin \operatorname{sp}\exp\{AT\}$, *then there exists a mild* T*-periodic solution of the differential inclusion* (6.3.1).

Proof. First of all, we note that from the assumptions of the theorem it follows that the translation multioperator P_T is defined on the whole space E. From assumptions (A1) and (F3″) it follows that there exists a number $N > 0$ such that

$$\| q(T) \| \leq N$$

for every function $q(\cdot) \in C([0, T]; E)$ of the form

$$q(t) = \int_0^t \exp\{A(t-s)\} f(s)\, ds$$

where $f \in \mathcal{P}_F(y)$ for any $y(\cdot) \in C([0, T]; E)$.

From the assumptions of the theorem it follows also that $\exp\{AT\}$ is a continuous $(Ce^{-\gamma T}, \chi)$-condensing linear operator and that we can choose $\eta > 0$ such that $\| x - \exp\{AT\}x \| > N$, for all $x \in E$, $\| x \| = \eta$.

Let $B_\eta \subset E$ be a ball of radius η with the center at the origin. Consider the family $H : [0, 1] \times \overline{B}_\eta \multimap E$ defined by

$$H(\lambda, x) = \lambda P_T(x) + (1 - \lambda) \exp\{AT\}x.$$

It is easy to see that the multimap H is quasi R_δ and (k, χ)-condensing where $k = \max\{Ce^{-\gamma T}, C \cdot C_1 e^{-\mu T}\}$.

To prove that H induces a homotopy between P_T and $\exp\{AT\}$ it remains only to show that $\mathrm{Fix}\, H(\lambda, \cdot) \cap \partial B_\eta = \emptyset$ for all $\lambda \in [0, 1]$. Suppose to the contrary that there exist $\lambda_0 \in [0, 1]$ and $x_0 \in \partial B_\eta$ such that

$$x_0 \in \lambda_0 P_T(x_0) + (1 - \lambda_0) \exp\{AT\}x_0.$$

Then

$$x_0 - \exp\{AT\}x_0 \in \lambda_0(P_T(x_0) - \exp\{AT\}x_0).$$

This implies that $x_0 - \exp\{AT\}x_0$ has the form

$$\lambda_0(y(T) - \exp\{AT\}x_0)$$

where $y(\cdot) \in \Sigma_{x_0}^F$, $y(T) = \exp\{AT\}x_0 + \int_0^T \exp\{A(T-s)\}f(s)\,ds$, $f \in \mathcal{P}_F(y)$. Hence

$$\| x_0 - \exp\{AT\}x_0 \| = \left\| \lambda_0 \Big(\int_0^T \exp\{A(T-s)\}f(s)\,ds \Big) \right\| \leq \lambda_0 N$$

giving a contradiction.

It was shown in [4] that $|\deg(i - \exp\{AT\}, \overline{B}_\eta)| = 1$ and hence

$$|\deg(i - P_T, \overline{B}_\eta)| = 1,$$

concluding the proof. □

We give one more result on the existence of periodic solutions based on one-sided estimates of the nonlinear term. For differential equations a result of such type was obtained by F. E. Browder [54].

Let $E = (H, \langle \cdot \rangle)$ be a Hilbert space.

Theorem 6.3.4. *Suppose that a linear operator $A : D(A) \subseteq H \to H$ and a multimap $F : \mathbb{R}_+ \times H \to Kv(H)$ satisfy the conditions* (A), (A1), (F_T), (F1), (F2), (F3′), ($F4_g$), (H1), (H2). *Assume in addition that*

(i) *$\langle Ax, x \rangle \leq 0$ for each $x \in D(A)$ and*

(ii) *there exists $R > 0$ such that every $x \in H$, $\| x \| = R$ has a neighborhood $U = U(x)$ with the property that, for all $x' \in U$ and a.e. $t \in [0, T]$ we have $\langle z, x' \rangle < 0$ for all $z \in F(t, x')$.*

Then there exists $x_0 \in H$, $\| x_0 \| < R$ such that a corresponding mild solution of the problem (6.3.1), (6.3.2) *is T-periodic.*

Proof. Note that above conditions guarantee that the translation multioperator P_T is defined on all of H and that it is condensing. Take an arbitrary x_*, $\| x_* \| \leq R$ and let $y(\cdot)$ be a mild solution of the differential inclusion (6.3.1) emanating from $y(0) = x_*$. Then

$$y(t) = \exp\{At\}x_* + \int_0^t \exp\{A(t-s)\}f(s)\,ds$$

where $f \in \mathcal{P}_F(y)$. If we suppose that $\| y(t_*) \| > R$ for some $t_* \in (0, T]$ then there exist $t_1 \in (0, t_*)$ and $\delta > 0$ such that $\| y(t_1) \| = R$ and $\| y(t) \| > R$, $y(t) \in U(y(t_1))$ for all $t \in (t_1, t_1 + \delta)$. Approximating x_* and f by sequences $\{x_n\} \subset D(A)$ and $\{f_n\}$, $f_n(t) \in D(A)$ a.e. $t \in [0, T]$ and using Proposition 5.2.1 and conditions (i), (ii), we have for a corresponding function y_n the following:

$$\begin{aligned}\frac{1}{2}\frac{d}{dt}\| y_n(t) \|^2 &\leq \langle Ay_n(t), y_n(t)\rangle + \langle f_n(t), y_n(t)\rangle \\ &\leq \langle f_n(t), y_n(t)\rangle.\end{aligned}$$

Integrating the last inequality and taking the limit as $n \to \infty$ we obtain for $t \in (t_1, t_1 + \delta)$:

$$\| y(t) \|^2 - \| y(t_1) \|^2 \leq \int_{t_1}^t \langle f(s), y(s)\rangle\,ds \leq 0,$$

giving a contradiction.

Hence the translation multioperator P_T along the solutions of the differential inclusion (6.3.1) maps the ball $\overline{B}_R$ into itself, giving the existence of a fixed point. □

Remark 6.3.2. If the multimap F is u.s.c., the condition (ii) can be simplified to the condition

(ii′) there exists $R > 0$ such that for every $x \in H$, $\| x \| = R$ and almost every $t \in [0, T]$ we have $\langle z, x\rangle < 0$ for all $z \in F(t, x)$.

The condition of condensivity of the translation multioperator can play an important role in solving certain controllability problems. We describe briefly one example using our result on the solvability of a system of inclusions (Theorem 3.6.5).

Consider the following problem. Let, as before, E be a separable Banach space; Z a Banach space of controls. It will be supposed also that the linear operator $A : D(A) \subseteq E \to E$ satisfies the conditions (A), (A1). Consider a control system whose dynamics is described by the relations

$$y'(t) = Ay(t) + f(t, x(t), u(t)), \quad t \in [0, T] \tag{6.3.4}$$

$$u(t) \in U(t, x(t)) \tag{6.3.5}$$

where $f : [0, T] \times E \times Z \to E$ is a continuous map and $U : [0, T] \times E \to K(Z)$ is an u.s.c. feedback multimap that maps bounded sets into bounded sets. A mild solution of the system (6.3.4), (6.3.5) is a pair $(y(\cdot), u(\cdot)) \in C([0, T]; E) \times L^\infty([0, T]; Z)$ such that

$$y(t) = \exp\{At\}y(0) + \int_0^t \exp\{A(t-s)\} f(s, y(s), u(s))\, ds$$
$$u(t) \in U(t, y(t)).$$

We will assume that every set

$$F(t, x) = f(t, x, U(t, x))$$

is convex for all $(t, x) \in [0, T] \times E$ and the u.s.c. multimap F besides the obviously fulfilled conditions (F1) and (F2) satisfies the conditions (F3′) and (F4$_g$). Finally we assume conditions (H1) and (H2), so we know from Theorem 6.3.1, that the translation multioperator along a mild solution of the differential inclusion

$$y'(t) \in Ay(t) + F(t, y(t)), \quad t \in [0, T] \tag{6.3.6}$$

is well defined, quasi R_δ and (k, χ)-condensing on bounded sets.

Consider the following controllability problem. Suppose $\widehat{\mathcal{R}} : E \times E \to K(E)$ is a compact acyclic multimap.

Theorem 6.3.5. *Under the above conditions, if in addition* $1 \notin \operatorname{sp} e^{AT}$ *then there exists a ball* $\overline{B}_r(0) \subset E$ *and a mild solution* $(y(\cdot), u(\cdot))$ *of the system* (6.3.4), (6.3.5) *which satisfies the boundary conditions*

$$y(0) \in \overline{B}_r(0) \tag{6.3.7}$$
$$y(T) \in y(0) + \widehat{\mathcal{R}}(y(0), y(T)) \tag{6.3.8}$$

and the optimality condition

$$j(y) = \inf_{z \in Q} j(z), \tag{6.3.9}$$

where Q is the set of all mild trajectories of the system (6.3.4), (6.3.5) *satisfying the conditions* (6.3.7), (6.3.8) *and j is a given lower semicontinuous functional.*

Proof. We consider the Banach spaces $X = E$ and $Y = C([0, T]; E)$ and apply Theorem 3.6.5 on the solvability of a system of inclusions. Let $P_T : X \to K(E)$ be the translation multioperator along the trajectories of the inclusion (6.3.6) and Σ_x^F, as usual, the set of all mild trajectories of the inclusion (6.3.6) with the initial value x.

Let us consider the multioperator $\mathcal{F} : X \times Y \multimap Y$,

$$\mathcal{F}(x, y) = \{z \in Y : z(t) = \exp\{At\}x + \int_0^t \exp\{A(t-s)\}g(s)\, ds, \quad g(\cdot) \in \mathcal{P}_F(y)\}.$$

It is easy to verify that $\mathcal{F}$ is closed and ν-condensing in the second variable on bounded sets of Y (see Theorem 5.1.4).

Define the multioperator $\mathcal{G}(x, y) : X \times Y \multimap X$ as

$$\mathcal{G}(x, y) = y(T) - \widehat{\mathcal{R}}(x, y(T)).$$

From Proposition 3.4.1 it follows that $\mathcal{G}$ is a Vietoris multimap.

Moreover, for any fixed $x \in X$ the set $S(x) = \{y : y \in \mathcal{F}(x, y)\}$ is nothing but Σ_x^F and hence is nonempty and R_δ.

Finally, the multioperator $\mathcal{Q}(x) = \mathcal{G}(x, S(x))$ can be represented as

$$\mathcal{Q}(x) = P_T(x) - \widehat{\mathcal{R}}(x, P_T(x))$$

and hence it is Vietoris (see Proposition 3.1.4) and (k, χ)-condensing on bounded sets (recall that the multimap $\widehat{\mathcal{R}}$ is compact and the Hausdorff MNC χ is algebraically semiadditive and regular).

Using the compactness of the range of the multioperator $\widehat{\mathcal{R}}$ and following the lines of the proof of Theorem 6.3.3 we can find a ball $\overline{B}_r(0) \subset X$ such that the Vietoris multimap $\mathcal{Q}$ will be homotopic on the boundary $\partial B_r(0)$ to the χ-condensing map e^{AT} and therefore $|\deg(\mathcal{Q}, \partial B_r(0)| = 1$ (the details are left to the reader).

Now we see that all the conditions of Theorem 3.6.5 are fulfilled and hence there exists a solution $(x, y) \in \overline{B}_r(0) \times Y$ of the system of inclusions

$$\begin{cases} y \in \mathcal{F}(x, y), \\ x \in \mathcal{G}(x, y). \end{cases}$$

Moreover, the (k, χ)-condensivity of the multimap $\mathcal{Q}$ and the ν-condensivity in the second variable of the multioperator $\mathcal{F}$ imply that the set of all solutions of the above system in $\overline{B}_r(0) \times Y$ is compact. This means that the set of all mild trajectories of the inclusion (6.3.6) satisfying the boundary conditions (6.3.7), (6.3.8) is compact. From the Filippov implicit function lemma (Theorem 1.3.3) we know that this set coincides with the set of all mild solutions of the control system (6.3.4), (6.3.5) satisfying the boundary conditions (6.3.7), (6.3.8). Therefore, there exists a mild trajectory $y(\cdot)$ with corresponding admissible control $u(\cdot)$ satisfying the optimality property (6.3.9). □

In conclusion of this section we consider the question of the condensivity of the translation multioperator when E is the Cartesian product of separable Banach spaces

$$E = E_1 \times \cdots \times E_N.$$

In this situation we assume that the linear operator A in E has a diagonal form

$$Ay = (A_1 y_1, \ldots, A_N y_N),$$

the multivalued nonlinearity $F : \mathbb{R}_+ \times E \to Kv(E)$ has the form of a Cartesian product

$$F(t, y) = F_1(t, y) \times \cdots \times F_N(t, y)$$

and the conditions (A_N) and (F_N) of Section 6.1 hold.

Then the inclusion (6.3.1) takes the form of a system

$$y_i'(t) \in A_i y_i(t) + F_i(t, y_1(t), \dots, y_N(t)), \quad i = 1, \dots, N. \tag{6.3.10}$$

In the space E consider the MNC κ defined by the formula

$$\kappa(\Omega) = q^{k-1} \| X(\Omega) \|_{R^N} + q^{k-2} \| LX(\Omega) \|_{R^N} + \cdots + \left\| L^{k-1} X(\Omega) \right\|_{R^N}, \tag{6.3.11}$$

where

$$\Omega \subseteq E = E_1 \times \cdots \times E_N, \quad X(\Omega) = (\chi_1(\Omega_1), \dots, \chi_N(\Omega_N)),$$

$\Omega_1, \dots, \Omega_N$ are the projections of the set Ω on the corresponding components of the space E and L is the translation operator for the time T along the trajectories of the system of ordinary differential equations (6.1.5). The norm in R^N is given by

$$\| x \|_{R^N} = \max_i |x_i|$$

and we assume that the constant q satisfies the relations $r(L) < q < 1$ where $r(L)$ is the spectral radius of L and the integer k is chosen so that $\left\| L^k \right\| < q^k$. Note that such a choice of constants q and k is possible by the exponential stability of the trivial solution of the system (6.1.5).

It is easy to see that the MNC κ is monotone, algebraically semiadditive, regular and invariant with respect to union with a compact set.

Theorem 6.3.6. *Under conditions* (A_N) *and* (F_N), *let the translation multioperator* P_T *for the system* (6.3.10) *be defined on a bounded set* $D \subset E = E_1 \times \cdots \times E_N$. *Then* P_T *is* (q, κ)*-condensing.*

Proof. Let us denote by P_τ, $\tau \le T$ the translation multioperator along the mild solutions of the system (6.3.10). Note that

$$P_t(x) \subseteq \left\{ \exp\{At\}x + \int_0^t \exp\{A(t-s)\}g(s)\,ds \ : \ g \in \mathcal{P}_F(\Sigma_x^F) \right\}. \tag{6.3.12}$$

Take $\Omega \subset D$ and set $P_{ti}(\Omega) = \mathrm{Proj}_{E_i} P_t(\Omega)$. By the monotonicity property of the Hausdorff MNC, estimates (6.1.3) and the χ-regularity property $(\widetilde{F}4_N)$ we have

$$\begin{aligned}
\chi_i\big(\big\{\exp\{A_i(t-s)\}g_i(s) \,:\, g \in \mathcal{P}_F(\Sigma_x^F),\ x \in \Omega\big\}\big)&\\
\le \chi_i\big(\big\{\exp\{A_i(t-s)\}F_i(s, \Sigma_x^F(s)) \,:\, x \in \Omega\big\}\big)&\\
\le \chi_i\,(\exp\{A_i(t-s)\}F_i(s, P_{s1}(\Omega), \dots, P_{sN}(\Omega)))&\\
\le \|\exp\{A_i(t-s)\,\|\}^{(\chi)} \cdot \chi_i(F_i(s, P_{s1}(\Omega), \dots, P_{sN}(\Omega)))&\\
\le e^{-\gamma_i(t-s)} \sum_{j=1}^{N} m_{ij}(s)\chi_j(P_{sj}(\Omega)).&
\end{aligned}$$

Hence

$$\begin{aligned}
\chi_i(P_{ti}(\Omega)) &\le \|\exp\{A_i(t-s)\}\,\|^{(\chi)} \cdot \chi_i(\Omega_i)\\
&\quad + \int_0^t e^{-\gamma(t-s)} \sum_{j=1}^{N} m_{ij}(s)\chi_j(P_{sj}(\Omega))\,ds\\
&\le e^{-\gamma_i t}\chi_i(\Omega_i) + \int_0^t e^{-\gamma_i(t-s)} \sum_{j=1}^{N} m_{ij}(s)\chi_j(P_{sj}(\Omega))\,ds
\end{aligned}$$

and therefore

$$X_i(P_t(\Omega)) \le e^{-\gamma_i t} X_i(\Omega) + \int_0^t e^{-\gamma_i(t-s)} M(s)X(P_s(\Omega))_i\,ds.$$

Applying Gronwall's lemma we obtain

$$X(P_T(\Omega)) \le LX(\Omega)$$

in the sense of component-wise comparison of vectors.

Then,

$$\begin{aligned}
\kappa(P_T(\Omega)) &= q^{k-1}\|\,X(P_T(\Omega))\,\| + q^{k-2}\big\|LX(P_T(\Omega))\big\| + \dots + \big\|L^{k-1}X(P_T(\Omega))\big\|\\
&\le q^{k-1}\big\|LX(\Omega)\big\| + q^{k-2}\big\|L^2X(\Omega)\big\| + \dots + \big\|L^kX(\Omega)\big\|\\
&\le q^{k-1}\big\|LX(\Omega)\big\| + q^{k-2}\big\|L^2X(\Omega)\big\| + \dots + q^k\big\|X(\Omega)\big\|\\
&= q(q^{k-1}\big\|X(\Omega)\big\| + q^{k-2}\big\|LX(\Omega)\big\| + \dots + \big\|L^{k-1}X(\Omega)\big\|) = q\kappa(\Omega).
\end{aligned}$$

□

6.3.2 Dissipative differential inclusions: periodic solutions and attractors

In this section we use the properties of the translation multioperator to prove the existence of a periodic solution and a global attractor for a semilinear differential inclusion satisfying a dissipativity condition in a separable Banach space E.

We assume that the translation multioperator P_T is defined on the whole space E, is quasi-R_δ and χ-condensing on bounded sets (see the previous section).

The set Y_x of all mild solutions of the inclusion (6.3.1) on $\mathbb{R}_+$ with the initial value $y(0) = x$ will be called *a thickened solution of* (6.3.1).

Definition 6.3.1. The inclusion (6.3.1) is said to be *dissipative* if there exists $d > 0$ such that for every thickened solution Y_x we have that

$$\limsup_{t\to\infty} \| Y_x(t) \| < d$$

where $Y_x(t) = \{y(t) \ : \ y \in Y_x\}$.

In other words, the inclusion (6.3.1) is dissipative provided that for every initial value $x \in E$ there exists $t(x) \geq 0$ such that each mild solution $y(t) = y(t, x)$ of the inclusion (6.3.1) satisfies

$$\| y(t) \| < d \quad \text{for } t \geq t(x).$$

Theorem 6.3.7. *Let us assume that*

(i) *the operator A is the infinitesimal generator of a decreasing semigroup:*

$$\| \exp\{At\} \| \leq Ne^{-\rho t}, \quad N \geq 1, \ \rho > 0;$$

(ii) $\| F(t, x) \| \leq \alpha + \beta \| x \|$ *where* $\beta < \rho/N$.

Then the differential inclusion (6.3.1) *is dissipative and the set of all mild solutions emanating from any bounded set $D \subset E$ is bounded.*

Proof. Take $R > 0$ so large that $\alpha/R + \beta = \delta < \rho/N$.

Let $y(t)$ be any mild solution of the problem (6.3.1) with initial value $x \in E$. If we suppose that $\| y(s) \| > R$ for all s on some interval $[t_0, a)$ we have, for $t \in [t_0, a)$, the following estimate:

$$\| y(t) \| \leq Ne^{-\rho(t-t_0)} \| y(t_0) \| + \int_{t_0}^{t} Ne^{-\rho(t-s)}(\alpha + \beta \cdot \| y(s) \|)\, ds$$

$$< Ne^{-\rho t}(e^{\rho t_0} \cdot \| y(t_0) \| + \int_{t_0}^{t} e^{\rho s}\delta \cdot \| y(s) \|\, ds).$$

Denoting the last expression by $z(t)$ we have that for a.e. $t \in [t_0, a)$

$$z'(t) = -\rho z(t) + N\delta \cdot \| y(t) \| < -\rho z(t) + N\delta \cdot z(t)$$

and using a standard theorem on differential inequalities we obtain

$$\| y(t) \| < z(t) \leq z(t_0) \cdot e^{(-\rho+N\delta)(t-t_0)} = N \cdot \| y(t_0) \| \cdot e^{(-\rho+N\delta)(t-t_0)}$$

and therefore for any initial value $x \in E$ there exists a moment $t_1 = t_1(x) \geq 0$ before which every mild solution $y(t, x)$ emanating from x will reach the ball $\overline{B}_R(0) \subset E$. Moreover from the above estimate we can see that, if $\| y(t') \| \leq R$ for any $t' \geq 0$, then $y(t)$ never leaves the ball $\overline{B}_{NR}(0) \subset E$ for $t \geq t'$.

The second conclusion of the theorem follows from the fact that, if Y_x is a thickened solution of (6.3.1) then

$$\| Y_x(t) \| \leq N \cdot \max\{R, \| x \|\}, \quad t \geq 0.$$

□

We prove the following proposition on the existence of a periodic solution for a dissipative inclusion. Given a translation multioperator P_T and a bounded set $D \subset E$, the set

$$\Gamma(D) = \bigcup_{n \geq 0} P_T^n(D)$$

is called the *orbit of D under P_T*.

Theorem 6.3.8. *Let the differential inclusion* (6.3.1) *be dissipative and the orbits of bounded sets under P_T bounded. Then the inclusion* (6.3.1) *has a T-periodic solution.*

Proof. Let $\overline{B}_0 \subset E$ be the ball $\{x : \| x \| \leq d\}$ where d is the number appearing in the Definition 6.3.1. Take balls $B_1, B \subset E$ such that $\overline{B}_0 \subset B_1 \subset \overline{B}_1 \subset B$ and $\Gamma(\overline{B}_0) \subset B_1$, $\Gamma(\overline{B}_1) \subset B$.

Consider now the translation multioperator P_T on the ball $\overline{B}$. We know that P_T is condensing with respect to the Hausdorff MNC in E and hence there exists a compact fundamental set $\boldsymbol{M}$ of P_T such that $\boldsymbol{M} \cap B_0 \neq \emptyset$.

Applying the dissipativity condition we obtain that for every point $x \in \overline{B}_1$ there exists a number $m(x)$ such that $P_T^n(x) \subset B_0$ for all $n \geq m(x)$. From the upper semicontinuity of the multioperator P_T it follows that there exists a neighborhood $V(x)$ of x such that $P_T^n(x') \subset B_0$ for all $x' \in V(x)$ and $n \geq m(x)$. The neighborhoods $V(x)$ cover the compact set $\overline{B}_1 \cap \boldsymbol{M}$, so we can select a finite cover $V(x_1), \ldots, V(x_k)$. For $m = \max\{m(x_1), \ldots, m(x_k)\}$ we will have

$$P_T^m(\overline{B}_1 \cap \boldsymbol{M}) \subset B_0.$$

In the asymptotic fixed point Theorem 3.4.4 we dealt with the restriction of a multimap to its essential fundamental set. We can apply it to conclude that there exists a fixed point $x \in B_0$ of P_T. This fixed point is the initial value from which the mild T-periodic solution starts.

□

Corollary 6.3.1. *Under conditions* (i) *and* (ii) *of Theorem* 6.3.7 *there exists a mild T-periodic solution of the differential inclusion* (6.3.1).

Our aim now is to describe the structure of the set of mild solutions of a dissipative differential inclusion (6.3.1). We show that this set has a global compact attractor.

To be more precise we need the following definition which extends a concept from differential equations (see, e.g. [148]).

Let $\mathcal{A}, \mathcal{B} \subset E$ be bounded sets. The deviation ρ of the set A from the set B is

$$\rho(\mathcal{A}, \mathcal{B}) = \sup_{a \in \mathcal{A}} \operatorname{dist}(a, \mathcal{B}).$$

Definition 6.3.2. A set $\mathcal{M} \subset C([0, +\infty); E)$ of bounded mild solutions of the inclusion (6.3.1) is said to be a Λ-*center* of the inclusion (6.3.1) if: (i) the set $\mathcal{M}(0) = \{m(0) : m \in \mathcal{M}\} \subset E$ is closed and P_T-invariant in the following sense: $P_T(x) \cap \mathcal{M}(0) \neq \emptyset$ for every $x \in \mathcal{M}(0)$ and $P_T(\mathcal{M}(0)) \supseteq \mathcal{M}(0)$; (ii) for every thickened solution Y_x of the inclusion (6.3.1) we have

$$\lim_{n \to \infty} \rho(Y_x(nT), \mathcal{M}(0)) = 0$$

and

$$\lim_{t \to \infty} \rho(Y_x(t), \mathcal{M}(t)) = 0.$$

Theorem 6.3.9. *Every dissipative differential inclusion* (6.3.1) *has a Λ-center which is compact in the sense of uniform convergence on every finite interval.*

Before proving this theorem we need some preliminaries.

Lemma 6.3.1. *If the inclusion* (6.3.1) *is dissipative then the orbit $\Gamma(x)$ of every point $x \in E$ under P_T is relatively compact.*

Proof. From the dissipativity condition and the fact that P_T is χ-condensing (and hence maps bounded sets into bounded ones) it follows that the set $\Gamma(x)$ is bounded. Applying the χ-condensivity of P_T to the equality

$$\Gamma(x) = P_T(\Gamma(x)) \cup \{x\}$$

we obtain that $\chi(\Gamma(x)) = 0$. □

Definition 6.3.3. A point $z \in E$ is said to be an *ω-limit point of the orbit* $\Gamma(x)$ if there exist a sequence of numbers $n_k \to \infty$ and a sequence $\{p_{n_k}\} \subset \Gamma(x)$, $p_{n_k} \in P_T^{n_k}(x)$ such that $p_{n_k} \to z$. The set of all ω-limit points of $\Gamma(x)$ is said to be an ω-limit set of $\Gamma(x)$ and it is denoted $\Omega(x)$.

It is easy to see that

$$\Omega(x) = \bigcap_{n\geq 0} \overline{\bigcup_{k\geq n} P_T^k(x)}$$

hence Lemma 6.3.1 implies that $\Omega(x)$ is a nonempty compact subset of $\overline{\Gamma(x)}$ if the inclusion (6.3.1) is dissipative. It is clear also that $\Omega(x)$ is contained in the ball $B_0 = \{x : \| x \| < d\}$.

Lemma 6.3.2. *If the inclusion* (6.3.1) *is dissipative then the set* $\Omega(x)$ *has the following properties:* (i) $\Omega(x)$ *is* P_T*-invariant in the sense that* $P_T(z) \cap \Omega(x) \neq \emptyset$ *for every* $z \in \Omega(x)$ *and* $P_T(\Omega(x)) \supseteq \Omega(x)$*;* (ii) *for every thickened solution* Y_x *of the inclusion* (6.3.1) *we have*

$$\lim_{n\to\infty} \rho(Y_x(nT), \Omega(x)) = 0.$$

Proof. Let $z \in \Omega(x)$ and $\{p_{n_k}\} \subset \Gamma(x)$ be a sequence such that $p_{n_k} \in P_T^{n_k}(x)$ and $p_{n_k} \to z$. Take arbitrary points $w_{n_k} \in P_T(p_{n_k}) \subseteq P_T^{n_k+1}(x)$. From Lemma 6.3.1 it follows that the sequence $\{w_{n_k}\}$ is relatively compact hence we can find a subsequence $\{w_{n_{k_j}}\}$ convergent to $w \in \Omega(x)$. The u.s.c. multioperator P_T is closed, therefore the relations $w_{n_{k_j}} \in P_T(p_{n_{k_j}})$, $p_{n_{k_j}} \to z$, $w_{n_{k_j}} \to w$ imply $w \in P_T(z)$ yielding $P_T(z) \cap \Omega(x) \neq \emptyset$.

On the other hand, for the same point $z \in \Omega(x)$ we have $p_{n_k} \in P_T^{n_k}(x) = P_T(P_T^{n_k-1}(x))$ and hence we can choose the sequence $v_{n_k} \in P_T^{n_k-1}(x)$ such that $p_{n_k} \in P_T(v_{n_k})$. Again using the relative compactness of the sequence $\{v_{n_k}\}$ we can choose the subsequence $v_{n_{k_j}}$ converging to $v \in \Omega(x)$. The same argument as above implies that $z \in P_T(v)$ and therefore $P_T(\Omega(x)) \supseteq \Omega(x)$.

Property (ii) of $\Omega(x)$ clearly follows from Lemma 6.3.1. □

Proof of Theorem 6.3.9. Consider the set $M = \bigcup_{x\in E} \Omega(x)$. It is clear that $M \subset B_0$ and it is also P_T-invariant in the above sense. The relation $P_T(M) \supseteq M$ implies its relative compactness. It is easy to verify that the compact set $\overline{M}$ is also P_T-invariant.

Let $\mathcal{M} \subset C([0, +\infty); E)$ be the set of all mild solutions of the inclusion (6.3.1) originating from the set $\overline{M}$. This is the required set. In fact, its compactness in the topology of uniform convergence on finite intervals follows from the compactness of $\overline{M}$ and the u.s.c. dependence of the solutions set on initial data, see Corollary 5.2.2.

Furthermore, from the same u.s.c. dependence it follows that, since the set $\overline{M}$ is compact, for given $\varepsilon > 0$ there exists $\delta > 0$ such that $\operatorname{dist}(x, \overline{M}) < \delta$, $x \in E$ implies $\rho(\Sigma(x)(t), \Sigma(\overline{M})(t)) < \varepsilon$ for $t \in [0, T]$ where $\Sigma(x) = \Sigma_x^F[0, T]$. Therefore for any thickened solution Y_x we have that the relation

$$\rho(Y_x(nT), \overline{M}) \leq \rho(Y_x(nT), \Omega(x)) \underset{n\to\infty}{\longrightarrow} 0$$

implies

$$\lim_{t\to+\infty} \rho(Y_x(t), \mathcal{M}(t)) = 0.$$

□

Remark 6.3.3. It is easy to see that the constructed Λ-center is minimal in the sense that it is contained in any other Λ-center of the dissipative differential inclusion (6.3.1).

Bibliographic notes

The method of integral multioperators in periodic problems for differential inclusions in a finite-dimensional space appears in [159], [93], [198], [176], [177]. In the infinite-dimensional case conditions under which the integral multioperator of a periodic problem for SDIBS is condensing were suggested in the papers of M. I. Kamenskii and V. V. Obukhovskii [133], [134]. For the case of systems of inclusions we refer the reader to the work of M. I. Kamenskii, P. Nistri and P. Zecca [132]. The method of integral multioperators was applied to boundary problems for SDIBS's on infinite intervals in the works of P. Zecca and P. Zezza [219] and Z. Ding and A. G. Kartsatos [77]. In Section 6.1 we study the case when the linear part of a SDIBS generates a C_0-semigroup. The techniques of analysis of the integral multioperators are those developed in Chapter 5.

The scheme of presentation of the averaging principle for a periodic problem of SDIBS follows the paper of M. I. Kamenskii, P. Nistri and P. Zecca [132].

The method of the translation multioperator plays an important role in the study of periodic problems (see, e.g. [145], [148]). However, its application is often restricted by the assumption of uniqueness of the solution. On the other hand, the extension of the method to the case of differential equations in infinite-dimensional spaces is complicated by the fact that the translation operator possesses the compactness property only in special cases (see [145]). In recent years the methods of topological degree theory for acyclic multimaps were applied to the study of the multivalued translation operator along the trajectories of differential equations and inclusions in finite-dimensional spaces (see the works of L. Górniewicz [103], [104] and the references therein). For control processes governed by systems of differential equations in Banach spaces the multivalued translation operator was studied in [131] and [135]. Section 6.3.1 is based on the paper of M. I. Kamenskii, V. V. Obukhovskii and P. Zecca [137]. The example of application of the translation multioperator to the controllability problem is taken from the paper [131].

The statements of Section 6.3.2 extend some results known for dissipative ordinary differential equations (see, e.g. [97], [96], [148], [113], [114]). They are based on the paper of V. V. Obukhovskii and P. Zecca [183].

Bibliography

[1] V. N. Afanasiev, V. B. Kolmanovskii, and V. R. Nosov. *Mathematical Theory of Control Systems Design*. Math. Appl. 341, Kluwer Academic Publishers, Dordrecht 1995.

[2] N. U. Ahmed. *Optimization and Identification of Systems Governed by Evolution Equations on Banach Spaces*. Pitman Res. Notes Math. Ser. 184, Longman Sci. Tech., Harlow 1988.

[3] N. Ahmed and K. Teo. *Optimal Control of Distributed Parameter Systems*. North Holland, New York 1981.

[4] R. R. Akhmerov, M. I. Kamenskii, A. S. Potapov, A. E. Rodkina, and B. N. Sadovskii. *Measures of Noncompactness and Condensing Operators*. Birkhäuser, Boston–Basel–Berlin 1992.

[5] R. R. Akhmerov, M. I. Kamenskii, A. S. Potapov, and B. N. Sadovskii. Condensing Operators. *J. Soviet Math.* **18** (1982), 551–592.

[6] W. Allegretto and P. Nistri. Periodic solutions and optimization problems for a class of semilinear parabolic control systems. *Topol. Methods Nonlinear Anal.* **5** (1995), 345–356.

[7] A. Andreini, M. Kamenskii, and P. Nistri. A result on the singular perturbation theory for differential inclusions in Banach spaces. *Topol. Methods Nonlinear Anal.* **15** (2000), 15–32.

[8] G. Anichini. Approximate selections for non convex set valued mappings. *Boll. Un. Mat. Ital.* B (7) **4** (1990), 313–326.

[9] G. Anichini, G. Conti, and P. Zecca. Approximation of nonconvex set-valued mappings. *Boll. Un. Mat. Ital.* C (6) **4** (1985) , 145–154.

[10] G. Anichini, G. Conti, and P. Zecca. A further result on the approximation of nonconvex set–valued mappings. *Boll. Un. Mat. Ital.* C (6) **4** (1985), 155–171.

[11] G. Anichini, G. Conti, and P. Zecca. Un teorema di selezione per applicazioni multivoche a valori non convessi. *Boll. Un. Mat. Ital.* C (6) **5** (1986), 315–320.

[12] G. Anichini and P. Zecca. Multivalued differential equations in Banach spaces. An application to control theory. *J. Optim. Theory Appl.* **21** (1977), 477–486.

[13] J. Appell. Multifunctions of two variables: Examples and counterexamples. In: *Topology in Nonlinear Analysis*, Banach Center Publ. 35, Inst. Math. Polish Acad. Sci., Warszawa 1996, 119–128.

[14] J. Appell, E. De Pascale, H. T. Nguyen, and P. P. Zabreiko. Multi-valued superpositions. *Dissertationes Math.* **345** (1995), 1–97.

[15] J. P. Aubin. *Applied Functional Analysis*. Wiley-Interscience, New York 1979.

[16] J. P. Aubin. *Viability Theory*. Birkhäuser, Boston–Basel–Berlin 1991.

[17] J. P. Aubin. *Optima and Equilibria*. Springer-Verlag, Berlin–Heidelberg–New York 1993.

[18] J. P. Aubin and A. Cellina. *Differential Inclusions. Set-Valued Maps and Viability Theory*. Springer-Verlag, Berlin–Heidelberg–New York–Tokyo 1984.

[19] J. P. Aubin and I. Ekeland. *Applied Nonlinear Analysis*. Wiley-Interscience, New York 1984.

[20] J. P. Aubin and H. Frankowska. *Set-Valued Analysis*. Birkhäuser, Boston–Basel–Berlin 1990.

[21] R. Bader. On the semilinear multi-valued flow under constraints and the periodic problem. In: *Differential Inclusions and Optimal Control* (J. Andres, L. Górniewicz and P. Nistri, eds.), Lecture Notes in Nonlinear Anal. 2, 1998, 51–55.

[22] R. Bader and W. Kryszewski. Fixed-point index for compositions of set-valued maps with proximally ∞-connected values on arbitrary ANR's. *Set-Valued Anal.* **2** (1994), 459–480.

[23] A. V. Balakrishnan. *Applied Functional Analysis*, 2nd ed., Springer-Verlag, New York–Heidelberg–Berlin 1981.

[24] J. N. Ball. Strongly continuous semigroups, weak solutions, and the variation of constants formula. *Proc. Amer. Math. Soc.* **63** (1977), 370–373.

[25] J. Banas and K. Goebel. *Measures of Noncompactness in Banach spaces*. Marcel Dekker, New York 1980.

[26] V. Barbu. *Nonlinear Semigroups and Differential Equations in Banach Spaces*, Noordhoff, Leyden 1976.

[27] C. Berge. *Théorie Générale des Jeux à n Personnes*, Gauthier-Villars, Paris 1957.

[28] C. Berge. *Espaces Topologiques. Fonctions Multivoques*, 2nd ed., Dunod, Paris 1966.

[29] V. I. Blagodatskih and A. F. Filippov. Differential inclusions and optimal control. *Trudy Mat. Inst. AN SSSR* **169** (1985), 194–253 (in Russian).

[30] N. A. Bobylev and S. K. Korovin. *Topological Methods in Variational Problems*. Publ. House of the Fac. of Math. of Comp. and Cybern., Moscow 1997 (in Russian).

[31] H. Bohnenblust and S. Karlin. On a theorem of Ville. In: *Contributions to the Theory of Games* (H. W. Kuhn and A. W. Tucker, eds.), Ann. of Math. Stud. 24, Princeton Univ. Press, Princeton 1950, 155–160.

[32] Yu. G. Borisovich. An application of the concept of rotation of a vector field. *Soviet Math. Dokl.* **4** (1963), 1584–1587.

[33] Yu. G. Borisovich. On the relative rotation of compact vector fields in linear spaces. *Trudy Sem. Funktsion. Anal. Voronezh Gos. Univ.* **12** (1969), 3–27 (in Russian).

[34] Yu. G. Borisovich. The relative rotation of compact vector fields and the Lefschetz number. *Trudy Sem. Funktsion. Anal. Voronezh Gos. Univ.* **12** (1969), 28–42 (in Russian).

[35] Yu. G. Borisovich. Modern approach to the theory of topological characteristics of nonlinear operators. I. In: *Global Analysis – Studies and Applications II*, Lecture Notes in Math. 1334, Springer-Verlag 1988, 199–220.

[36] Yu. G. Borisovich. A modern approach to the theory of topological characteristics of nonlinear operators. II. In: *Global analysis – studies and applications IV*, Lecture Notes in Math. 1453, Springer-Verlag 1990, 21–50.

[37] Yu. G. Borisovich. Global analysis of operator equations appearing in singular problems of optimizations. *Dokl. Ross. Akad. Nauk* **355** (1997), 11–13 (in Russian).

[38] Yu. G. Borisovich. A generalized degree of multivalued mappings and its applications to nonlinear problems. *Nonlinear Anal.* **30** (1997), 101–109.

[39] Yu. G. Borisovich, B. D. Gelman, E. Mukhamadiev and V. V. Obukhovskii. On the rotation of multi–valued vector fields. *Soviet Math. Dokl.* **10** (1969), 956–958.

[40] Yu. G. Borisovich, B. D. Gelman, A. D. Myshkis, and V. V. Obukhovskii. Topological methods in the fixed-point theory of multi-valued maps. *Russian Math. Surveys* **35** (1980), 65–143.

[41] Yu. G. Borisovich, B. D. Gelman, A. D. Myshkis, and V. V. Obukhovskii. Multivalued mappings. *J. Soviet Math.* **24** (1984), 719–791.

[42] Yu. G. Borisovich, B. D. Gelman, A. D. Myshkis, and V. V. Obukhovskii. *Introduction to the theory of multivalued maps.* Publishing House of the Voronezh University, Voronezh 1986 (in Russian).

[43] Yu. G. Borisovich, B. D. Gelman, A. D. Myshkis, and V. V. Obukhovskii. Multivalued analysis and operator inclusions. *J. Soviet Math.* **39** (1987), 2772–2811.

[44] Yu. G. Borisovich, B. D. Gelman, A. D. Myshkis, and V. V. Obukhovskii. New results in the theory of multivalued mappings. I. Topological characteristics and solvability of operator relations. *J. Soviet Math.* **49** (1990), 800–855.

[45] Yu. G. Borisovich and Yu. E. Gliklikh. On the Lefschetz number for a class of multi–valued maps. *Seventh Math. Summer School* (Katsiveli 1969), Izd. Akad. Nauk Ukrain. SSR, Kiev 1970, 283–294 (in Russian).

[46] Yu. G. Borisovich and V. V. Obukhovskii. On the optimization problem for controlled systems of parabolic type. *Proc. Steklov Inst. Math.* **211** (1995), 85–91.

[47] Yu. G. Borisovich and Yu. I. Sapronov. Topological theory of compactly restrictible maps. *Trudy Sem. Funktsional. Anal.* Voronezh Gos. Univ. **12** (1969), 43–68 (in Russian).

[48] K. Borsuk. *Theory of Retracts.* Monografie Mat. 44, PWN, Warszawa 1967.

[49] D. Bothe. Semilinear differential inclusions with application to hybrid system. Submitted to SIAM J. Math. Anal.

[50] A. Bressan. On the qualitative theory of lower semicontinuous differential inclusions, *J. Differential Equations* **77** (1989), 379–391.

[51] A. Bressan. Upper and lower semicontinuous differential inclusions: a unified approach, In: *Nonlinear Controllability and Optimal Control* (H. Sussmann, ed.), Marcel Dekker, 1990, 21–31.

[52] A. Bressan and V. Staicu. On nonconvex perturbations of maximal monotone differential inclusions, *Set-Valued Anal.* **2** (1994), 415–437.

[53] F. E. Browder. On a generalization of the Schauder fixed point theorem. *Duke Math. J.* **26** (1959), 291–303.

[54] F. E. Browder. Existence of periodic solutions for nonlinear equations of evolution. *Proc. Nat. Acad. Sci. USA* **53** (1965), 1100–1103.

[55] J. Bryszewski and M. Serbinowski. On the minimal displacement under admissible maps. *Bull. Acad. Polon. Sci. Ser. Sci. Math.* **27** (1979), 201–207.

[56] C. Castaing and M. Valadier. *Convex Analysis and Measurable Multifunctions*. Lecture Notes in Math. 580, Springer-Verlag, Berlin–Heidelberg–New York 1977.

[57] A. Cellina. The role of approximation in the theory of multivalued mappings. In: *Differential games and related topics*, Proc. Internat. Summer School, Varenna 1970, North-Holland, Amsterdam 1971, 209–220.

[58] A. Cellina and A. Lasota. A new approach to the definition of topological degree for multi-valued mappings. *Atti Accad. Naz. Lincei Rend. Cl. Sci. Fis. Mat. Natur.* (8) **47** (1969), 434–440.

[59] S. S. Ceron, O. Lopes. α-Contractions and attractors for dissipative semilinear hyperbolic equations and systems. *Ann. Mat. Pura Appl.* **160** (1991), 193–206.

[60] K. C. Chang. Free boundary problems and the set-valued mappings. *J. Differential Equations* **49** (1983), 1–28.

[61] G. Conti, W. Kryszewski, and P. Zecca. On the solvability of systems of non-convex inclusions in Banach spaces. *Ann. Mat. Pura Appl.* (IV) **160** (1991), 371–408.

[62] G. Conti, and P. Nistri. A fixed point theorem for noncompact acyclic-valued maps. *Rend. Accad. Sci. Fis Nat. Napoli (4)* **42** (1975), 510–517.

[63] G. Conti, P. Nistri, and P. Zecca. Systems of set-valued equations in Banach spaces. In: *Delay Differential Equations and Dynamical Systems* (S. Busenberg and M. Martelli, eds.) Lecture Notes in Math. 1475, Springer-Verlag 1991, 98–109.

[64] G. Conti, V. Obukhovskii, and P. Zecca. On the topological structure of the solution set for a semilinear functional-differential inclusion in a Banach space. In: *Topology in Nonlinear Analysis*, Banach Center Publ. 35, Inst. of Math. Polish Acad. of Sci., Warszawa 1996, 159–169.

[65] G. Conti, V. Obukhovskii, and P. Zecca. The topological degree theory for a class of noncompact multimaps. *Ann. Math. Pura Appl.* (IV) **178** (2000), 103–113.

[66] G. Conti and J. Pejsachowicz. Fixed points theorems for multivalued weighted maps. *Ann. Math. Pura Appl.* (IV) **126** (1980), 319–341.

[67] J. F. Couchouron and M. Kamenskii. A unified topological point of view for integro-differential inclusions. In: *Differential Inclusions and Optimal Control* (J. Andres, L. Górniewicz and P. Nistri, eds.), Lecture Notes in Nonlinear Anal. 2, 1998, 123–137.

[68] J. F. Couchouron and M. Kamenskii. An abstract topological point of view and a general averaging principle in the theory of differential inclusions. *Nonlinear Anal.* **42** (2000), 625–651.

[69] J. Daneš. Some fixed point theorems. *Comment. Math. Univ. Carolin.* **9** (1968), 223–235.

[70] G. Darbo. Punti uniti in trasformazioni a codominio non compatto. *Rend. Sem. Mat. Univ. Padova* **24** (1955), 84–92.

[71] F. S. De Blasi. Characterizations of certain classes of semicontinuous multifunctions by continuous approximations. *J. Math. Anal. Appl.* **106** (1985), 1–18.

[72] K. Deimling. *Nonlinear Functional Analysis.* Springer-Verlag, Berlin–Heidelberg–New York–Tokyo 1985.

[73] K. Deimling. *Multivalued Differential Equations.* De Gruyter Ser. Nonlinear Anal. Appl. 1, Walter de Gruyter, Berlin–New York 1992.

[74] E. De Pascale and R. Guzzardi. On the boundary-values dependence for the topological degree of multi-valued noncompact maps. *Boll. Un. Mat. Ital.* A (5) **13** (1976), 110–116.

[75] J. Diestel. Remarks on weak compactness in $L_1(\mu, X)$. *Glasgow Math. J.* **18** (1977), 87–91.

[76] J. Diestel, W. M.Ruess and W. Schachermayer. Weak compactness in $L_1(\mu, X)$. *Proc. Amer. Math. Soc.* **118** (1993), 447–453.

[77] Z. Ding and A.G. Kartsatos. Nonresonance problems for differential inclusions in separable Banach spaces. *Proc. Amer. Math. Soc.* **124** (1996), 2357–2365.

[78] Do Hong Tan. On continuity of fixed points of multivalued collectively condensing mappings. *Indian J. Pure Appl. Math.* **15** (1984), 631–632.

[79] A. Dold. *Lectures in Algebraic Topology.* Springer-Verlag, Berlin 1972.

[80] R. Dragoni, J. W. Macki, P. Nistri, and P. Zecca. *Solution Sets of Differential Equations in Abstract Spaces.* Pitman Res. Notes Math. Ser. 342, Longman Sci. Tech., Harlow 1996.

[81] N. Dunford and J. T. Schwartz. *Linear Operators, Part I.* Interscience, New York 1958.

[82] Z. Dzedzej. Fixed point index theory for a class of nonacyclic multivalued maps. *Dissertationes Math.* **253** (1985), 1–53.

[83] Z. Dzedzej. Equivariant selections and approximations. In: *Topological Methods in Nonlinear Analysis* (Z. Kucharski, ed.), Gdańsk Sci. Soc., Gdańsk 1997, 25–31.

[84] I. Ekeland and R. Teman. *Convex Analysis and Variational Problems*. North Holland, Amsterdam, 1976.

[85] V. V. Fedorchuk and V. V. Filippov. *General Topology. Basic Constructions*. Moscow University Press, Moscow 1988 (in Russian).

[86] O. P. Filatov and M. M. Hapaev. *The Averaging of Systems of Differential Inclusions*. Moscow Univ. Publ., Moscow 1998 (in Russian).

[87] A. F. Filippov. On some problems of the theory of optimal control. *Vestn. Moscow Univ.* **2** (1959), 25–32 (in Russian).

[88] P. M. Fitzpatrick and W. V. Petryshyn. Fixed point theorems for multivalued noncompact acyclic mappings. *Pacif. J. Math.* **54** (1974), 17–23.

[89] H. Frankowska. A priori estimates for operational differential inclusions, *J. Differential Equations* **84** (1990), 100–128.

[90] A. Fryszkowski. Continuous selections for a class of non-convex multivalued maps. *Studia Math.* **76** (1983), 163-174.

[91] M. Furi and M. Martelli. On the minimal displacement under acyclic-valued maps defined on a class of ANR's. *Boll. Un. Mat. Ital.* (4) **11** (1975) (Suppl. fasc.), 238–246.

[92] D. Gabor and W. Kryszewski, On the solvability of systems of noncompact and nonconvex inclusions. *Differential Equations Dynam. Systems* **6** (1998), 377–403.

[93] B. D. Gelman. Multivalued integral operators and ω-periodic solutions. *Voronezh Gos. Univ. Trudy Mat. Fak.* **4** (1970), 35–44 (in Russian).

[94] B. D. Gelman and V. V. Obukhovskii. Some fixed point theorems for multivalued maps of condensing type. In: *Algebraic Problems of Analysis and Topology*, Publishing House of the Voronezh University, Voronezh 1990, 110–115 (in Russian).

[95] B. D. Gelman and V. V. Obukhovskii. New results in the theory of multivalued mappings. II. Analysis and applications. *J. Soviet Math.* **64** (1993), 853–883.

[96] V. M. Gershtein. On the theory of dissipative differential equations in a Banach space. *Funktsional Anal. i Prolozhen.* **4** (1970), 99–100 (in Russian).

[97] V. M. Gershtein and M. A. Krasnoselskii. Structure of the set of solutions of dissipative equations. *Soviet Math. Dokl.* **9** (1968), 1356–1359.

[98] Yu. E. Gliklikh. Fixed points of multi-valued maps with non-convex images and the rotation of multi-vector fields. *Trudy Aspirantov Mat. Fak. Voronezh Gos. Univ.* **1** (1971), 30–38.

[99] K. Goebel and W. A. Kirk. *Topics in Metric Fixed Points Theory.* Cambridge Univ. Press, Cambridge 1990.

[100] S. Goldberg. *Unbounded Linear Operators: Theory and Applications.* McGraw-Hill, New York 1966.

[101] L. S. Goldenstein, I. Gohberg and A. S. Markus. Investigation of some properties of bounded linear operators in the connection with their g-norm. *Uchen. Zap. Kishinev. Gos. Univ.* **29** (1957), 29–36 (in Russian).

[102] L. Górniewicz. Homological methods in fixed point theory of multivalued maps. *Dissertationes Math.* **129** (1976), 1–71.

[103] L. Górniewicz. Topological approach to differential inclusions. In: *Proc. Conference on Topological Methods in Differential Equations and Inclusions* Kluwer Academic Publishers, Dordrecht 1994.

[104] L. Górniewicz. *Topological Fixed Point Theory of Multivalued Mappings.* Kluwer Academic Publishers, Dordrecht–Boston–London 1999.

[105] L. Górniewicz, A. Granas, and W. Kryszewski. On the homotopy method in the fixed point index theory of multi-valued mappings of compact absolute neighborhood retracts. *J. Math. Anal. Appl.* **161** (1991), 457–473.

[106] L. Górniewicz and Z. Kucharski. Coincidence of k-set contraction pairs. *J. Math. Anal. Appl.* **107** (1985), 1–15.

[107] L. Górniewicz and M. Lassonde. Approximation and fixed points for compositions of R_δ-maps. *Topology Appl.* **55** (1994), 239–250.

[108] L. Górniewicz, P. Nistri, and V. Obukhovskii. Differential inclusions on proximate retracts of Hilbert spaces. *Internat. J. Non-Linear Differential Equations, Theory, Methods, Appl.* **3** (1997), 13–26.

[109] L. Górniewicz, P. Nistri, and P. Zecca. Control problems on closed subsets of $\mathbb{R}^n$ via feedback controls. *Topol. Methods Nonlinear Anal.* **2** (1993), 163–178.

[110] I. N. Gourova. A statement of the type of the relatedness principle and the second theorem of N. N. Bogolyubov in the averaging principle of parabolic equations. *Qualitative and Appl. Meth. of the Study of Operator Equat.*, Yaroslavl 1982, 47–58 (in Russian).

[111] O. Hadžić. *Fixed Point Theory in Topological Vector Spaces.* Univ. of Novi Sad, Inst. of Math., Novi Sad 1984.

[112] J. K. Hale. Continuous dependence of fixed points of condensing maps. *J. Math. Anal. Appl.* **46** (1974), 388–394.

[113] J. K. Hale. *Theory of Functional Differential Equations*. Springer-Verlag, New York–Heidelberg–Berlin 1977.

[114] J. K. Hale. *Asymptotic Behavior of Dissipative Systems*. Math. Surveys Monogr. 25, Amer. Math. Soc., Providence, RI, 1988.

[115] J. K. Hale, S. M. Verduyn Lunel. *Introduction to Functional Differential Equations*. Springer-Verlag, New York–Heidelberg–Berlin 1993.

[116] E. Hille and R. S Phillips. *Functional Analysis and Semigroups*. 3rd ed. Amer. Math. Soc., Providence, RI, 1974.

[117] C. J. Himmelberg, J. R. Porter and F. S. Van Vleck. Fixed point theorems for condensing multifunctions. *Proc. Amer Math. Soc.* **23** (1969), 635–641.

[118] S. Hu and N. S. Papageorgiou. *Handbook of Multivalued Analysis, Vol. I: Theory*. Kluwer Academic Publishers, Dordrecht–Boston–London 1997.

[119] S. Hu and N. S. Papageorgiou. *Time-Dependent Subdifferential Evolution Inclusions and Optimal Control*. Mem. Amer. Math. Soc. 133 (632), Amer. Math. Soc., Providence, RI, 1998.

[120] D. M. Hyman. On decreasing sequences of compact absolute retracts. *Fund. Math.* **64** (1969), 91–97.

[121] A. D. Ioffe and V. M. Tihomirov. *Theory of Extremal Problems*. North Holland, Amsterdam 1979.

[122] Ya. A. Izrailevich. On the concept of relative rotation of multi-valued vector field. *Trudy Sem. Funktsional. Anal. Voronezh Gos. Univ* **12** (1969), 111–115. (in Russian).

[123] Ya. A. Izrailevich and V. V. Obukhovskii. On equivariant multi-valued maps. *Soviet Math. Dokl.* **13** (1972), 864–867.

[124] Ya. A. Izrailevich and V. V. Obukhovskii. Some topological characteristics of equivariant multi-valued maps. *Voronezh Gos. Univ. Trudy Mat. Fak.* **10** (1973), 52–61 (in Russian).

[125] M. Q. Jacobs. Remarks on some recent extensions of Filippov's implicit functions lemma. *SIAM J. Control Optim.* **5** (1967), 622–627.

[126] M. Q. Jacobs. Measurable multivalued mappings and Lusin's theorem. *Trans. Amer. Math. Soc.* **134** (1968), 471–481.

[127] M. I. Kamenskii. Applications des operateurs condencents au problemes des solutions periodiques pour les equations paraboliques quasi-lineaires. *C. R. Acad. Sci. Paris Sér. I Math.* **318** (1994), 719–722.

[128] M. I. Kamenskii. On the averaging principle for quasilinear parabolic equations with delay. *Russian Acad. Sci. Dokl. Math.* **50** (1995), 66–69.

[129] M. I. Kamenskii. On a modificattion of the averaging principle for degenerated equations. *Dokl. Ross. Akad. Nauk.* **347** (1996), 151–153 (in Russian).

[130] M. I. Kamenskii. On a the investigation of the stability of periodic solutions for a new class of systems of quasilinear equations in a Banach space. *Dokl. Ross. Akad. Nauk.* **353** (1997), 13–16 (in Russian).

[131] M. I. Kamenskii, P. Nistri, V. V. Obukhovskii, and P. Zecca. Optimal feedback control for a semilinear evolution equation. *J. Optim. Theory Appl.* **82** (1994), 503–517.

[132] M. I. Kamenskii, P. Nistri, and P. Zecca. On the periodic solutions problem for a parabolic inclusions with a large parameter. *Topological Meth. Nonlin. Anal.* **8** (1996), 57–77.

[133] M. I. Kamenskii and V. V. Obukhovskii. On periodic solutions of differential inclusions with unbounded operators in Banach spaces. *Zb. Rad. Prirod.-Mat. Fak. Ser Mat.* **21** (1991), 173–191.

[134] M. I. Kamenskii and V. V. Obukhovskii. Condensing multioperators and periodic solutions of parabolic functional-differential inclusions in Banach spaces. *Nonlinear Anal.* **20** (1993), 781–792.

[135] M. I. Kamenskii and V. V. Obukhovskii. The operator of translation along trajectories of semilinear controlled systems. *Differential Equations* **32** (1996), 755–762.

[136] M. I. Kamenskii, V. V. Obukhovskii, and P. Zecca. Method of the solutions sets for a quasilinear functional-differential inclusion in a Banach space. *Differential Equations Dynam. Systems* **4** (1996), 339–350.

[137] M. I. Kamenskii, V. V. Obukhovskii, and P. Zecca. On the translation multioperator along the solutions of semilinear differential inclusions in Banach spaces. *Canad. Appl. Math. Quart.* **6** (1998), 139–155.

[138] M. I. Kamenskii, V. V. Obukhovskii and P. Zecca. On semilinear differential inclusions with lower semicontinuous nonlinearities. *Ann. Math. Pura Appl.* (IV) **178** (2000), 235–244.

[139] T. Kato. *Perturbation Theory for Linear Operators*. Springer-Verlag, New York 1966.

[140] M. Kisielewicz. Multivalued differential equations in separable Banach spaces. *J. Optimiz. Theory Appl.* **37** (1982), 231–249.

[141] M. Kisielewicz. *Differential Inclusions and Optimal Control.* PWN-Polish Scientific Publishers, Warszawa; Kluwer Academic Publishers, Dordrecht–Boston–London 1991.

[142] E. Klein and A. Thompson. *Theory of Correspondences.* Wiley, New York 1984.

[143] V. B. Kolmanovskii and A. D. Myshkis. *Applied Theory of Functional Differential Equations.* Kluwer Academic Publishers, Dordrecht 1992.

[144] V. B. Kolmanovskii and A. D. Myshkis. *Introduction to the Theory and Applications of Functional Differential Equations.* Kluwer Academic Publishers, New York 1998.

[145] M. A. Krasnoselskii. *The Operator of Translation along the Trajectories of Differential Equations.* Amer. Math. Soc., Providence, RI, 1968.

[146] M. A. Krasnoselskii and S. G. Krein. On the averaging principle in nonlinear mechanics. *Uspehi Mat. Nauk.* **10** (1955), 147–152 (in Russian).

[147] M. A. Krasnoselskii and A. V. Pokrovskii. *Systems with Hysteresis.* Springer-Verlag, Berlin–Heidelberg–New York 1989.

[148] M. A. Krasnoselskii and P. P. Zabreiko. *Geometrical Methods of Nonlinear Analysis.* Springer-Verlag, Berlin 1984.

[149] M. A. Krasnoselskii, P. P. Zabreiko, E. I. Pustyl'nik and P. E. Sobolevskii. *Integral Operators in Spaces of Summable Functions.* Noordhoff International Publishing, Leyden 1976.

[150] S. G. Krein. *Linear Differential Equations in Banach Spaces.* Amer. Math. Soc., Providence, RI, 1971.

[151] W. Kryszewski. Topological and approximation methods of degree theory of set–valued maps. *Dissertationes Math.* **336** (1994), 1–101.

[152] W. Kryszewski. *Homotopy Properties of Set-Valued Mappings*, Univ. N. Copernicus Publishing, Torun 1997.

[153] Z. Kucharski. A coincidence index. *Bull. Acad. Polon. Sci. Ser. Math. Astronom. Phys.* **24** (1976), 245–252.

[154] Z. Kucharski. Two consequences of the coincidence index. *Bull. Acad. Polon. Sci. Ser. Math. Astronom. Phys.* **24** (1976), 437–443.

[155] K. Kuratowski. Sur les espaces complets. *Fund. Math.* **15** 1930, 301–309.

[156] K. Kuratowski. Les fonctions semi-continues dans l'espace des essembles fermes. *Fund. Math.* **18** (1932), 148–159.

[157] K. Kuratowski. *Topology*, Vols. I and II, 4th. ed.. Panstowowe Wyd Nauk., Warszawa; Academic Press, New York 1966.

[158] A. Lasota and Z. Opial. An application of the Kakutani–Ky Fan theorem in the theory of ordinary differential equations. *Bull. Acad. Polon. Sci. Ser. Sci. Math. Astronom. Phys.* **13** (1965), 781–786.

[159] A. Lasota and Z. Opial. Fixed-point theorems for multivalued mappings and optimal control problems. *Bull. Acad. Polon. Sci. Ser. Sci. Math. Astronom. Phys.* **16** (1968), 645–649.

[160] A. Lasota and Z. Opial. An approximation theorem for multi-valued mappings. *Podst. Sterow.* **1** (1971), 71–75.

[161] T. W. Ma. Topological degrees of set-valued compact fields in locally convex spaces. *Dissertationes Math.* **92** (1972), 1–47.

[162] J. W. Macki, P. Nistri, and P. Zecca. Measurable and directionally continuous selections for the control of uncertain systems. *Dynamics Contin. Discrete Impuls. Systems* **2** (1996), 397–409.

[163] V. Makarov and A. M. Rubinov. *Mathematical Theory of Economic Dynamics and Equilibria*. Springer-Verlag, New York 1977.

[164] S. A. Marano. Controllability of partial differential inclusions depending on a parameter and distributed parameter control processes. *Matematiche* **45** (1990), 283–300.

[165] G. Marino. Nonlinear boundary value problems for multivalued differential equations in Banach spaces. *Nonlinear Anal.* **14** (1990), 545–558.

[166] R. Martin. *Nonlinear Operators and Differential Equations in Banach Spaces*. Wiley, New York 1976.

[167] I. Massabo, P. Nistri, and J. Pejsachowicz. On the solvability of nonlinear equations in Banach spaces. In: *Fixed Point Theory* (E. Fadell and G. Fournier, eds.), Lecture Notes in Math. 886 (1980), 270–289.

[168] E. Michael. Continuous selections I. *Ann. Math.* **63** (1956), 361–382.

[169] E. Mitidieri and I. I. Vrabie. Differential inclusions governed by non convex perturbations of m-accretive operators. *Differential Integral Equations* **2** (1989), 525–531.

[170] A. D. Myshkis. Generalizations of the theorem on a stationary point of a dynamical system inside a closed trajectory. *Mat. Sb.* **34** (1954), 525–540 (in Russian).

[171] H. Nikaido. *Convex Structures and Economic Theory.* Academic Press, New York 1968.

[172] P. Nistri, V. V. Obukhovskii, and P. Zecca. On the solvability of systems of inclusions involving noncompact operators. *Trans. Amer. Math. Soc.* **342** (1994), 543–562.

[173] P. Nistri, V. V. Obukhovskii, and P. Zecca. Viability for feedback control systems in Banach spaces via Carathéodory closed-loop controls. *Differential Equations Dynam. Systems* **4** (1996), 367–378.

[174] R. D. Nussbaum. Degree theory for local condensing maps. *J. Math. Anal. Appl.* **37** (1972), 741–766.

[175] V. V. Obukhovskii. On some fixed point principles for multivalued condensing operators. *Trudy Mat. Fac. Voronezh Univ.* **4** (1971), 70–79 (in Russian).

[176] V. V. Obukhovskii. Periodic solutions of control systems. *Trudy Mat. Fac. Voronezh Univ.* **7** (1972), 68–76 (in Russian).

[177] V. V. Obukhovskii. On periodic solutions of differential equations with multivalued right-hand side. *Trudy Mat. Fac. Voronezh Univ.* **10** (1973), 74–82 (in Russian).

[178] V. V. Obukhovskii. Asymptotic fixed point theorems and stationary points of dynamical systems without uniqueness. *Trudy Aspirantov Voronezh Gos. Univ.* **2** (1974), 30–38 (in Russian).

[179] V. V. Obukhovskii. On the topological degree for a class of noncompact multivalued mappings. *Funktsionalnyi Analiz (Ulyanovsk)* **23** (1984), 82–93 (in Russian).

[180] V. V. Obukhovskii. Semilinear functional–differential inclusions in a Banach space and controlled parabolic systems. *Soviet J. Automat. Inform. Sci.* **24** (1991), 71–79.

[181] V. V. Obukhovskii and E. V. Gorokhov. On the definition of the rotation for a class of compactly restrictible multivalued vector fields. *Trudy Mat. Fac. Voronezh Univ.* **12** (1974), 45–54 (in Russian).

[182] V. Obukhovskii and P. Rubbioni. On a controllability problem for systems governed by semilinear functional differential inclusions in Banach spaces. *Topol. Methods Nonlinear Anal.* **15** (2000), 141–151.

[183] V. V. Obukhovskii and P. Zecca. On some properties of dissipative functional differential inclusions in a Banach space. *Topol. Methods Nonlinear Anal.* **15** (2000), 369–384.

[184] V. V. Obukhovskii and P. Zecca. On a property of the solutions set for a semilinear differential inclusion with an almost lower semicontinuous nonlinearity. To appear.

[185] V. I. Opoitsev. *Equilibrium and Stability in Models of Collective Behavior.* Nauka, Moscow 1977 (in Russian).

[186] N. S. Papageorgiou. Representation of set valued operators. *Trans. Amer. Math. Soc.* **292** (1985), 557–572.

[187] N. S. Papageorgiou. On multivalued evolutions equations and differential inclusions in Banach spaces. *Comment. Math. Univ. San. Paul.* **36** (1987), 21–39.

[188] N. S. Papageorgiou. On quasilinear differential inclusions in Banach spaces. *Bull. Pol. Acad. Sci. Math.* **35** (1987), 407–415.

[189] N. S. Papageorgiou. On multivalued semilinear evolution equations. *Boll. Un. Mat. Ital.* B (7) **3** (1) (1989), 1–16.

[190] N. S. Papageorgiou. Optimal control of nonlinear evolution inclusions. *J. Optim. Theory Appl.* **67** (1990), 321–354.

[191] N. S. Papageorgiou. Evolution inclusions with nonconvex valued orientor fields. *Bull. Math. Soc. Sci. Math. Roumanie* **34** (1990), 161–172.

[192] N. S. Papageorgiou. A minimax optimal control problem for evolution inclusions. *Yokohama Math. J.* **39** (1991), 1–19.

[193] N. S. Papageorgiou. Boundary value problems and periodic solutions for semilinear evolution inclusions. *Comment. Math. Univ. Carolin.* **35** (1994), 325–336.

[194] N. H. Pavel. *Differential Equations, Flow Invariance and Applications.* Pitman Res. Notes Math. Ser. 113, Longman Sci. Tech., Harlow 1984.

[195] A. Pazy. *Semigroups of Linear Operators and Applications to Partial Differential Equations.* Springer-Verlag, Berlin 1983.

[196] W. V. Petryshyn and P. M. Fitzpatrick. A degree theory, fixed point theorems, and mapping theorems for multivalued noncompact mappings. *Trans. Amer. Math. Soc.* **194** (1974), 1–25.

[197] V. A. Plotnikov, A. V. Plotnikov, and A. N. Vityuk. *Differential Equations with Multivalued Right-Hand Side.* Asymptotic Methods. Astro Print, Odessa 1999 (in Russian).

[198] A. I. Povolotskii and E. A. Gango. Periodic solutions of differential equations with a multi-valued right hand-side. *Leningrad. Gos. Ped. Inst. Uchen. Zap.* **464** (1970), 235–242 (in Russian).

[199] R. T Rockafellar. Integral functionals, normal integrands and measurable multifunctions. In: *Nonlinear Operators and of Variations* (L. W. Waelbroek, ed.), *Lecture Notes in Math.* 543, Springer-Verlag, Berlin 1976, 157–207.

[200] Yu. I. Sapronov. On the homotopy classification of condensing maps. *Trudy Mat. Fac. Voronezh Univ.* **6** (1972), 78–80 (in Russian).

[201] B. N. Sadovskii. On a fixed point principle. *Funktsional. Anal. i Prilozhen.* **1** (1967), 74–76 (in Russian).

[202] B. N. Sadovskii. Measures of noncompactness and condensing operators. *Problemy Mat. Anal. Slozhn. Sistem* **2** (1968), 89–119 (in Russian)

[203] B. N. Sadovskii. Limit-compact and condensing operators. *Russian Math. Surveys* **27** (1972), 85–155.

[204] L. Schwartz. *Analyse Mathematique*. Vol. I, Hermann, Paris 1967.

[205] Shi Shuzhong. Théorèmes de viabilité pour les inclusions aux dérivées partiélles, *C. R. Acad. Sci. Paris Sér. I Math.* **303** (1986), 11–14.

[206] Shi Shuzhong. Nagumo type condition for partial differential inclision, *Nonlinear Anal.* **12** (1988), 951–967.

[207] Shi Shuzhong. Viability theorems for a class of differential-operator inclusions, *J. Differential Equations* **79** (1989), 232–257.

[208] E. Spanier. *Algebraic Topology*. McGraw-Hill, New York 1966.

[209] E. U. Tarafdar and H. B. Thompson. On the solvability of nonlinear noncompact operator equations. *J. Austral Math. Soc.* (Ser. A) **43** (1987), 103–126.

[210] A. A. Tolstonogov. *Differential inclusions in a Banach space*. Nauka, Novosibirsk 1986 (in Russian).

[211] A. A. Tolstonogov. On solutions of evolution inclusions. I. *Siberian Math. J.* **33** (1992), 500–511.

[212] A. A. Tolstonogov and I. A. Finogenko. On solutions of a differential inclusions with lower semicontinuous nonconvex right hand-side in a Banach space. *Math. USSR Sbornik* **53** (1986), 203–231.

[213] A. A. Tolstonogov and Ya. I. Umanskii. On solutions of evolution inclusions. II. *Siberian Math. J.* **33** (1992), 693–702.

[214] Ya. Umanskii. On a property of the solutions set of differential inclusions in a Banach space. *Differ. Uravn.* **28** (1992), 1346–1351 (in Russian).

[215] Ya. Umanskii. On the existence of solutions for a class of evolution inclusions on a closed set. I. *Differ. Uravn.* **30** (1994), 1139–1147 (in Russian).

[216] Ya. Umanskii. On the existence of solutions for a class of evolution inclusions on a closed set. II. *Differ. Uravn.* **30** (1994), 2188–2190 (in Russian).

[217] A. Vanderbauwhede. On a modified degree theory for multivalued mappings. *Simon Stevin* **50** (1976–1977), 65–86.

[218] J. R. L. Webb. On the degree theory for multi-valued mappings and applications. *Boll. Un. Mat. Ital.* (4) **9** (1974), 137–158.

[219] P. Zecca and P. Zezza. Nonlinear boundary value problems in Banach spaces for multivalued differential equations on a non–compact interval. *Nonlinear Anal.* **3** (1979), 374–352.

Index

Bei Fragen zur Produktsicherheit wenden Sie sich bitte an:
If you have any questions regarding product safety,
please contact:

Walter de Gruyter GmbH
Genthiner Straße 13
10785 Berlin
productsafety@degruyterbrill.com